INNOVATIONS IN RECYCLING TECHNOLOGIES FOR PLASTIC WASTE

INNOVATIONS IN RECYCLING TECHNOLOGIES FOR PLASTIC WASTE

Edited by

Dr. Kailash Malode and
Saiyed Sakil ahemad

Delve PUBLISHING

www.delvepublishing.com

Innovations in Recycling Technologies for Plastic Waste

Dr. Kailash Malode and Saiyed Sakil ahemad

Delve Publishing

4164 Lakeshore Road

Burlington, ON L7L 1A4

Canada

www.delvepublishing.com

Email: orders@arclereducation.com

ISBN: 978-1-77956-425-2 (Hardcover)

ABOUT THE EDITORS

Dr. Kailash Malode (1980) is presently serving as Assistant Professor, Department of Soil Science and Agriculture Chemistry, college of Agriculture, Parul University, Vadodara, Gujarat. He obtained his B. Sc. (Ag.) in 2006 from Marathwada Agriculture University, Parbhani. M. Sc. (Ag.)2009 in Dr. Panjabrao Deshmukh Krishi Vidyapeeth,Akola and Ph. D. Soil Science And Agriculture Chemistry in 2014 from Vasantrao Naik Marathwada Krishi Vidhyapeeth, Parbhani. He started his career as Assistant professor in 2014. His field of specialization is soil fertility, Remote Sensing and Organic Farming. He has published 02 books and 15 research paper in National and international journals of reputed. He has participated many National, State Seminar and symposiums. He has written 02 practical manuals for under graduate course for the benefit of students besides several Radio talk, Lectures in farmer traning programme. Dr. Kailash has also written 21 popular articles in Shati Bharti, Shati pragati and other reputed Magazines. He has Best writer award on 2015 for Krishi dooth popular Magazines of Maharashtra and One Young scientist award in 2020.

Saiyed Sakil ahemad is an eminent scholar with an Honors Bachelor's degree in Forestry from the College of Forestry, Navsari Agricultural University (NAU) Navsari, and a Master's in Environmental Science and Technology from the Institute of Science & Technology for Advanced Studies & Research (ISTAR), Anand. With over 5 years of teaching experience as an Assistant Professor at the College of Agriculture, Parul University in Vadodara, Gujarat, his scholarly contributions extend globally, with over six publications illuminating the intersection of Agriculture, Forestry, and Environment.

TABLE OF CONTENTS

LIST OF FIGURES

LIST OF TABLES

ABSTRACT

Plastics, celebrated for their versatility, affordability, and durability, have become indispensable across industries. However, their extensive use and poor disposal practices have resulted in severe environmental challenges. A significant portion of global plastic production is dedicated to single-use items, leading to the accumulation of waste in landfills and natural environments. This persistent issue is exacerbated by the reliance on finite resources, such as oil and gas, for plastic manufacturing. Recycling has emerged as a critical strategy to address these challenges, offering the potential to reduce resource consumption, carbon emissions, and waste volumes. Despite its importance, recycling rates vary widely depending on the type of plastic, its application, and regional capabilities, highlighting the need for technological advancements and efficient practices.

This book, Recycling Technologies for Plastic Waste, provides a comprehensive exploration of cutting-edge innovations in plastic recycling. The initial chapters lay the groundwork by analyzing the types, sources, and environmental impact of plastic waste, while emphasizing the critical need for sustainable recycling solutions. Subsequent sections delve into specific aspects of recycling: recovery of polymeric materials, recycling of engineering and thermosetting plastics, synthetic fibers, and waste rubbers. Special attention is given to emerging solutions for marine plastic debris and the thermal analysis of recycled materials to enhance quality and performance.

The book also examines mechanical recycling processes for packaging plastics and anticipates future advancements in recycling technologies and applications. By combining insights from environmental science, materials engineering, and waste management, it offers a multidisciplinary perspective on tackling plastic waste sustainably. Designed for researchers, engineers, policymakers, and educators, this book bridges knowledge gaps and provides actionable insights into modern recycling techniques. It also serves as a valuable resource for students and academics, equipping them with the tools needed to advance the field of plastic waste management. Through its detailed analysis and forward-looking approach, this book contributes to global sustainability efforts by promoting innovative solutions for one of the planet's most pressing environmental issues.

PREFACE

Plastics, known for their affordability, lightweight, and durability, can be molded into a wide array of products and are used extensively across various sectors. This versatility has led to a significant increase in plastic production over the past 60 years. However, the widespread use and disposal of plastics have resulted in notable environmental challenges. Approximately 4% of global oil and gas production, which are finite resources, is utilized as raw material for plastics, with an additional 3-4% used in their production process. A considerable portion of the plastic produced annually is designated for single-use packaging or short-lived products that are discarded shortly after manufacture, highlighting the unsustainable nature of current plastic usage. Furthermore, due to the resilience of the polymers, large amounts of end-of-life plastics are accumulating in landfills and natural environments worldwide.

Recycling has emerged as a crucial strategy to mitigate these environmental issues and is one of the most rapidly evolving sectors within the plastics industry. By recycling, it is possible to decrease oil consumption, lower carbon dioxide emissions, and reduce the volume of waste that needs to be managed. Recycling is one among several waste-reduction strategies, which also include minimizing material use through product downsizing or reuse, exploring alternative biodegradable materials, and recovering energy through incineration.

Although plastic recycling began in the 1970s, the rate of recycling varies by region, type of plastic, and application. Recent decades have seen significant growth in the recycling of packaging materials in several countries. Technological advancements in collection, sorting, and reprocessing of recyclable plastics are creating new recycling opportunities. With continued efforts from the public, industry, and governments, there is potential to substantially increase the diversion of plastic waste from landfills to recycling facilities in the coming decades.

Recycling Technologies for Plastic Waste is to explore and elucidate the latest advancements in recycling technologies for various types of plastic waste. The book seeks to provide comprehensive insights into innovative methods and technologies that enhance recycling efficiency and sustainability.

Chapter 1 lays the foundation by providing a comprehensive overview of plastic waste and recycling. It covers the different types and sources of plastic waste, including their environmental impact and the critical need for effective recycling practices. This chapter establishes the context for the subsequent discussions by detailing the challenges and opportunities in managing plastic waste.

Chapter 2 deals with the recovery and recycling of polymeric and plastic materials. It explores various techniques and technologies employed to reclaim and repurpose these materials, emphasizing the importance of innovative approaches in improving recycling efficiency and sustainability.

Chapter 3 focuses on the recycling of engineering plastics such as polycarbonate (PC), high-impact polystyrene (HIPS), acrylonitrile butadiene styrene (ABS), and polyamide (PA), along with their blends and composites. This chapter addresses the specific challenges associated with recycling these robust plastics and discusses advancements in technology that aim to enhance the recycling processes for these materials.

Chapter 4 examines the recycling of waste rubbers, presenting methods to transform used rubber into value-added products. It highlights the economic and environmental benefits of recycling rubber, as well as the technological innovations that make this process more effective and efficient.

Chapter 5 covers the recycling of thermosetting plastics, which are known for their durability and resistance to high temperatures. This chapter discusses the unique challenges these materials present and the innovative methods developed to overcome these challenges and facilitate their recycling.

Chapter 6 addresses the issue of marine plastic debris, focusing on the sources and impact of plastics found in marine environments. It explores technologies and strategies for recycling marine plastic debris, aiming to mitigate the environmental damage caused by plastic pollution in the oceans.

Chapter 7 discusses the recycling of synthetic fiber plastics, which are commonly found in textiles and other products. This chapter explores the methods for processing these fibers and the potential for repurposing them into new products, highlighting advancements in recycling technology for synthetic fibers.

Chapter 8 provides an analysis of the thermal properties of recycled plastics. It details how thermal analysis can be used to evaluate the quality and performance of recycled materials, offering insights into the potential improvements in recycling processes through thermal evaluation.

Chapter 9 focuses on the mechanical recycling of packaging plastics. It covers the processes involved in breaking down used packaging materials and repurposing them into new products, discussing the challenges and advancements in mechanical recycling techniques.

Chapter 10 looks towards the future of recycling and recycled plastics, exploring emerging technologies and new applications. It discusses how innovations in recycling can contribute to sustainability efforts and the broader implications for the future of plastic waste management.

This book is written for researchers, engineers, and professionals in environmental science, materials science, and waste management. It is also valuable for policymakers and educators interested in the latest advancements in plastic waste recycling. Additionally, the book serves as a resource for students and academics studying related fields, offering in-depth knowledge and insights into the latest recycling technologies and practices.

Chapter 1

Introduction to Plastic Waste and Recycling

Contents

INTRODUCTION

Plastic waste has grown to be a significant problem over the last ten years, both on land and at sea. Plastic has gone from being praised as a scientific marvel to being despised as an environmental blight in just over a century. Over the past year, the media has focused a lot of attention on river and ocean plastics, particularly on the major issues in Asia and the amount of floating plastic that is discovered in the middle of the ocean. In addition, the media frequently shows us images of plastic and plastic bags engulfing marine life. Every ocean has plastic products and packaging in it. In fact, a recent report revealed the discovery of a plastic bag at the bottom of the Pacific Ocean's Mariana Trench, which is located 11 kilometers below the surface of the water. There doesn't seem to be much of an impact from cleaning up plastic debris on coastal beaches. Additionally, less than 20% of plastic is recycled globally currently, so it is hoped that this book will spur new innovations in waste disposal and recycling techniques.

In 1850, the first plastic material was created, and in 1907, Bakelite was created. The invention of nylon, which was used in place of silk in parachutes and ropes, during World War II marked the start of the plastic revolution. The 1950s and 1960s saw the real advancement of manufactured goods and new polymers. Even though the public was aware of environmental issues as early as the 1960s (Silent Spring, Rachel Carson, 1962), at the time, as with many new and inventive inventions, little thought was given to the problems associated with plastic waste. Currently, the annual production of plastic exceeds 300 million tonnes, of which 8 million find their way into the ocean. The mass of plastic in the oceans is predicted to surpass that of fish by 2050 if we keep polluting the seas in this manner.

Additionally, it has been documented that 90% of seabirds have plastic in their bellies and that one in three marine life species have been discovered entangled in marine debris. With nearly 40% of all plastic usage occurring globally, the largest end-use market segment is plastic packaging. An average plastic bag lasts 15 minutes, and over 500 billion plastic bags are used globally each year. (Geyer et al, 2017)

1.1 PLASTIC MATERIALS: COMPOSITION AND CLASSIFICATION

In most cases, plastics are made up of fillers, binders, plasticizers, pigments, and other ingredients. The binder determines most of the properties of plastic, and binder molecules are often the source of the plastic's name. Both natural and synthetic binder types, such as casein, milk protein, or cellulose derivatives, are possible. Notably, synthetic resins make up the majority of binders. Polyethylene is the primary ingredient in plastics. An ethylene polymer with the molecular and empirical formulas CH_2–CH_2 and $(–CH_2–CH_2–)_n$, respectively, can be used to characterize it in accordance with the necessary characteristics of the final product. Polyethylene is unaffected by most organic solvents, acids, alkalis, and water. Depending on their chemistry and physical characteristics, thermosets and thermoplastics are two types of plastic that can be differentiated. Thermoplastics are plastics that can be melted, heated, and molded; once cooled, they solidify. For thermoplastics, these three procedures can also be repeated. This property of the plastic also qualifies them for mechanical recycling, which is a useful way to control waste. The internal structure of thermoplastics, which includes chemical bonding, can be used to classify them along with other structural features and attributes.

The largest application of chlorine gas worldwide is in the manufacturing of polyvinyl chloride (PVC). An estimated 16 million tons of chlorine, or 40% of the world's annual production, are consumed by human activity. Organochlorine is most frequently produced in PVC. It can be referred to as a massive class of compounds that have recently been the subject of scientific and regulatory investigation due to their widespread use and detrimental effects on both the environment and public health. More hazardous to the community than plastic waste made of plastics are most plastic wastes with chemical compositions free of chlorine. The production of vinyl, the production of hazardous compounds, and the excessive use of energy and resources at different stages of the production process all have an adverse effect on the environment.

Natural gas, oil, or chlorine gas—which is primarily produced from sea salt via high-energy electrolysis—can be used to make ethylene. These are the two main components that go into making vinyl. Ethylene dichloride (EDC), scientifically known as 1,2- dichloroethane, is created when chemical reactions combine chlorine gas with the organic molecule ethylene. This manufacturing process is known as "chlorination.". Organic HCl, a by-product of this process, is combined with more ethylene to produce more EDC through the chemical process of oxychlorination. The produced EDC is concurrently further converted into chloroethylene (VCM, vinyl chloride monomer), via a procedure called pyrolysis. When the VCM monomers produced during the pyrolysis process are joined, a long chain of PVC known as white powder is produced. Pure PVC is added along with stabilizers, plasticizers, colorants, and other necessary additives that can supply any specific attribute for the intended plastic working. PVC in its pure form is not very useful due to its stiffness, brittleness, and capacity to gradually accelerate its disintegration with intensity from UV radiation. By incorporating additives into the polymer to increase its flexibility and moldability, PVC is rendered usable. Vinyl records, sewage and water pipes, clothing, water bottles, and medical containers are just a few examples of items that commonly use PVC. In addition, it's used in flooring, electrical conductors, furniture, and other practical wires.

Unlike HDPE, which is transparent and semirigid and has a long chain of identical subunits, LDPE is a long chain of long-chain monomers with a large branching structure. LDPE is made by free radical polymerization, which calls for extremely specific conditions like high pressure and temperatures between 80 and 300 degrees Celsius. The synthesis of LDPE requires between 4000 and 40,000 carbon atoms in total, along with a large number of short branches and subbranches. Stirred autoclaving and tubular methods are two other processes that can be utilized to produce LDPE. Because of the advantages that tubular reactors offer—such as a higher rate of ethylene transformation—they are currently used more often than autoclaving. A few common applications for LDPE include laundry bags, bin bags, drink cartons, work tables, drink ring holders, machine parts, lids, trays, protective shells, computer hardwires, playground fixtures, and containers.

Polypropylene (PP) is a thermoplastic polymer that is used in various applications; its empirical formula is $(C_3H_6)_n$. Propylene is transformed into a continuous chain of the polymer through a polymerization reaction to create PP, a semi-nonpolar, partially crystalline chemical molecule. The invention of polypropylene (PP) in 1954 was motivated by its advantages as a less dense polyolefin than other commodities. One of the many benefits of using polypropylene (PP) in a variety of applications and conversion processes, such as injection and extrusion molding, is its resistance to chemicals. Its chemical and physical characteristics are linked to chemical branching and high-temperature resistance. PP can fabricate a wide range of household items, including bottles, instrument jars (which may need to be cleaned frequently for use in a clinical setting), funnels, pails, and trays. In many applications, polypropylene (PP) is a better option than polyethylene due to its superior mechanical properties and colorlessness. In many applications, polyethylene is a better option than polyethylene because of PP's colorlessness and superior mechanical properties. Packing tape, food containers, crisp bags, straws, hobby model supplies, lunch boxes, bottle caps, clothing, and surgical instruments and tools are just a few of the industries that use PP extensively.

1.1.2 Thermosets

Thermosets are polymers that go through several physicochemical conversion processes under different heat treatments. The formation of a three-dimensional network is facilitated by a cross-linking reaction that materializes the chemical linkage between macromolecular chains. These thermoset molecules cannot be reconstituted or remelted after undergoing the heating treatment, and the transformation process itself is irreversible. The fact that thermosets can transition from a liquid with a low viscosity to a solid with a high melting point shows how versatile thermosets are in producing a wide range of materials with distinct chemical and physical properties. The viscosity of thermosetting monomers or subunits is usually low, allowing for customization and consumer-friendly simplicity. The application of various additives can optimize and maximize the performance of thermosets, thereby enabling these materials to be utilized for a wide range of specific applications.

The production of polyurethane, also referred to as a carbonate in the commercial context, is the result of the polymerization of organic monomers known as urethane. Another name for many thermoset polyurethanes is thermoplastic polyurethanes. Polyurethane's versatility and physical and chemical properties make it a popular material for use in many products, such as paints, coatings, foams, furniture, adhesives, and insulators. Like many other kinds of polymers, polyurethanes are mostly made of petrochemicals, either as an essential component or subunit or as the main building block of their main structural elements.

1.2 TECHNIQUES OF PLASTIC PROCESSING

Plastic processing is the series of operations that transform raw materials such as plastic or polymer into refined products that have the power to improve or lower one's standard of living in many areas, including financial, health, and developmental ones. The food and beverage processing industries use plastics extensively. By adding certain synthetic materials, also known as additives, plastics can have their durability, applications, and modifiability increased. Plasticizers containing bisphenols and phthalates are two examples of additives that can aid in the altering processes. High-quality plastics can be made from polymer using a variety of methods. High-quality plastics can be made from polymer in a variety of ways, which can be divided into three groups. For example, primary processing methods include transfer molding, compression, extrusion, and injection; secondary methods include thermoforming, coating, calendaring and fabrication, roto-coating, and casting; and tertiary methods include welding, briquetting, and drilling. (Geyer et al, 2017)

1.2.1 Primary Techniques

Thermosets, sometimes referred to as thermoplastics, are produced at a temperature that is controlled by using an injection process that uses a screw pump or plunger to reduce the viscosity of the polymer that is heated and injected under controlled pressure by compressing it into runners through a nozzle, molds cavities, and gates. The mold injection method is used to make many different products, such as those used in the car industry, bottle caps, spools, gem clips, bobbins, buckets, and crates. Another popular processing method is blow molding, which involves heating a region to the point where plastic melts and can take on deformities from the raw material of plastic pellets. This is done with the use of electricity and band heaters. Gas tanks, drinking bottles, armrests, portable toilets, and air ducts are just a few of the products that are made using the blow molding technique. In the course of the extrusion process, raw thermoplastic materials or resins are loaded into the top-mounted hopper, where the force of gravity causes them to fall into the extruder's barrel. The processing of extruding plastics can be completed by adding chemical additives, such as colorants and UV inhibitors, to the resin before it reaches the hopper. These chemical additives can be in liquid or pellet forms. Plastic films and sheeting, strapping, thermoplastic coatings, multilayer films, pipe, and tubing are a few items that can be produced by extrusion. Compression molding is another method that uses heat to process plastics. During the processing of plastic materials, a heated polymer is injected into a hot mold cavity. After the plug or closure has sealed the mold completely, the material is compressed to fill the mold cavities from top to bottom. A material with complex thickness and length patterns can be produced more easily using this compression molding technique. Users from a wide range of industries and individuals find the products made using this technology appealing due to their high strengths, hardness, and durability. Engine handles, cisterns, plugs, electrical sockets, and switches for engine casings are just a few examples of the many useful items that can be made by compression and molding processes. Transfer molding is yet another popular plastic processing technique used by numerous experts to create various rubber component types. Calculating, placing, and adding molding quantities into the pot is necessary during the processing process. The material is then heated and compressed, forcing it into the mold cavity.

1.2.2 Secondary Techniques

Hollow objects can be made with great success using the rotational molding process for plastic. Unlike earlier methods, there is no pressure applied at any point during the process. A short production process is beneficial from an economic perspective since casting techniques shorten the production process and lower production costs. A sort of plastic molding called thermoforming is useful for creating a wide range of useful plastic instruments. Little plastic sheets are heated during the manufacturing process to make handling easier. To make the necessary products, the sheets are heated to a temperature that is malleable, and the production process is then completed by cooling the finished product. Calendaring is a kind of secondary processing method used to make high-volume plastics and various high-quality plastic sheet and film products. It is widely employed in the production of PVA and other polymers with various characteristics.

After passing through an extruder that applies pressure and heat to the molten polymer, the resulting sheets are shaped by calendaring rolls. The casting method, which involves pouring a liquid state into a mold with cavities that mimic the shape of the finished product, is another intriguing and useful way to process plastic. The liquid assumes the shape of the plastic required to make the intended product once it has solidified. The mold needs to be removed or cast out of the product for the solidified component to finish the process. (Geyer et al, 2017)

1.3 GLOBAL PLASTIC PRODUCTION

Due to their outstanding physicochemical characteristics (e.g. durability, availability, hygienic, lightweight, and flexibility) and being cost-effective, plastic has become a primary product around the world and has diverse applications in industrial and commercial commodities. The global production of plastic has surged dramatically in response to the expanding demand for these goods. The amount of plastic produced annually worldwide has increased over the past ten years, from 2 million tons in the 1950s to 359 million tons in 2018 and 368 million tons in 2020. Global plastic production has historically increased by about 9% annually. According toscientific reports, the global production of plastic reached 311 million tons annually by 2014.

This suggests that, although global plastic production has grown dramatically to 20,000 percent in 65 years, global plastic production has increased by about 25 percent annually in just 5 years. China leads the world in plastic production, accounting for 26%, 20%, and 19% of global plastic production, respectively (Figure 1.1). North America and Europe follow in second place. Furthermore, current long-term forecasts suggest that the production of plastic goods is expected to continue growing without showing any signs of slowing down. Based on scientific research, it is estimated that by 2050, the global annual plastic production will range from 850 million to 1124 million tons, with an additional 33 billion tons of plastic materials produced. Nonetheless, the unprecedented consumption of materials containing plastic, such as PPE and packaging, can exacerbate these projections.

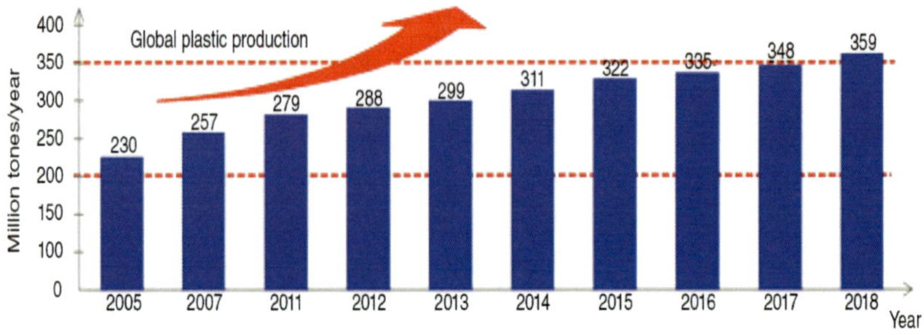

Figure 1.1. Global plastics production.

Source: *Hong Y. Study on the Maximum Level of Disposable Plastic Product Waste. Sustainability. 2023; 15(12):9360. https://doi.org/10.3390/su15129360*

Table 1.1 lists the most widely used and abundant polymers, which include low-density polyethylene (LDPE), PVC, PP, PET, HDPE, and polystyrene (PS) (Figure 1.2).. Taken as a whole, they account for almost 90% of global plastic production. Most plastic products, especially those used in food and drink packaging, have internationally recognized codes that identify the type of polymer used to make the commodity. This allows specific types of plastic materials to be distinguished from other types. The current system of coding was released by the American Society for Testing and Materials (ASTM). In order to facilitate reprocessing procedures and facilitate the easy identification of plastic materials, the Society of Plastics Industry (SPI) standardized the most widely used commodity plastics in 1998. Still, ASTM International assumed responsibility for the codes' supervision in 2008. In 2013, ASTM International decided to update the ASTM D7611 system by changing the well-known three mutually chasing arrows, revising the symbols, and substituting solid equilateral triangles for them. The initial symbols bore a striking resemblance to the global recycling symbol, which is why this action was taken. The fact that many recycling facilities at the time would only accept plastics with specific codes and would not accept any other plastic sorts suggests that was a source of confusion, even in spite of the mutually chasing arrows As a result, customers couldn't understand why the plastics were turned away despite having a recycling symbol. ASTM International therefore aimed to ensure that the introduced symbols and abbreviations only identified the type of plastic, independent of the plastic's potential for recycling. Consequently, the solid equilateral triangle system was introduced to give the stakeholder society an effective and reliable way to use the resin recognition coding system. (Geyer et al, 2017)

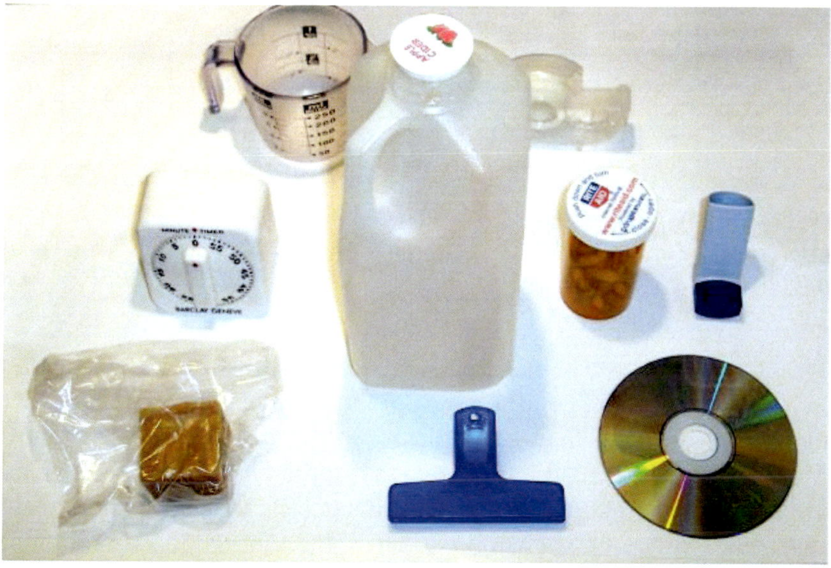

Figure 1.2. Plastic items.

Source: *By ImGz - Own work, CC BY-SA 3.0, https://commons.wikimedia.org/w/index.php?curid=3781690*

Table 1.1. *The world's top 10 exports of plastic waste (10,000 t) from 2016 to 2018, 2020, and 2022*

2016		2017		2018		2020		2022	
Country	**Export**	**Country**	**Export**	**Country**	**Export**	**Country**	**Export**	**Country**	**Export**
Hong Kong	212.60	Hong Kong	128.56	Japan	61.69	Germany	85.37	Germany	73.43
USA	205.61	USA	107.31	USA	51.41	Japan	82.07	Netherlands	68.42
Japan	88.53	Japan	84.24	Germany	39.12	USA	62.45	Japan	59.77
Germany	46.85	Germany	41.87	Mexico	23.38	UK	56.10	Belgium	47.74
Mexico	31.88	Mexico	27.91	Hong Kong	22.57	Nether-lands	41.32	UK	47.21
Nether-lands	29.04	Vietnam	27.97	France	19.89	Blegium	34.62	USA	43.75
Vietnam	24.98	Thailand	25.83	Thailand	18.63	France	33.37	France	34.25
France	24.71	Nether-lands	22.12	Nether-lands	18.49	Italy	21.72	Sri Lanka	23.90
Spain	22.42	Spain	19.41	Italy	16.83	Slovenia	18.19	Italy	20.65

2016		2017		2018		2020		2022	
Country	**Export**	**Country**	**Export**	**Country**	**Export**	**Country**	**Export**	**Country**	**Export**
Indone-sia	18.11	Italy	19.19	Belgium	16.67	Austria	17.93	Canada	18.35

Source: *Macheca AD, Mutuma B, Adalima JL, Midheme E, Lúcas LHM, Ochanda VK, Mhlanga SD. Perspectives on Plastic Waste Management: Challenges and Possible Solutions to Ensure Its Sustainable Use. Recycling. 2024; 9(5):77. https://doi.org/10.3390/recycling9050077*

Plastic waste, in spite of its exceptional qualities, has become a major global concern. About one-third of all plastics produced each year are anticipated to be disposable, meaning they are typically thrown away after 12 months of manufacture. Furthermore, it is estimated that approximately 10% of all plastics produced have been dumped into the world's oceans. According to the United Nations Joint Group of Experts on the Scientific Aspects of Marine Pollution (GESAMP) assessment, only 20% of waste in the marine environment is a result of activities occurring at sea; the remaining 80% comes from land-based sources (Figure 1.3). Scientific estimates indicate that in 2015, about 8 million tons of plastic waste were reached by the ocean, compared to assessments that between 4 and 12 points 7 million tons of plastic litter ended up in the world's ocean in 2010 alone. By 2050, it's expected that this quantity will increase to roughly 32 million tons yearly. As a result, there are a number of environmental and health-related issues raised by the growing amount of marine plastic debris.

Figure 1.3. Plastic litter can substantially end up in the global ocean.

Source: *By MichaelisScientists - Own work, CC BY-SA 4.0, https://commons.wikimedia.org/w/index.php?curid=48418056*

1.4 THE HEALTH AND ENVIRONMENTAL EFFECTS OF PLASTIC DEBRIS

Increasing urbanization, fast industrialization, and population growth have all contributed to the different environmental issues brought on by human activity. Solid waste management is regarded as one of the most important environmental issues and has become one of the most urgent issues facing our planet, especially in metropolitan areas and megacities. Municipal solid waste (MSW) generation is expected to increase from its current 2 billion tons per year to 3 billion tons by 2025. (Van Cauwenberghe et al, 2015; Shen et al, 2020)

MSW is made up of a variety of waste materials, such as inorganic wastes like plastic, glass, and metal as well as organic residues like fruits, vegetables, and food scraps (Figure 1.4).

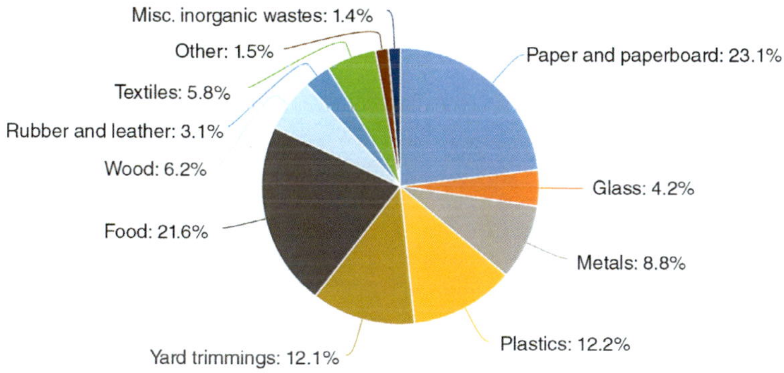

Figure 1.4. MSW components.

Source: *https://www.epa.gov/facts-and-figures-aboutmaterials-waste-and-recycling/guide-facts-and-figures-report-about.*

Plastic litter fractions make up a sizable portion of the MSW's inorganic components. The majority of the plastic waste found in MSW is made up of bottles, bags, packing materials, lids, containers, and cups. Compared to other types of MSW, plastic wastes have received a lot more attention due to their stability and durability, which stem from their polymeric nature. Plastic waste has become a significant issue in the modern world due to the accelerating rate at which plastic is produced and the dearth of suitable management, treatment, and disposal methods. Approximately 16 per cent of plastic waste is produced annually in India, more than 10 per cent in China, and 2.5 per cent in the United Kingdom. It takes centuries for them to completely degrade because they are resistant and nonbiodegradable. As a result, plastic waste in landfills and natural settings tends to accumulate rather than break down. There are several negative health and environmental consequences that can arise from the environment's increasing concentration of plastic waste. Figure 1.5 shows the fate and harmful effects of plastic particles.

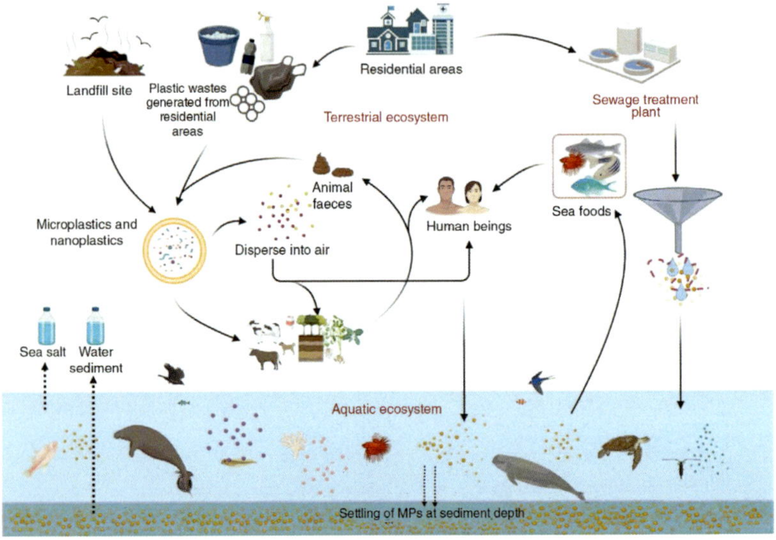

Figure 1.5. Fate and detrimental effects of plastic products on the ecosystem.

Source: *Ziani K, Ioniță-Mîndrican C-B, Mititelu M, Neacșu SM, Negrei C, Moroșan E, Drăgănescu D, Preda O-T. Microplastics: A Real Global Threat for Environment and Food Safety: A State of the Art Review. Nutrients. 2023; 15(3):617. https://doi.org/10.3390/nu15030617*

1.4.1 Plastic Debris and Microplastics

As scientific communities struggle to manage the massive amount of improperly disposed plastic waste, the introduction of microplastics has presented a serious new threat to global security. Polymer particles ranging in size from 1 to 5 mm are known as microplastics. Based on their sources, microplastics are divided into primary and secondary categories. "Primary microplastic" describes mainly micro-sized plastic particles. They can be found in body washes, lipstick, toothpaste, face cleansers, and cosmetics. On the other hand, "secondary microplastic" describes tiny plastic particles that are created when larger plastic items, like face masks and clothing, break down because of exposure to harsh environmental elements like UV rays and mechanical stresses. Thus, potential sources of microplastics include road marking and tiers, construction sites, sports arenas, landfills, littering, plastic production industries, mulching in agricultural fields, and the use of cosmetics and healthcare products. As part of the standardized size and color sorting system (SCS), microplastics are divided into ten categories based on their appearance and size: pellets, which are plastic spheres with diameters ranging from 1 to 5 mm; microbeads, which are small spherical pieces of plastic less than 1 mm to 1 μm in diameter; fragments, which are irregularly shaped pieces of plastic less than 5–1 mm in length along their longest dimension; microfiber, as well as film, which are thin sheets of plastic less than 1 mm–1μm in size along their longest dimension; foam, which are plastics less than 5 mm to 1 mm in size along their longest dimension; and microfoam, which are foam-like plastics less than 1–1μm in size along their longest dimension. These plastic particles could disperse either horizontally or vertically. The density of polymers is almost entirely responsible for the vertical transportation of (micro)plastics in the water column. While hydrodynamic processes like wind, river flow, and ocean currents can cause them to spread horizontally, animals may also transfer them after ingesting them. Microplastics may find their way into acceptor ecosystems through this distribution process, which could lead to them ending up in freshwater, oceans, soil, groundwater, Arctic snow, the atmosphere, human food sources like fish, and eventually the human body. Examples of primary and secondary microplastics are shown in Figure 1.6.

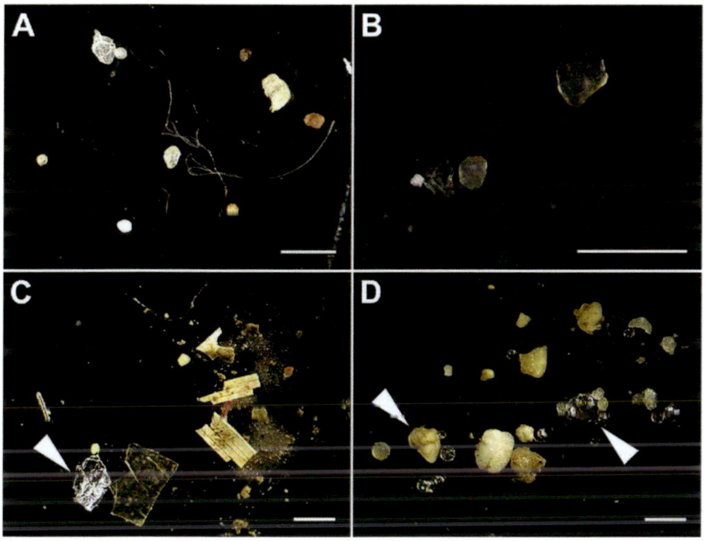

Figure 1.6. Microplastics.

Source: *By Martin Wagner et al., - Wagner et al.,: Microplastics in freshwater ecosystems: what we know and what we need to know. In: Environmental Sciences Europe. 26, 2014, doi:10.1186/s12302-014-0012-7, CC BY 4.0, https://commons.wikimedia.org/w/index.php?curid=39507778*

Therefore, it is obvious that the widespread persistence and accumulation of (micro)plastics in ecosystems around the globe endangers public health. However, due to ethical constraints, a dearth of reliable detection methods, and strict biosecurity regulations for handling human samples, our understanding of the potential risks that microplastics may pose to public health is fairly limited. Thus, the relationship between the increased incidence of related human disorders and the environmental effects of (micro)plastics is still up for debate. The section that follows discusses the various pathways through which (micro) plastics can enter the human body as well as any potential harm to health. Next, we'll talk about how plastic waste affects the environment. By assisting readers in comprehending the complex environmental health concerns associated with (micro)plastic contamination, the chapter hopes to suggest future directions for (micro)plastics research. (Van Cauwenberghe et al, 2015; Shen et al, 2020)

1.4.2 Plastic Effects on Human Health

Three important routes exist for microplastics to enter the human body: ingesting food and water tainted with microplastics, breathing in microplastics from the surroundings, and coming into touch with microplastics on the skin from dust, objects, or textiles. The toxicity routes associated with possible exposure pathways to microplastics in the human body are depicted in Figure 1.7. It has been demonstrated that foods consumed by humans, including commercial fish, mussels, sugar, table salt, and drinking water, contain microplastics.

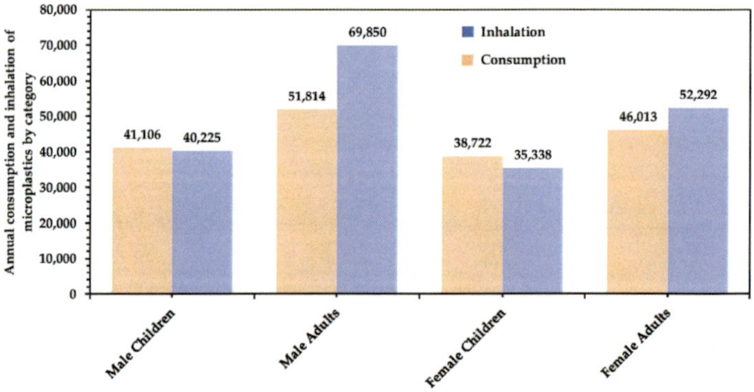

Figure 1.7. Potential pathways of microplastic exposure in the human body and their toxicity routes.

Source: *Macheca AD, Mutuma B, Adalima JL, Midheme E, Lúcas LHM, Ochanda VK, Mhlanga SD. Perspectives on Plastic Waste Management: Challenges and Possible Solutions to Ensure Its Sustainable Use. Recycling. 2024; 9(5):77. https://doi.org/10.3390/recycling9050077*

As a result, eating is thought to be the main way that people are exposed to microplastics. It is estimated that each person consumes between 39,000 and 52,000 particles of microplastic annually based on food consumption. Foodstuffs can introduce microplastics into the gastrointestinal tract, where they may cause alterations in gut microbial composition, permeability, and inflammation.

Microplastics have the ability to settle in aquatic or terrestrial environments, stay suspended in the atmosphere, and then return to the atmosphere. These microplastics in the air that are the result of improperly managed plastic waste can be categorized as particulate matter (PM) components of air pollution. Since the smallest size of microplastics that can be found using conventional methods is 5 μm, microplastics could be present in an area of the PM that is currently unknown. However, a significant amount of airborne microplastics in urban, suburban, isolated mountain, and indoor environments have been measured and described by recent research. According to the findings of a scientific study, indoor microplastic concentrations in private apartments or offices ranged from 3 to 15 MP particles/m^3. Also, González-Pleiter et al, reported the concentration of microplastics in Spain's outdoor environment is 13.5 particles per millimeter in a rural area and 13.9 particles per millimeter in Madrid. Adults are thought to breathe about 15 m^3 of air on average each day. Because of this, each person is exposed to varying amounts of airborne microplastics every day, depending on factors like employment, the population density of the city in which they live, and the amount of time they spend indoors. Adults who breathe in airborne microplastics are thought to inhale between 26 and 130 microplastics daily. Moreover, another route through which plastic particles are exposed to the human body is through dermal contact, which is caused by dust, synthetic fibers, and microbeads in cosmetics. This route primarily involves nanoplastics (<100nm).

Microplastics are toxic in a variety of ways after exposure. Microplastics have the potential to be the source of oxidative stress in the human body because of their large surface area, which can release oxidizing pollutants like metals that have been adsorbed to it or reactive oxygen species (ROS) that are released during the inflammatory response. For example, exposure to microplastics has been linked to oxidative stress in mice and zebrafish (*Danio rerio*). Cytotoxicity can also result from oxidative stress, plastic particle toxicity, and inflammation. Schirinzi et al have shown that exposure to PS and polyethylene at concentrations ranging from 0.05 to 10 mg/l enhanced ROS to high levels in human epithelial and cerebral cells, which contributed to cytotoxicity. Energy homeostasis is the balance between intake, expenditure, and the quantity of energy available from reserves. Microplastics may alter human energy homeostasis by decreasing nutrient (energy) intake, raising energy expenditure, and modifying metabolism, according to recent research. However, it might be challenging to see these effects because humans need more energy than the species under study and are exposed to lower exposure concentrations.

Other tissues may become exposed as a result of localized accumulation of plastic particles or their translocation within the body. For example, swallowed microplastics could find their way into the ileum and break through the mucus in the intestinal lumen. Microplastics are transported to the lymph and blood vessels by M-cells, despite the fact that T-cells, B-cells, and macrophages absorb plastic particles. Microplastics thus re-release and migrate to different textiles before ending up in feces and urine. The size, charges, surface area, and hydrophobicity of microplastics determine whether they accumulate or move. On the other hand, it has been demonstrated that the inflammation of the tissues increases the permeability and translocation of epithelial boundaries. Furthermore, it has been observed that unhealthful diets high in fructose sugar and saturated fats also make the gastrointestinal mucosa more permeable, which raises the concentration of MPs in blood vessels or the lymphatic system.

Depending on how widely they are dispersed, particles after exposure may trigger either local or systemic immune responses. In certain cases, such as those involving genetic predisposition, environmental exposure may be adequate to compromise immune function and encourage autoimmune or immunosuppressive disorders. Microplastics in the air have the potential to impair immune function. Mytilus species have been shown to exhibit tissue-dependent immune response modulation and immunosuppression following microplastic exposure. This hasn't been seen in humans yet, though. Therefore, more investigation into the effects of microplastics on the immune system is necessary.

Notwithstanding the inherent toxicity of microplastics, numerous studies reveal that they have the ability to ensnare and transport a wide variety of chemical pollutants, such as pathogens, heavy metals, toxic chemicals, antibiotics, and persistent organic pollutants (POPs). After exposure, microplastics may expose organisms to higher concentrations of the previously mentioned potentially harmful pollutants, or they may even make those pollutants more toxic. But pathogens' adsorption to microplastics is largely dependent on several factors, including the negative effects of vector microplastics' translocation, the types of ingested particles, the release rate, the clearance time, the amount of contamination, its toxic effects, and its translocation in body cells.

Neurotoxicity, which is connected to neurological disorders, can result from exposure to micropollutants. Tests for microplastic toxicity in vivo have demonstrated that microplastics can significantly affect neuronal behavior. Microplastics in the brain of *Dicentrarchus labrax* (European seabass) have been linked to a decrease in acetylcholinesterase (AChE), an increase in anaerobic energy generation, and oxidative stress with an increase in the amount of lipid peroxidation. Furthermore, it has been noted that PS exposure affects mouse neurotransmission by changing neurotransmitter levels in the blood and raising AChE activity. Regarding the evidence of neurotoxicity, it is important to consider how microplastics may be linked to neurotoxicity in humans, increasing the chance of developing neurological disorders. This is especially true when assessing microplastics in cells or other living things. (Van Cauwenberghe et al, 2015; Shen et al, 2020)

1.4.3 Plastic and Climate Change

Prior studies on the environmental effects of plastic waste primarily focused on its sources, distribution, fate, toxicity, and behavior; only a limited amount of attention was given to the undeniable role that plastic materials play in the increase of greenhouse gases in the atmosphere.

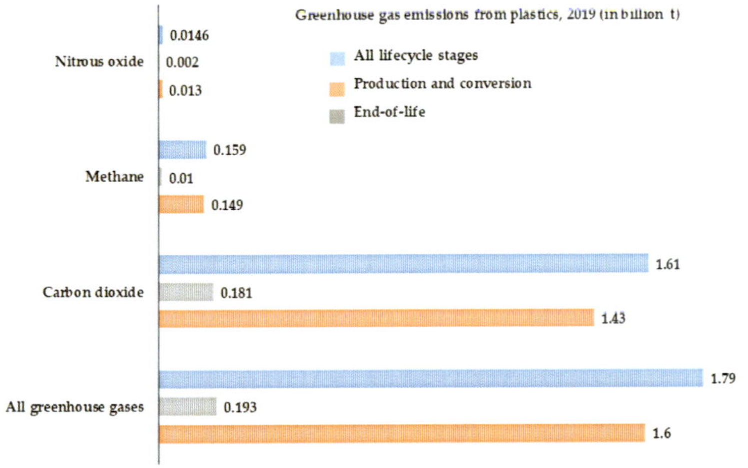

Figure 1.8. Greenhouse gas emissions (in billion t) from plastics in 2019 (emissions measured in t of carbon dioxide-equivalents (CO_2-e). The numbers that appear in the figure mean billions of t of greenhouse gas emitted.

Source: *Macheca AD, Mutuma B, Adalima JL, Midheme E, Lúcas LHM, Ochanda VK, Mhlanga SD. Perspectives on Plastic Waste Management: Challenges and Possible Solutions to Ensure Its Sustainable Use. Recycling. 2024; 9(5):77. https://doi.org/10.3390/recycling9050077*

The increase in plastic waste has had a markedly negative impact on the planet's climate. Scientific research indicates that every phase of the plastic life cycle—including the extraction and transportation of raw materials, manufacturing, disposal of plastic waste, and even the plastics' entry into the environment—contributes to greenhouse gas emissions. Thus, the data suggest that GHS emissions from the cradle to the grave of plastic are inevitable and significant, despite the limited information on the role of plastics in increasing atmospheric GHG. Figure 1.8 illustrates the direct and indirect contributions of plastic to climate change. Among the common approaches to dealing with plastic litter are incineration, recycling, landfilling, and sanitation. These methods of managing plastic waste have a direct impact on greenhouse gas emissions.

Recycling plastic waste refers to the physical process of extracting material without modifying the molecular makeup of the polymer. Comparing recycled plastic to other plastic waste management techniques now in use, recycling plastic considerably lowers greenhouse gas emissions. In theory, more recycling could lessen the demand for raw materials and stop emissions from producing the same amount of raw materials. Reusing 3.17 million tons of plastic waste in 2014 could have prevented 3.2 million tons of carbon dioxide emissions, the equivalent of taking 670,000 cars off the road for a year, according to the US Environmental Protection Agency. Recycling plastic packaging may also be able to stop 12.4 million tons of carbon dioxide emissions. Since recycled plastic materials are being used to make new plastic instead of raw materials, the process is more than three times more efficient in terms of greenhouse gas emissions. This is primarily because original products are being replaced and renewable energy is being conserved. Nevertheless, very little of the "recyclable" plastic waste is converted into the main products because of numerous restrictions and difficulties. Recycling is therefore regarded as the main strategy with a long-term solution to the plastic problem.

In recent times, burning has been seen as a fast solution to the pervasive contamination of plastics derived from land. It may also produce heat and energy for human consumption in addition to effectively lowering plastic pollution. During the burning process of incineration, plastic waste is transformed into combustion gas, fly ash, and bottom ash. In urban areas, collected plastic waste is burned with biomass or fossil fuels in waste incinerators, power plants, and other industrial facilities, mostly paper mills, cement kilns, and utility boilers. However, burning plastic waste can produce greenhouse gases (GHGs), most notably CO_2. According to scientific data, each ton of plastic packaging waste normally contains 79 percent flammable carbon, releasing 790 kg of carbon dioxide equivalent into the atmosphere every ton. Plastic waste accounted for over half of the 11 million tons of CO_2e emissions from waste incineration in the US in 2015 (5.9 million tons). Burning plastic waste has an environmental impact equivalent to driving 1.26 million cars annually in the United States.

A sanitary landfill typically uses liners or clay to keep waste away from groundwater and adds a layer of soil to reduce the amount of waste that is exposed to the air. The main source of greenhouse gas emissions from sanitary landfills is organic waste, which includes decomposing paper, wood, and food waste. There isn't a story about greenhouse gas emissions from plastic landfills yet. The pre-landfill debris is treated and classified, and fossil fuels are used to transport the debris from collection points to landfills, which contributes to the emissions linked to landfill plastic packing litter. However, this does not completely remove the possibility of landfill-related GHG emissions.

Apart from the aforementioned management strategies, an extra thirty-two percent of plastic packaging waste remains untreated. Dumping, littering, and burning are some of the different methods used to deal with unmanaged plastic debris, and these practices are common in areas lacking in waste management infrastructure. However, the full impact of unmanaged plastic waste on global warming remains unclear. Open burning, which is the practice of burning flammable waste outdoors without any precautions to reduce air pollution, has a negative impact on both human health and the climate because it takes place at lower temperatures than in a debris incinerator. Scientific data indicates that during open burning, one ton of plastic debris releases 2.9 million tons of greenhouse gas emissions. In general, it's unclear how the open disposal of plastic waste affects global warming. Current research indicates that the breakdown of plastic in non-aquatic environments exposed to sunlight can release GHGs at a faster rate than in oceanic environments. However, the magnitude and annual rate of these emissions remain unknown. Despite a dearth of data on waste management techniques, examining the greenhouse gas emissions from unmanaged plastic waste can reveal the entire extent of the threat that plastic packing debris poses to global warming. Unmanaged or improperly managed plastic waste contributes significantly to global warming, and this effect is largely correlated with the percentage of open burning.

The effects of microplastics on the environment and their release of greenhouse gases will continue even after they are disposed of. It was found that several of the most widely used plastic types release detectable levels of two greenhouse gases (ethylene and methane) after being exposed to radiation. The range of emission rates for CH_4 is 10–4100 pmol/day/g, while the range for C_2H_6 is 20–5100 pmol/day/g. Furthermore, compared to recycled plastics, virgin plastics had substantially higher greenhouse gas emissions. Furthermore, while the greenhouse gas emissions from virgin plastic increase over time, those from aged plastic remain constant. This is probably caused by anti-ultraviolet plasticizers, which lessen the effects of UV light and slow down the deterioration process. The rate at which greenhouse gas emissions from plastic materials are produced may be deemed negligible when compared to other sources of emissions, such as vehicle transportation, industrial processes, and agricultural activities. However, emissions of greenhouse gases linked to plastic degradation will likely rise along with an increase in plastic manufacturing and the improper disposal of waste plastics, which may be a bigger concern.

When plastics are mined, transported, refined, and manufactured, greenhouse gases are unavoidably released. The estimated global greenhouse gas emissions in 2015 from wells to refineries were 1.07 gigatons CO_2e. Given that about 4% of crude oil is used as a raw material for plastics, the global oil industry is estimated to have contributed 68 million tons of CO_2e to plastic manufacturing emissions in 2015. Up until now, new facilities for producing natural gas have either been completed or strongly recommended, and in the ensuing decades, more developments are planned. These facilities are motivated by the rapidly growing plastics industry as well as the necessity for natural gas. Because oil, coal, and gas extraction are major industries that are only the beginning of the plastic manufacturing process, their impact on greenhouse gas emissions is concerning. It is unlikely that these emissions would be reduced without a significant reduction in these large industries. Furthermore, the production of plastics releases greenhouse gases into the atmosphere. The process of turning petrochemical raw materials, such as propylene and ethylene, into usable commodities is what causes these emissions. Manufacturing facilities frequently control greenhouse gas emissions during production based on the product's effectiveness, control system, and service life. In 2014, approximately 17 million tons of CO_2e, or 46,324 tons per day, were produced by 72 plastic manufacturing facilities in the United States. It is extremely difficult to attribute greenhouse gas emissions from industry to the production of plastics because there are numerous industrial processes involved in turning fossil fuels into plastic materials, in addition to a vast number of manufacturing steps. There is a dearth of data regarding greenhouse gas emissions throughout the plastic production process, but mounting evidence suggests that plastic manufacturing is linked to greenhouse gas emissions. Both the production of plastic and the consequences of greenhouse gas emissions are global.

Aside from the direct release of greenhouse gases, plastic waste, particularly that found in maritime environments, has the potential to indirectly but eventually more significantly contribute to global warming through its effects on the organisms that form the foundation of the ocean food web. The ocean is thought to be the largest naturally occurring carbon dioxide pool and plays a vital role in removing carbon from the atmosphere through adsorption. The fundamental needs for human survival are in jeopardy since the ocean's capacity to absorb carbon is uncertain, which will drastically alter the earth's carbon cycle. Is it possible for oceanic (micro)plastic to prevent the ocean from sequestering carbon? Research has shown that (micro)plastics have a detrimental effect on phytoplankton growth and photosynthesis. The oceanic ecosystem greatly depends on phytoplankton. As the primary producer in marine ecosystems, phytoplankton uses atmospheric carbon that has been sequestered to produce organic matter and oxygen through photosynthesis (Figure 1.9). However, the presence of microplastics at the ocean's surface that reflect and block sunlight can lessen phytoplankton's ability to absorb sunlight and carry out photosynthesis. According to research conducted in laboratories, phytoplankton is negatively impacted by microplastic exposure; the more tiny the microplastic, the more harmful the effect. (Van Cauwenberghe et al, 2015; Shen et al, 2020)

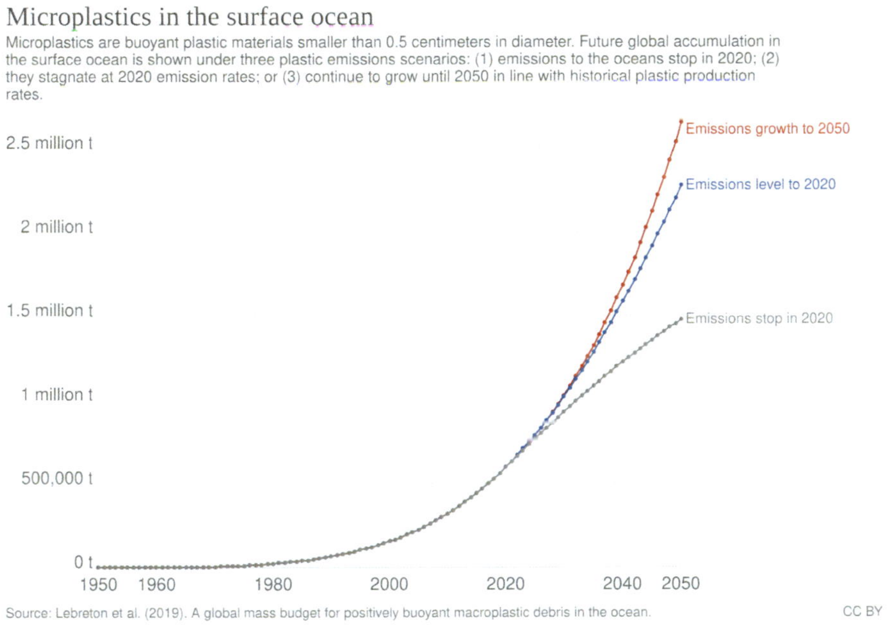

Figure 1.9. Microplastics.

Source: *https://commons.wikimedia.org/wiki/File:Microplastics_in_the_surface_ocean,_OWID.svg*

Furthermore, because microplastics are toxic to them and interfere with their development and reproduction, they may harm zooplankton. The main and most important consumers of phytoplankton are zooplanktons. These are essential for the movement of mass, the renewal of ocean energy and nutrients, the cycling of biogenic elements, the transfer of genetic information up the food chain, and the breakdown of pollutants in the water. Through their ability to respire, zooplanktons can break down particulate organic carbon (POC) in the ocean, which can have an impact on the depths at which POC remineralizes in the ocean and the capacity of marine organisms to absorb carbon from the atmosphere. Should zooplanktons remain unentangled during the OCS procedures, the sequestered carbon will promptly re-enter the atmosphere and water. On the other hand, the abundance of microplastics in the oceans may negatively affect the oceans' capacity to absorb CO_2. Scientific studies have acknowledged that microplastics may have negative effects on zooplankton (copepods) and reduce their carbon consumption because they cause satiety. Copepods that had consumed (micro)plastic consumed forty percent less food, and as time went on, their eggs became smaller and less likely to hatch, increasing the copepods' overall mortality rate. Furthermore, exposing zooplankton to more (micro)plastics over time may drastically decrease the amount of carbon-containing material they consume. Furthermore, fecal particles carry the carbon that phytoplankton consumes to the ocean floor because zooplankton eats it first. Subsequently, these particles gradually sink to the bottom of the ocean and settle in the mud. Cole et al.,observed that microplastics may be transported to the seabed by fecal pellets. According to scientific investigations, fecal particles containing microplastics exhibited a descending rate that was 1.35 times lower and significantly smaller comparable spherical diameters. Furthermore, less carbon settles on the ocean floor because microplastic-polluted pellets descend and degrade more quickly than unpolluted pellets. But our understanding of the behavior and effects of (micro)plastics in the deep ocean is still in its infancy. Therefore, more research is needed to understand the issue's likely dimensions, scope, and key players. (Van Cauwenberghe et al, 2015; Shen et al, 2020)

1.5 RECYCLING

Roughly 9% of all plastic ever produced is thought to have been recycled; the remaining 12% has been burned, and the remainder ends up in landfills, seas, and soils. Nearly half of all plastic produced to date has reportedly been made since 2000, and the plastics industry is expanding quickly. The United States recycles less than 9% of its plastic production, while the global average is currently less than 20%.

The term "recycle" has been used to encompass a wide range of beneficial methods of handling plastic. It truly ought to be reserved for waste plastic that is recycled or processed again to create new or different kinds of products. In fact, breaking down the polymeric material is one way to address the issue of plastic waste. Real recycling is rare, and even the plastic bottles that we in the UK separate for collection each week might not actually become plastic bottles at all. Instead, they might hydrolyze and become monomers, which would then be repolymerized into polyethylene terephthalate (PET), which is used to make a variety of products, including carpets. The plastic that is recycled the best is PET. The challenge of sorting different types of plastic is one of the issues with recycling plastic waste. Symbols are often noted on plastic packaging to aid in the separation. A list of international labeling numbers is provided below:

- 1dPET (polyethylene terephthalate)
- 2dHDPE (high-density polyethylene)
- 3dPVC (polyvinyl chloride) .
- 4dLDPE (low-density polyethylene)
- 5dPP (polypropylene) .
- 6dPS (polystyrene)

Water and beverage bottles are made of PET, of which 110 billion are produced annually; shampoo, milk, and freezer bags are made of HDPE; plastic bags and food packaging film are made of LDPE; bottle caps and plastic bags are made of PP; plastic cups and cutlery are made of PS; hot drink cups and protective packaging are made of expanded PS. It is difficult to recycle mixed plastic packaging (trays, tubs, and pots) composed of polyurethane and PS. Before being turned into new products, plastic that is collected for recycling is first sorted according to the type of polymer. These products may include food trays, drink bottles, wheel bins, flower pots, garbage sacks, carrier bags, and even polyester fabric for clothing.

Our plastic waste is mostly the result of supermarkets. Approximately 40% of all plastics produced are reportedly used as packaging materials. The thin plastic bag from the grocery store is, in fact, one of the emblems of the issue of plastic waste. Supermarket managers tell us that to keep produce fresh for consumers, plastic packaging is required. In light of this, a Dutch campaign group demanded at the beginning of 2018 that supermarkets start using packaging free of plastic. As a result, one supermarket, EkoPlaza in Amsterdam, recently established a plastic-free aisle with over 700 items, including snacks, meat, cereals, and sources. The aisle serves as a testing ground for cutting-edge new cardboard packaging and compostable biomaterials. This is a critical milestone in the effort to cut down on plastic waste. Paper bags will soon replace plastic bags at stores all over the world, and consumers will bring their own reusable containers for fruit, vegetables, cereal, pasta, coffee, and even wine. This is because the idea is becoming more and more popular.

There doesn't seem to be any other option to deal with the current waste backlog except to burn the remaining material (and hopefully recover the energy) while taking into account how polluting the fumes are. This is mostly due to the challenge of distinguishing between the various kinds of plastic. There is a strong desire to take action as a result of the media's recent coverage of plastic waste. For instance, the Indian government outlawed the production, use, storing, selling, importation, and transportation of numerous plastic products. The courts are being asked to review this hurried and reckless decision. Some governments have adopted a more cautious stance; in the UK, initiatives are underway to increase the 5p carrier bag fee and phase out "avoidable" plastic waste by 2042. The UK is also investigating how levies or fees can contribute to waste reduction and force manufacturers of plastic packaging to cover the cost of disposing of waste. Additionally, the UK government has established additional funding for studies on plastic innovation. According to a directive from the European Commission, by 2030, all plastic packaging manufactured in the EU needs to be recyclable and reusable. It lays out several steps to address plastic waste in its "Plastic Strategy," such as reducing the amount of single-use plastic produced and limiting the use of microplastics. Only 30% of the 25 million plastic tons produced annually in the European Union are recycled; the majority are burned or landfilled. 95% of plastic packaging in the US is never collected or recycled, according to estimates, and 86% of plastic packaging worldwide. PET is the most recycled plastic-type in the US, with 19% of all plastic bottles made from it being recycled. The next two types of polyethylene are low density (used to make flimsy plastic bags) and high density (10.3%), which are used to make gallon-sized liquid containers. Recyclability rates for other plastics are between 1% and less, including polypropylene, polystyrene (used in food packaging), and polyvinyl chloride (used in hose pipes).

Recent reports have brought attention to yet another major issue pertaining to plastic waste. Rich industrialized and even developing nations, like Malaysia, Turkey, Indonesia, and India, have been exporting their garbage to developing nations like these that have inadequate or nonexistent waste management. Waste is frequently dumped in massive open piles rather than being recycled or treated. Germany, the US, and Japan are the biggest export violators. A TV program that aired on BBC1 on June 10, 2019, brought attention to this issue by showcasing the mountains of plastic waste—some of which originated in the UK—that are accumulating in Malaysia and posing a health risk, particularly when burned in open fires for disposal. China prohibited the import of plastic waste in 2018, most likely due to these factors.

1.5.1 What to do?

Plastic is a material that is essential to our future even though it is not perfect. We have to find solutions to the issue because we are unable to go back in time. It would be very expensive and release a lot of CO_2 and other greenhouse gases to replace every plastic item with one made of an alternative material. It will take experts from a variety of fields, including chemistry, engineering, material sciences, and biology, to create new plastics that are both biodegradable and made from nonfossil feedstock.

Maybe this is a good place to start when it comes to cutting back on the amount of plastic we use, since packaging accounts for over a third of all plastic produced. Following the example set by the Dutch supermarket, a few UK supermarkets have recently introduced plastic-free aisles.

Many fundamental actions need to be taken, and these actions should serve as the foundation for governments and local authorities: reducing the amount of plastic waste produced by increasing plastic recycling and possibly limiting the kinds of plastic that can be produced; limiting or outright outlawing the use of mixed plastic or multilayer plastic products (e.g. simplify the use of plastic material); make sure that plastics are kept within a circular economy; put in place incentives with plastic producers and the public to lessen the environmental impacts of plastics; design packaging with recycling in mind; prevent plastic littering by putting in place incentive infrastructure; create tracing systems to make it easy to identify different types of plastic to ensure better recovery and recycling of plastics; educate the public about reducing plastic waste; and improve the facilities that local authorities provide for collecting plastic waste. In summary, it should be standard practice to reduce, redesign, reuse, and recycle.

Reduce the amount of single-use items you own, and make recycling plastics a top priority for all local government agencies. Reducing the quantity of plastic bottles—the majority of which are filled with water—that are sold each day could be a good place to start. In 2020, there will be half a trillion (or 0.5×10^{12}) plastic bottles sold worldwide, according to estimates. Most of these are single-use plastic bottles, and tap water is just as healthy for humans as expensive bottled water in the developed world. It will simply mean that rather than purchasing bottled water, people should bring a reusable bottle to be filled. This would significantly reduce the quantity of plastic waste that is thrown away.

Packaging materials need to be redesigned, and multilayer packaging needs to be abandoned, in order to include more recycled plastic. It would entail lowering the variety of plastics produced so that only recyclable plastics are used for packaging and the production of essential goods. Since they don't break down in water like toilet paper does, wet wipes containing plastic (polyester or polypropylene) have grown to be a significant source of pollution in recent years. They serve as pain relief, baby wipes, cleaning pads, industrial wipes, and pet care products. Toilet blockages are caused by wet wipes (over 90% of clogged toilets in the UK are reportedly caused by wet wipes). They are also to blame for the formation of fatbergs in sewers. Every year, more than 11 billion wet wipes are used in the UK.

Microplastics are a new plastic issue that has recently come to light. This is especially crucial for the oceans, where 73 percent of deepwater fish in the Northwest Atlantic have microplastics in their stomachs, according to a recent National University of Ireland study. Although microplastics have been found to absorb human pathogens (bacteria), little is known about whether or how microplastics impact the health of humans or animals. This book includes several chapters on microplastic. It is a good idea to hold plastic manufacturers and producers accountable for gathering and recycling plastic waste. This will probably lead to a decrease in the total amount of plastic produced globally and will also restrict the types of plastic produced since only certain types of plastic can be recycled.

Sun-induced degradation of plastic materials, etc. is a hot subject for research these days, with oxobiodegradation at the top of the list. In conclusion, oxobiodegradable plastic does decompose, albeit only into tiny pieces of microplastic substance. In addition, these microplastic fragments should eventually decompose more quickly than traditional nonbiodegradable plastic material, which would reduce the amount of plastic that ends up in mid-ocean. Due to their larger surface area, this process may, regrettably, take years or even decades to chemically break down into water, carbon dioxide, and possibly methane. There is currently no conclusive evidence based on the ongoing research into the detrimental effects of microplastic material on human health. Given the microplastic problem with oxoplastics, it is advisable to restrict the production of all plastics to recyclable varieties. Biodegradable and compostable plastics are likely too costly to produce and, in most cases, degrade very slowly into their constituent chemical elements. This would entail limiting the production of plastics to about six different varieties, all of which are recyclable. So, mixed plastic products are no longer an option.

All of this indicates that perhaps there is a case to use oxoplastics in the interim period before new legislation and manufacturing processes come into play. There are many things that each one of us could do to make a difference:

- Always carry a reusable bag, coffee mug, and bottle
- Give up buying bottled water, use a reusable bottle filled with tap water, and persuade authorities to have filling points in public places.
- Reuse plastic bags that they can be washed, dried, and reused
- Avoid excessive food packaged food, where possible, from shops that sell products free of plastic packaging

- Purchase food from bulk bins and fill your own reusable container
- Stop using plastic drinking straws
- Avoid buying frozen food as the packaging is usually plastic
- Eat fresh fruit and squeeze fresh juice and do not buy juice that usually comes in plastic bottles
- Pack picnic lunches and school lunches in reusable containers
- Do not use disposable plastic cutlery
- Avoid items such as cosmetics that contain microbeads
- Petition governments to (a) limit the types of plastic made, (b) ban exporting plastic waste, (c) burn plastic waste, if really necessary, in plastic-to-energy plants fitted with filters to polluting gases and particulates, and (d) ensure no plastic is landfilled.
- Do not purchase or use wet wipes.
- Do not accept free plastic "kids" toys from retailers or food outlets.

(Van Cauwenberghe et al, 2015; Shen et al, 2020)

REFERENCES

1. Rivard, C., Moens, L., Roberts, K. et al., (1995). Starch esters as biodegradable plastics: Effects of ester group chain length and degree of substitution on anaerobic biodegradation. Enzyme Microb. Technol. 17: 848–852.
2. Ageyeva, T., Sibikin, I., and Kovacs, J. (2019). A review of thermoplastic resin transfer molding: Process modeling and simulation. Polym. J. 11: 1555.
3. Alam, M., Kaur, J., Khaira, H., and Gupta, K. (2015). Extrusion and extruded products: Changes in quality attributes as affected by extrusion process parameters: A review. Crit. Rev. Food Sci. Nutr. 56: 445–473.
4. Andrady, A.L. and Neal, M.A. (2009). Applications and Societal Benefits of Plastics. Philos. Trans. R. Soc. B: Biol. Sci. 364: 1977–1984.
5. Appiah, J.K., Berko-Boateng, V.N., Tagbor, T. A. (2017). Use of waste plastic materials for road construction in Ghana. Case Studies in Construction Materials, 6, 1-7.
6. Asgher, M., Afzal, M., Qamar, S.A., and Khalid, N. (2020). Optimization of biosurfactant production from chemically mutated strain of Bacillus subtilis using waste automobile oil as low-cost substrate. J. Environ. Sustain. 3: 405–413.
7. Asgher, M., Afzal, M., Qamar, S.A., and Khalid, N. (2020). Optimization of biosurfactant production from chemically mutated strain of Bacillus subtilis using waste automobile oil as low-cost substrate. J. Environ. Sustain. 3: 405–413.
8. Balogun Ayodeji, T. (2019). Pyrolysis of waste plastics: a case study of polyethylene. International Journal of Advanced Academic Research Sciences, Technology and Technology, 5(3), 1-16.
9. Bonhomme, S., Cuer, A., Delort, A.M., Lemaire, J., Sancelme, M., Scott, G. (2003). Environmental biodegradation of polyethylene. Polymer Degradation and Stability, 81, 441-452.
10. Brunner, I., Fischer, M., Rüthi, J., Stierli, B., Frey, B. (2018). Ability of fungi isolated from plastic debris floating in the shoreline of a lake to degrade plastics. PLoS ONE 13(8): e0202047.
11. Cai, Y., Mitrano, D.M., Heuberger, M. (2020). The origin of microplastic fiber in polyester textiles: The textile production process matters. J. Clean. Prod. 267: 121970.
12. Chai, B., Wei, Q., She, Y. (2020). Soil microplastic pollution in an e-waste dismantling zone of China. Waste Manag. 118: 291–301.

13. Charles, W., Walker, L., and Cord-Ruwisch, R. (2009). Effect of pre-aeration and inoculum on the start-up of batch thermophilic anaerobic digestion of municipal solid waste. Bioresour. Technol. 100: 2329–2335.
14. Chen, G., Fu, Z., Yang, H., and Wang, J. (2020). An overview of analytical methods for detecting microplastics in the atmosphere. TrAC Trends Anal. Chem. 130: 115981.
15. Coyle, R., Hardiman, G., and Driscoll, K.O. (2020). Microplastics in the marine environment: A review of their sources, distribution processes, uptake and exchange in ecosystems. Case Stud. Chem. Environ. Eng 2: 100010.
16. Das, A. and Mahanwar, P. (2020). A brief discussion on advances in polyurethane applications. Adv. Ind. Eng. Polym. Res 3: 93–101.
17. Evangeliou, N., Grythe, H., Klimont, Z.(2020). Atmospheric transport is a major pathway of microplastics to remote regions. Nat. Commun. 11: 1–11.
18. Evode, N., Qamar, S.A., Bilal, M. (2021). Plastic waste and its management strategies for environmental sustainability. Case Stud. Chem. Environ. Eng 4: 100142.
19. Fahrenfeld, N.L., Arbuckle-Keil, G., Naderi Beni, N., and Bartelt-Hunt, S.L. (2019). Source tracking microplastics in the freshwater environment. TrAC - Trends Anal. Chem. 112: 248–254.
20. Geyer, R., Jambeck, J. R., & Law, K. L. (2017). Production, use, and fate of all plastics ever made. Science Advances, 3(7), e1700782.
21. Gramentz, D. (1988). Involvement of loggerhead turtle with the plastic, metal, and hydrocarbon pollution in the Central Mediterranean Marine Pollution Bulletin, 19 (1988), 11-13.
22. Harris, M.E. and Walker, B. (2010). A novel, simplified scheme for plastics identification: JCE classroom activity 104. J. Chem. Educ. 87: 147–149.
23. Herbort, A.F., Sturm, M.T., Fiedler, S. (2018). Alkoxy-silyl induced agglomeration: A new approach for the sustainable removal of microplastic from aquatic systems. J. Polym. Environ. 26: 4258 4270.
24. Hirai, H., Takada, H., Ogata, Y., (2011). Organic micropollutants in marine plastics debris from the open ocean and remote and urban beaches. Mar. Pollut. Bull. 62: 1683–1692.
25. J. Thornton, "Environmental impacts of polyvinyl chloride (PVC) building materials," A Healthy Building Network Report, Washington D.C; 2002.
26. Jambeck, J. R., Geyer, R., Wilcox, C., Siegler, T. R., Perryman, M., Andrady, A., Narayan, N., & Law, K. L. (2015). Plastic waste inputs from

land into the ocean. Science, 347(6223), 768-771.

27. Järlskog, I., Strömvall, A.-M., Magnusson, K. et al., (2020). Occurrence of tire and bitumen wear microplastics on urban streets and in sweepsand and washwater. Sci. Total Environ. 729: 138950.
28. jha N, Pradhan N, Singh S, Barla A, Shrivastava A, Khatua P., Rai, V., Bose, S. (2017). Evaluation of HDPE and LDPE degradation by fungus, implemented by statistical optimization. Scientific Reports, 7: 39515.
29. Kathiresan, K. (2003). Polythene and Plastics-degrading microbes from the mangrove soil. Revista de Biología Tropical, 51(3-4), 629-633.
30. Kathiresan, K. (2003). Polythene and Plastics-degrading microbes from the mangrove soil. Revista de Biología Tropical, 51(3-4), 629-633.
31. Laglbauer, B.J.L., Franco-Santos, R.M., Andreu-Cazenave, M. et al., (2014). Macrodebris and microplastics from beaches in Slovenia. Mar. Pollut. Bull. 89: 356–366.
32. Laist, D.W. (1997). Impacts of marine debris: Entanglement of marine life in marine debris including a comprehensive list of species with entanglement and ingestion records, pp. 99-140 in: J.M. Coe and D.B. Rogers (eds.), Marine debris: Sources, impacts, and solutions. Springer-Verlag. New York, NY.
33. Landsecker, K. and Bonten, C. (2019). Thermoforming simulation of heat conductive plastic materials using the K-BKZ model. AIP Conf. Proceed. 030049.
34. Lebreton, L. C., Van der Zwet, J., Damsteeg, J. W., Slat, B., Andrady, A., & Reisser, J. (2017). River plastic emissions to the world's oceans. Nature Communications, 8, 15611.
35. Lebreton, L., Slat, B., Ferrari, F., Sainte-Rose, B., Aitken, J., Marthouse, R., & Noble, K. (2018). Evidence that the Great Pacific Garbage Patch is rapidly accumulating plastic. Scientific Reports, 8(1), 4666.
36. Lefebvre, C., Saraux, C., Heitz, O. et al., (2019). Microplastics FTIR characterisation and distribution in the water column and digestive tracts of small pelagic fish in the Gulf of Lions. Mar. Pollut. Bull. 142: 510–519.
37. Lerchner, J., Wolf, A., Buchholz, F., Mertens, F., Neu, T.R., Harms, H., Maskow, T. (2008) Miniaturized calorimetry-A new method for real-time biofilm activity analysis. Journal of Microbiological Methods, 74, 74-81.
38. Liu, Y., Zhou, C., Li, F. et al., (2020). Stocks and flows of polyvinyl chloride (PVC) in China: 1980-2050. Resour. Conserv. Recycl. 154: 104584.
39. Lucas, N., Bieniame, C., Belloy, C., Queneudec, M., Silvestre, F., Nava-Saucedo, J.E.. (2008). Polymer biodegradation: mechanisms and estimation techniques. Chemosphere, 73, 429-442.

40. Mendible, G., Rulander, J., and Johnston, S. (2017). Comparative study of rapid and conventional tooling for plastics injection molding. Rapid Prototyp. J. 23: 344–352.
41. Moser, M.L., Lee, D.S. (1992). A fourteen-year survey of plastic ingestion by western North Atlantic seabirds. Colonial Waterbirds, 15 (1992), 83-94.
42. Mukhamediev, M.G. and Bekchanov, D.Z. (2019). New anion exchanger based on polyvinyl chloride and its application in industrial water treatment. Russ. J. Appl. Chem. 92: 1499–1505.
43. Ngo, P.L., Pramanik, B.K., Shah, K., and Roychand, R. (2019). Pathway, classification and removal efficiency of microplastics in wastewater treatment plants. Environ. Pollut. 255: 113326.
44. Oehlmann J., Schulte-Oehlmann, U., Kloas, W., Jagnytsch, O., Lutz, I., Kusk, O., Wollenberger, L., Santos, E.M., Paull,G., Van Look, K., Tyler, CR. (2009). Critical analysis of the biological impacts of plasticizers on wildlife. Phil. Trans. R. Soc. B 364, 2047–2062.
45. Ogila, K., Shao, M., Yang, W., and Tan, J. (2017). Rotational molding: A review of the models and materials. EXPRESS Polym. Lett. 11: 778–798.
46. Prata, J.C., da Costa, J.P., Duarte, A.C., and Rocha-Santos, T. (2019). Methods for sampling and detection of microplastics in water and sediment: a critical review. TrAC - Trends Anal. Chem. 110: 150–159.
47. Qamar, S.A., Ashiq, M., Jahangeer, M. et al., (2020). Chitosan-based hybrid materials as adsorbents for textile dyes–A review. Case Stud. Chem. Environ. Eng. 2: 100021.
48. Raziyafathima, M., Praseetha, P. K., Rimal Isaac R.S. (2016). Microbial Degradation of Plastic Waste: A Review. J Pharm Chem Biol Sci, 4(2), 231-242.
49. Rhein, S. and Schmid, M. (2020). Consumers' awareness of plastic packaging: More than just environmental concerns. Resour. Conserv. Recycl. 162: 105063.
50. Rochman, C.M., Brookson, C., Bikker, J. et al., (2019). Rethinking microplastics as a diverse contaminant suite. Environ. Toxicol. Chem. 38: 703–711.
51. Romo-Uribe, A. and Lichtenhan, J.D. (2021). Melt extrusion and blow molding parts-per-million POSS interspersed the macromolecular network and simultaneously enhanced thermomechanical and barrier properties of polyolefin films. Polym. Eng. Sci. 61: 245–257.
52. Seneviranlne, G., Tennekoon, N.S., Weerasekara, MLMAW, Nandasena

K.A. (2006). Polyethylene biodegradation by a developed Penicillium-Bacillus biofilm. Current Science, 90, 20 – 21.

53. Shah A.A., Hasan, F., Hameed, A., Ahmed, S. (2008). Biological degradation of plastics: a comprehensive review. Biotechnol. Adv., 26 (3), 246-265.
54. Sharma, M., Dubey, A., Pareek, A. (2014). Algal flora on degrading polythene waste. CIBTech Journal of Microbiology, 3, 43-47.
55. Shen, L., Haufe, J.I., and Patel, M. (2009). Product overview and market projection of emerging bio-based plastics-PRO-BIP 2009. In: Report for European Polysaccharide Network of Excellence (EPNOE) and European Bioplastics, vol. Vol. 243, 1–245.
56. Shen, M., Huang, W., Chen, M. et al., (2020). (Micro)plastic crisis: Un-ignorable contribution to global greenhouse gas emissions and climate change. J. Clean. Prod. 254: 120138.
57. Sun, J., Dai, X., Wang, Q. et al., (2019). Microplastics in wastewater treatment plants: Detection, occurrence and removal. Water Res. 152: 21–37.
58. umar, R.V., Kanna, G.R., Elumalai, S. (2017). Biodegradation of Polyethylene by Green Photosynthetic Microalgae. J Bioremediat Biodegrad., 8: 381.
59. Urbanek, A.K., Rymowicz, W., Mironczuk, A. M. (2018). Degradation of plastics and plastic degrading bacteria in cold marine habitats. Appl. Microbiol. Biotechnol., 102(18), 7669-7678.
60. Van Cauwenberghe, L., Devriese, L., Galgani, F. et al., (2015). Microplastics in sediments: A review of techniques, occurrence and effects. Mar. Environ. Res. 111: 5–17.
61. Vinayagamoorthy, R. and Rajmohan, T. (2018). Machining and its challenges on bio-fibre reinforced plastics: A critical review. J. Reinf. Plast. Compos. 37: 1037–1050.
62. Wang, F., Pan, S., Zhang, P. et al., (2018). Synthesis and application of phosphorus- containing flame retardant plasticizer for polyvinyl chloride. Fibers Polym. 19: 1057–1063.
63. Wright, S.L., Thompson, R.C., Galloway, T.S. (2013). The physical impacts of microplastics on marine organisms: a review. Environ. Pollut. 178, 483–492.
64. Yang, L., Zhang, Y., Kang, S. et al., (2021). Microplastics in freshwater sediment: A review on methods, occurrence, and sources. Sci. Total Environ. 754: 141948.

65. Zettler, E.R., Mincer, T.J., Amaral-Zettler, L.A. (2013). Life in the plastisphere: microbial communities on plastic marine debris. Environmental Science and Technology, 47, 7137-7146.

66. Zobkov, M., Esiukova, E., Zyubin, A., and Samusev, I. (2019). Microplastic content variation in water column: The observations employing a novel sampling tool in stratified Baltic Sea. Mar. Pollut. Bull. 138: 193–205.

Chapter 2

Recovery and Recycling of Polymeric and Plastic Materials

Contents

INTRODUCTION

The recovery and recycling of polymeric and plastic materials have become increasingly critical in addressing the global environmental challenges posed by plastic waste. With the rapid growth of the plastic industry over the past century, plastics have become an integral part of modern life, used in everything from packaging and construction to electronics and medical devices. However, the durability and widespread use of plastics have also led to significant environmental issues, including pollution, resource depletion, and the accumulation of non-biodegradable waste in landfills and oceans.

The recycling of plastics is a key strategy in mitigating these environmental impacts. By recovering and reprocessing used plastic materials, the need for virgin plastic production is reduced, conserving natural resources such as petroleum, which is the primary raw material for most plastics. Recycling also helps to decrease greenhouse gas emissions associated with plastic manufacturing and reduces the volume of plastic waste that ends up in landfills or the natural environment, where it can persist for hundreds of years.

There are various methods for recycling plastic materials, including mechanical recycling, chemical recycling, and energy recovery. Mechanical recycling involves the physical processing of plastics to produce new materials, while chemical recycling breaks down plastics into their chemical components for reuse in manufacturing. Energy recovery, although less preferable, involves converting waste plastics into energy through incineration.

Despite the clear benefits, the recycling of plastics faces several challenges, such as the complexity of plastic waste streams, contamination, and economic viability. These issues necessitate ongoing research and innovation to improve recycling technologies and processes, as well as the implementation of policies and practices that encourage higher recycling rates and the development of sustainable materials. Advancing the recovery and recycling of polymeric and plastic materials is essential for reducing environmental impact and promoting a circular economy.

2.1 RECOVERY OF PLASTICS

In terms of applications, the three main pillars of modern human civilization are metals, polymers, and cements. Plastics have many uses because they are resource-efficient, non-toxic, and aesthetically pleasing. Even though plastics were first produced on a large scale in the 1950s, plastics are now an indispensable part of modern life. Strong, light, and moldable, plastic materials can be used to create a wide range of products, including toys, mobile phones, protective packaging for different foods, product packaging for consumer and industrial goods, lightweight and safe parts for cars and aircraft, medical devices, personal protective equipment, building insulation, materials for office and home furniture, pipeline transportation, and many more uses. Polyvinyl chloride (PVC), Polystyrene (PS), Polyethylene Terephthalate (PET), Polyethylene (PE), Low Density (LDPE), Linear Low Density (LLDPE), High Density (HDPE), Polypropylene (PP), and Polytetrafluoroethylene (PTFE or Teflon) are common plastics.

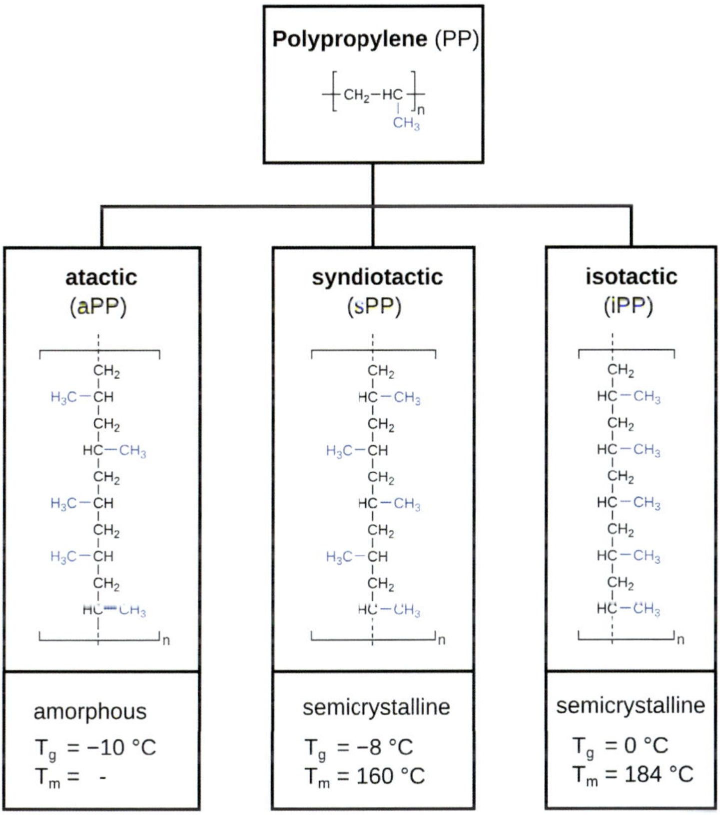

Figure 2.1. Chemical structures of common plastics.

Source. By User:Minihaa. (Adapted to English by User:Vis M) - File:Polypropylene_tacticity_de.svg, CC0, https://commons.wikimedia.org/w/index.php?curid=131584141

Plastics have revolutionized living standards by replacing metal, wood, and cement with their lightweight, highly durable alternatives. By swapping out metallic and wooden parts, transportation by rail, road, and air is now lighter and more fuel-efficient, reducing the need for non-renewable fossil fuels. Other significant contributions made by plastics include the creation of high-density data storage systems, laser systems for accurate medical procedures, plastic tubing to protect optical fibers, which enables the rapid transfer of large amounts of data, and compact disks to enhance recording quality for the music industry. Energy savings have resulted from laptop and desktop computer weight and size reductions made possible by the use of plastics. The unique requirements of electrical and electronic applications can be satisfied by the use of plastics. According to Fisher et al (2004)., refrigerators with plastic foam insulation can save 17 times as much energy. The production of plastics increases the demand for renewable polysaccharide-based feedstocks and hydrocarbon resources, resulting in a rise in the amount of solid plastic waste (SPW) on Earth.

2.1.1 Two Major Types of Plastics

- *Thermoplastics* which can be molded more than once with the application of heat: PVC, PE, LDPE, LLDPE, HDPE, PP, PET, PS, PTFE, High-impact polystyrene (HIPS), Acrylonitrile–butadiene–styrene (ABS), Styrene–acrylonitrile copolymer (SAN), Polymethyl methacrylate (PMMA), Thermoplastic elastomer (TPE), Polyoxymethylene (POM), Polybutylene terephthalate (PBT), Polycarbonate (PC), Polyamide (PA), Ethylene–vinyl alcohol copolymer (EVOH), Copolymer from ethylene and vinyl acetate (EVA), Polyphenylene oxide (PPO), Polyphenylenesulfide (PPS), etc., are widely used.

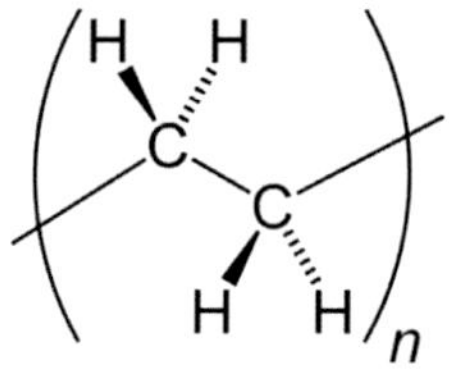

Figure 2.2. Chain structure of polyethylene.

Source. https://en.m.wikipedia.org/wiki/File:Polyethylene-repeat-2D.png

- *Thermosets* which can be molded only once: Phenol–formaldehyde resin (PF), Melamine–formaldehyde resin (MF), Urea–formaldehyde resin (UF), Unsaturated Polyester (UP), Epoxy, Vinyl Esters, Silicone, Acrylic Resins, etc., are widely used. Engineering plastics like ABS, PC, PA, PET, polysulphone (PSU), polyetherketone (PEK), polyetheretherketone (PEEK) and polyimides (PI) have high heat resistance, rigidity, chemical stability, flame resistance with high mechanical resistance. The above materials are widely used in electrical and electronic equipment due to the above properties. Again, fast-changing technologies in the above two areas, a huge amount of disposable materials are produced (Tarantili *et al.*,, 2010). Engineering plastics market value is predicted to reach above $90 billion by 2020 (Chauhan *et al.*,, 2019). In 2018, worldwide production of plastics crossed 359 million metric tons, where the production from Europe was 62 million metric tons. China produces nearly more than one-fourth of the global production of plastics with a steady growth of its imports to the United States (Plastics Europe, 2019).

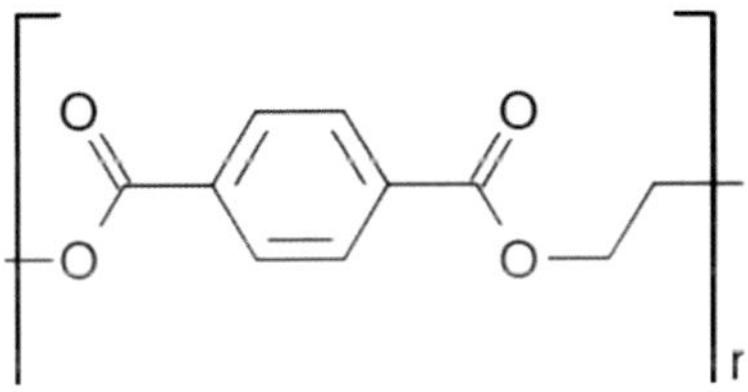

Figure 2.3. Repeating unit of PET.

Source: Di Lorenzo ML. Crystallization of Poly(ethylene terephthalate): A Review. Polymers. 2024; 16(14):1975. https://doi.org/10.3390/polym16141975

Advantages and Disadvantages of Plastic Materials (Siddique *et al.*,, 2008)

Advantages:

- Plastics can meet specific technical needs and have many applications.
- Low density of the plastics helps in reducing the cost of transportation.
- Safe and hygienic for food packaging.
- Durable and long-lasting.
- Water, chemical and impact resistant.
- Non-conductive to heat and electricity
- Production cost is comparatively cheaper than other materials for similar use.
- May be used in combination with aluminum foil, paper and adhesives.
- Excellent esthetic appeal.

Disadvantages:

- Plastic manufacturing uses potentially harmful chemicals like colorizing materials.

- Phthalates may be released when plastic toys are sucked by children.
- The presence of toxic chemicals and exposure to plastics may result in birth defects, tumors, cancer and immune system disorders.
- Thermosetting plastics can be molded only once.
- Widely used plastic items such as food packaging become waste within a short time after purchase.
- Most of the plastics are non-degradable, long time is required to degrade, which may take up to hundreds of years.
- The disposal of plastics may have environmental impacts on plankton through to whales. Plastics need to undergo an environmental risk assessment before their use in some cases.
- Space requirement for landfills by plastic waste is a major environmental problem.

2.1.2 Regeneration and Recovery of Plastics

Regeneration and plastic recovery are more desirable for the following reasons, in addition to lowering the quantity of plastic waste that needs to be disposed of:

- Conservation of non-renewable fossil fuels: plastic production uses 8% of the world's oil production, 4% as feedstock and 4% during manufacture.
- Reduced consumption of energy.
- Reduced amounts of solid waste going to landfill can also reduce the municipal solid wastes.
- Reduced emissions of carbon dioxide (CO_2), nitrogen oxide (NO) and sulfur dioxide (SO_2).

Regeneration of plastics refers to the conversion of plastics via the synthesis of new petrochemicals, oligomers, or monomers. Recovery is the process of extracting valuable materials from waste for further use or processing. It involves gathering, classifying, and combining different materials. The most common uses of pyrolysis techniques are in the production of synthesis gas through partial oxidation or gasification. With the development of contemporary technologies, the exchequer may be able to save a significant sum of money through the collection and separation of different waste kinds, appropriate regeneration, and recovery of plastics. It is impossible to overlook the direct economic impact that recovered and regenerated plastics have.

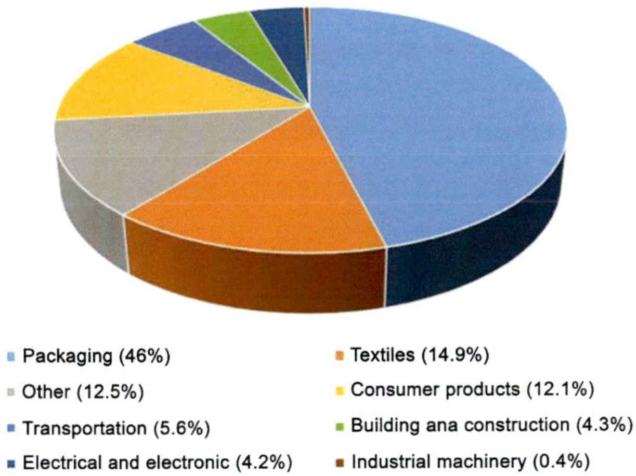

Figure 2.4. Waste hierarchy.

Source: Macheca AD, Mutuma B, Adalima JL, Midheme E, Lúcas LHM, Ochanda VK, Mhlanga SD. Perspectives on Plastic Waste Management: Challenges and Possible Solutions to Ensure Its Sustainable Use. Recycling. 2024; 9(5):77. https://doi.org/10.3390/recycling9050077

Currently, after just one use, almost 95% of plastic packaging is thrown away. It takes a lot of work to separate plastic from non-plastic contaminants like paper, glass, metal, cloth, and wood because plastic waste is typically contaminated by these materials. Using separate bins for plastics, papers, metals, and food waste will help solve the issue. Apart from the aforementioned, heterogeneous polymeric components are only possible to find in plastic waste. The amount of plastic solid waste can be decreased by using newly developed processes like direct incineration, which produces monomers through one or two stages of combustion. These procedures lessen the amount of gas and oil used in the process of making virgin plastics (Al-Salem et al., 2009, Singh et al.,2017).

When direct mechanical crush methods are used for recovery, physical methods do not alter the chemical structures and properties of the plastics. Plastics are broken down chemically or undergo pyrolysis to yield monomers as regenerated products during chemical recovery processes. Landfilling is the more convenient option because chemical methods are more expensive. Plastics are also burned frequently, although this practice is outlawed in many locations due to environmental pollution. The regeneration of thermosetting phenolic plastics is accomplished through mechanochemical techniques. According to Wu et al., (2014), using the aforementioned method results in the production of reactive groups following the breakdown of the molecular structure, a decrease in crosslinking density, and an improvement in the plasticity of the regenerated products. It has been discovered that the generation of monomers can efficiently regenerate plastics. Chemical regeneration is one method of producing the same parent plastics from different pathways of plastic waste regeneration. Additionally, this procedure may lessen the market's need for plastic feedstocks (Oku, 2005). Numerous plastic packaging items may contain identification codes, as depicted in the following figure:

Figure 2.5. Packaging codes on plastics.

Source: By Z22 - Own work, CC BY-SA 3.0, https://commons.wikimedia.org/w/index.php?curid=33928886

The quality of products made from mixed plastics is significantly decreased due to differences in melting and processing temperatures as well as incompatibility.

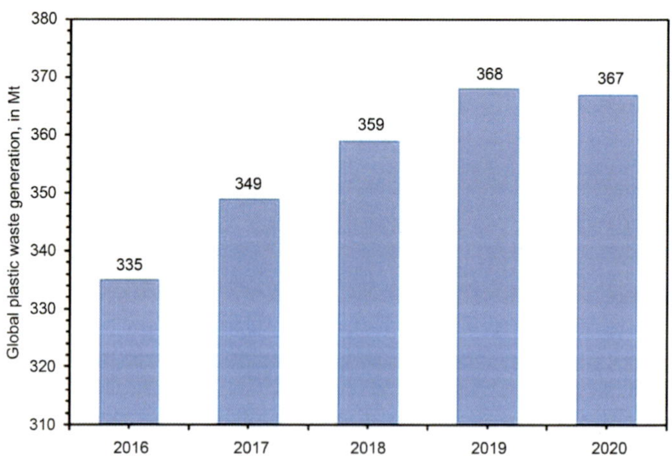

Figure 2.6. Waste generated.

Source. The global percentage of the top five companies' branded plastic waste generated throughout their value chain.

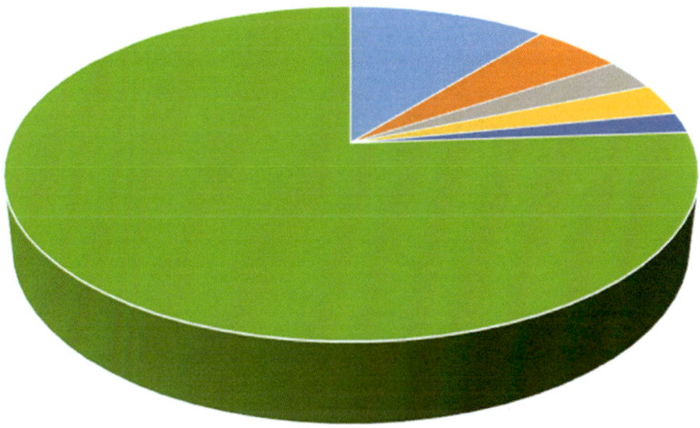

Figure 2.7. The global percentage of the top five companies' branded plastic waste generated throughout their value chain.

Source: Macheca AD, Mutuma B, Adalima JL, Midheme E, Lúcas LHM, Ochanda VK, Mhlanga SD. Perspectives on Plastic Waste Management: Challenges and Possible Solutions to Ensure Its Sustainable Use. Recycling. 2024; 9(5):77. https://doi.org/10.3390/recycling9050077

Table 2.1. *Case study: plastic waste dumping in Indonesia as reported by Phelan et al, 2020; PMCID: PMC7386615*

Villages studied—Selayar.

		Total Population	**Male**	**Female**	**Total Households**
	Bontosikuyu sub-district	**15,170**	**7,381**	**7,789**	**3,848**
No.	**Villages**				
1.	Bahuluan	318	157	161	84
2.	Tambalongan	1328	654	674	358
3.	Appatanah	870	442	428	207

Villages studied—Wakatobi.

		Total Population	**Male**	**Female**	**Total Households**
	Kaledupa sub-district	**10,964**			**2,847**
No.	**Villages**				
1.	Sombano	930	500	430	170
2.	Balasuna	930	480	450	260
3.	Mantigola	935	465	470	350
4.	Lewuto	640	350	290	180
5.	Langge	1150	545	605	288
6.	Tampara	1076	542	534	244
7.	Kaswari	681	336	345	157

Sample characteristics.

	n	**% of sample**
Total Sample	**473**	-
Female	263	56%
Male	210	44%
Education–Primary and Middle School	269	57%
Education–High School	134	28%
Education–University	70	15%
Long term residents of the region (+15 years)	421	89%
Fishers and/or farmers	194	41%
Housewife	132	28%

Individual knowledge scale questions with percent of correct responses.

Single response questions	Correct answer(s)	N	% correct		
1. Organic waste (e.g. food waste, plant litter) thrown on the ground will quickly break down and disappear (become part of the soil).	Agree	473	335 (70.8%)		
2. Snack food wrappers and other plastic packaging thrown on the ground will quickly break down or disappear.	Disagree	472	300 (63.6%)		

3. The closest recycling facility (plastic processing factory) is located in?	Makassar	473	21 (4.4%)		
4. Do fish and other marine animals eat plastic waste?	Yes	471	55 (11.7%)		
5. What effect does plastic waste have on the environment?	Negative effect	471	273 (58.0%)		
6. Does burning rubbish, including plastic, affect human health?	Yes	471	350 (74.3%)		
7. Rubbish left on the ground will eventually make its way into the ocean.	Agree	470	223 (47.4%)		
8. In the ocean, how long does the plastic that makes up a plastic bag last?	Up to 20 years	469	143 (30.5%)		
9. In the ocean, how long does the plastic that makes up a plastic bottle last?	UP to 400 years	469	66 (14.1%)		
10. In the ocean, how long does discarded fishing line does last?	Up to 600 years	467	59 (12.6%)		
11. Have you heard about 'microplastics'—tiny pieces of plastic floating in the ocean?	Yes	470	51 (10.9%)		
Multi-response questions		N	1 correct	2 correct	3 correct
12. Which of the following items is ok to throw (discard) on the ground?	Food waste, waste paper	473	238 (50.3%)	79 (16.7%)	NA
13. Which of the following items can be re-processed in a factory?	Plastic water bottle/ sealed cup, coca-cola can, glass bottle	473	223 (47.1%)	94 (19.9%)	87 (18.4%)

14. Do any of the following types of rubbish affect marine life?	Plastic litter, cigarette butts, discarded fishing line	473	315 (66.6%)	42 (8.9%)	6 (1.3%)

Summary of focus group participants and summary statistics.

No	Location	Group	No. of participants
1	Bahuluang1	female	12
2	Bahuluang2	male	10
3	Tambolongan1	female	12
4	Tambolongan2	male	12
5	Appatanah–mixed	female	5
		male	7
6	Tampara1	female	12
7	Tampara2	male	10
8	Sombano1	female	10
9	Sombano2	male	10
10	Mantigola1	female	12
11	Mantigola2	male	10
	Total		122
	female		63
	male		59
	Summary statistics		
	Gender:		
	Female (n = 63): 54.64%		
	Male (n = 59): 48.36%		
	Age:		
	Range: 21–62; Mean = 35		
	Education: >50% finished primary school		
	Occupations:		
	Female: Over 75% housewives; others were local government officers, fishers, kiosk owners, or fish traders.		
	Male: Over 50% fishers; others were local government officers, farmers, kiosk owners or small business owners		

Estimated household level waste flows per week.

	kg per week
Average waste from a typical household (4–5 people) based on 2kg/day per household	14
Waste from recreational activities produced by members of household	1.125
Waste from community gatherings, weddings and other special events; Aggregated per household	1.125
Fishing and boat travel–estimated per household	1.125
Total household waste	17.375 kg/household/week

Village activities contributing to plastic in the ocean.

Boat travel (10 groups)	Participants explained that as the communities live on small islands, the need to commute between islands is unavoidable. Increasing household income, which makes travelling affordable and increases the number of children able to attend high school on the main island, and increasing population, are among factors affecting the frequency of boat travel. Many passengers bring food and drinks for the trip and dump their plastic packaging into the ocean.
Dumping household rubbish directly into the ocean (10 groups)	This is commonly practised by many members of coastal communities. The amount of plastic wastes continues to increase as the population increases. Participants also mentioned that rubbish brought in by the currents made them less hesitant to dump their household rubbish in the ocean.
Fishing activities (7 groups)	Discarded or lost fishing lines, nets, and other broken fishing gear also contribute to the amount of plastic in the ocean.
Growth in the number of kiosks (6 groups)	Population growth creates more demand for snacks and drinks, leading to more kiosks opening.
Seasonal rubbish washed up from the ocean (6 groups)	Although this phenomenon occurs mainly in the monsoon months, the amount of the rubbish is overwhelming and many orders of magnitude larger than that produced by the village. Accounts describe it as a 'landfill dumped into the ocean'. Rubbish remains on the beaches year after year, or is washed back into the water, if not addressed.

Community gatherings and meetings (3 groups).	Growing population means more community gatherings such as betrothals, weddings, and picnics, and meetings, at which water is commonly served in single use plastic cups or bottles.
Seaweed farming (2 groups in one village).	The people in one village in Wakatobi are mostly seaweed farmers. Seaweed farming uses floats made from fragments of polystyrene and used plastic bottles, which can break away and remain in the ocean.

Source: Phelan AA, Ross H, Setianto NA, Fielding K, Pradipta L. Ocean plastic crisis-Mental models of plastic pollution from remote Indonesian coastal communities. PLoS One. 2020 Jul 28;15(7):e0236149. doi: 10.1371/journal.pone.0236149. PMID: 32722705; PMCID: PMC7386615.

In order to be treated, 29.1 million tonnes of plastic waste were collected in the EU28+NO/CH in 2018. Exports of plastic waste outside the EU fell by 39% between 2016 and 2018.

2.1.3 Different Paths for Plastic Recovery

In the plastics industry, SPW recycling, treatment, and recovery are fairly common. Re-extrusion is the primary recycling method; the advantages and disadvantages of mechanical recycling have been discussed for this secondary method. It has been noted that in order to maximize the recovery of plastic waste, the solid plastic waste used in mechanical recycling processes must possess qualities comparable to commercial-grade plastics. The more environmentally friendly tertiary chemical or thermo-chemical treatments of waste plastic are pyrolysis-based processes that yield monomers or feedstocks. One of the energy recovery processes is the quaternary treatment technologies, which are the incineration or combustion technologies that burn SPW in addition to burning municipal solid waste (Ragaert et al., 2017; Singh et al, 2017).

- ***Primary recovery*:** This is the reprocessing of the scraps and by-products of the same plastic material produced during the processing of finished products or extrusion process. This method of recycling is also known as re-extrusion process.
- ***Secondary recovery or mechanical recovery*:** Reprocessing scrap plastics by physical means into plastic recyclates or products. This process was commercialized in 1970s. In this process, the SPW is converted into powders, or flakes, or pellets resulting in volume reduction which depends on the components present in the SPW. This process also involves size reduction (pellets, flakes or powders), contaminant separation (separation of paper, metals and dusts using

cyclone), floating (separation based on density differences in a floating tank) and finally washing, drying and palletization. This process is the best for a single component SPW like industrial scraps and homogeneous post-consumer plastics recovery. Mechanical recovery mostly follows the methods of recovery of plastics where no change in chemical structures and properties takes place and plastics are used directly after mechanical crush or adhesive bonding. Besides low operational costs and easy operational methods, the mechanical recovery rate is low with limited applications (Al-Salem et al.,, 2009; Hopewell et al.,, 2009; Gu et al.,, 2017; et al.,Ragaert et al.,, 2017, Singh et al.,, 2017et al.,).

- ***Tertiary recovery or chemical feedstock recovery:*** Also known as chemical recycling where the chemical structures of the SPW components are depolymerized by pyrolysis to produce oil and gas fractions in the presence of air at high temperatures (Ragaert *et al.,*, 2017). Catalytic cracking, steam degradation and liquid–gas hydrogenation are also commonly used in tertiary recycling of SPW for the production of various monomers. Condensation-polymerized plastics like nylons polyesters, and polyurethanes are suitably converted to monomers by glycolysis, methanolysis, and hydrolysis processes. Whereas the addition of polymerized plastics like vinyls, acrylics, fluoroplastics, and polyolefins are difficult to reprocess. These wastes may be sorted out and mixed with respective virgin resins after they are powdered and blended with virgin plastics for remolding into finished goods or using suitable compatibilizers to make recycled products. Fluidized bed pyrolysis provides excellent heat and mass transfer between the particles and has been found to be most suitable for the chemical processing of the mixed plastic wastes. It has been reported that using fluidized bed pyrolysis, 25%–45% of product gas with a high heating value and 30%–50% of an oil rich in aromatics could be recovered (Kaminsky, 1995). Compared with virgin polyester, chemically recovered fibers provide lower desired physicochemical characteristics. Although, chemically recycled fibers can be applied in a wider range of applications than mechanically recycled fibers, mechanically recycled plastics provide better environmental impacts than chemically recycled plastics (Shen *et al.,*, 2010).
- ***Quaternary recovery or process engineered fuel*:** This process consists of the conversion of pre-consumer or post-consumer plastics into an industrial fuel that meets market specifications. When all

the material recovery processes from the SPW fail due to economic reasons, they are burned to produce heat energy converted to hot water, steam and electricity. During the thermal recycling of plastics, energy derived from plastics along with other high-caloric waste streams is used as fuel. Thermal decomposition is a costlier process with environmental pollution when some toxic gaseous components are liberated into the atmosphere (Al-Salem *et al.*,, 2009).

- ***Landfilling with waste plastic*:** This may be the last option with relatively less environmental pollution (Rebeiz and Craft, 1995, Siddique et al.,, 2008).

2.1.4 Recent Developments in Plastic Regeneration and Recovery

Value recovery from plastics waste by pyrolysis using molten salts:

Melted salts have been used in a low temperature (420–480°C) pyrolysis that was done in a lab. Light oils, aromatics, paraffin waxes, and highly valuable monomers are the end products of pyrolysis. Despite the process's high capital requirements, one tonne per day of capacity is a manageable amount of output (Bertolini and Fontaine, 1987).

Thermal degradation of mixed plastic waste:

Mixed plastic waste with less than 1% chlorine content can be processed using the Hamburg Pyrolysis Process to yield approximately 48% weight percentage of oil that contains benzene, toluene, xylenes, and styrene and has a calorific value of about 50 MJ/Kg. The goods can be fed into petrochemical processes as feedstock. According to Kaminsky et al.,, the residue contains heavy metals and chlorine with a calorific value of roughly 40 MJ/Kg. 1996).

Recovery of plastics using froth flotation:

Froth flotation has been employed to extract plastics from municipal waste. The high-energy surfaces of inorganic minerals facilitate their robust adsorption by solutes, thereby strengthening the electrical forces necessary for flotation. Conversely, the low-energy surfaces of plastics impede solute adsorption, resulting in a lack of electrical forces for flotation (Alter, 2005).

Rapid characterization of plastics using laser-induced plasma spectroscopy (LIPS):

A small laser-induced plasma spectrometer has been created so that post-consumer plastics can be quickly identified using LIPS. Additionally, a software package including data processing and acquisition has been created. For online, real-time recycling material analysis, nearly 90–99 percent reliable identification of nearly all analyzed plastics has been reported (Anzano et al., 2006).

Alkaline hydrothermal treatment of brominated high impact polystyrene (HIPS-Br) for bromine and bromine-free plastic recovery:

When high impact polystyrene (HIPS-Br) is debrominated, Br can be extracted as KBr from an aqueous solution following the process. With a molecular weight comparable to the original HIPS-Br, the plastic can be recovered as individual pellets. When very little HIPS-Br is present, debromination using only water is done (Brebu et al., 2006).

Recycling of cable plastics:

The life cycle assessment of various recycling options for waste cable plastic includes energy recovery or incineration, chemical recycling, and mechanical recycling. The methods for recycling Swedish plastics in China include cable stripping, open burning in granulation, and energy recovery. According to Lindahl and Winsnes (2005), recycling plastics from Sweden in China is worse for the environment.

A goal programming approach for the plastic recycling system in Thailand:

To properly plan for the recycling of solid plastic in Thailand, a multi-objective recycling program called mixed integer goal programming, or MIGP, has been developed. The main goals of the aforementioned study are to lower overall costs, raise the amount of plastic recovered, and increase the desired plastic materials in the recycling process. The entire budgeted costs are necessary in order to obtain the desired plastic materials and the targeted amount of recycled plastic (Wongthatsanekorn, 2009).

Plastic waste recovery using free fall triboelectric separator:

A straightforward triboelectric separator for separating PE and PVC waste. According to laboratory experiments, the characteristics of the materials to be separated, the applied potential difference between the electrodes, the separator tube's material, diameter, and inclination, as well as its rotational speed, all affect how pure the separated materials are (Bendimerad et al., 2009).

Plastic waste management and uncertainties in a life cycle perspective:

Reusing and recycling materials to increase material efficiency is the goal of EU waste management policies. For the environmentally beneficial processes with life cycle assessment studies, several waste management options are discussed. The primary methodological factors, including uncertainties, for managing plastic waste are identified and discussed (Lazarevic et al., 2010).

Separation of packaging plastics by froth flotation:

It has been reported that a froth flotation process can be used to separate post-consumer polyethylene terephthalate (PET) from other packaging plastics with a similar density, which include 11.9 percent PS (polystyrene), 2.5 percent PVC (polyvinyl chloride), and 85 percent PET. Treatment with an alkali-surfactant mixture, followed by froth flotation in a pilot plant that is continuously run, can enhance separation, yielding a concentrate grade that contains 97.2 percent PET, 1.1 percent PVC, and 1.1 percent PS (Carvalho et al., 2010).

Recycling of high-density polyethylene to fuel:

Plastic waste is pyrolyzed to yield hydrocarbon fractions such as diesel, kerosene, and gasoline. HDPE, LDPE, PET, PA, PUR, PP, PS, PVC, and so forth. are employed as feedstocks for the pyrolysis process, which successfully converts the polyolefins and polystyrenes into hydrocarbon fractions. With the right catalysts, HDPE can be successfully pyrolyzed in sophisticated reactors to produce middle distillates (Kumar et al., 2011).

Recycling of plastics from cable waste:

There have been reports of a commercial process called PlastSep that can produce polyolefin and PVC-rich fractions from cable plastic waste. With a 99.5% yield, copper and aluminum can be recycled after being separated from the plastic fractions (Boss et al., 2011).

Catalytic pyrolysis of plastic wastes:

Synthetic zeolite (ZSM-5, Zeolyst International) that is sold commercially and leftover red mud have been utilized as catalysts for the pyrolysis of solid waste plastic that includes both recycled and virgin plastic. The quality and yields of recycled products are impacted by the use of catalysts. At 440°C, ZSM-5 encourages the production of low molecular weight gases and more aromatic-containing liquids from feedstock that contains polyolefinic acid. However, at 500°C, red mud that is primarily composed of Fe2O3, Al2O3, and TiO2 yields more gases and increases the amount of aromatics in the liquids. Another cost-effective method of recovering and regenerating waste made of solid plastic could be the catalytic pyrolysis described above (López et al., 2011).

Waste plastic recycling in Japan:

Since 2000, Japan has implemented the Packaging and Recycling Law. In addition to the life cycle assessment, mechanical and thermal recycling techniques are used for PET plastic waste. The best way to lessen the environmental impact on sustainable development has been determined to be the thermal decomposition of waste plastic or the gasification of waste plastic for the production of ammonia (Kamo, 2011).

Dissolved air flotation on separation of waste:

To separate waste plastics containing acrylonitrile butadiene styrene (ABS) and polystyrene (PS), the dissolved air flotation method has been employed. Tannic acid and terpinol are used as frothers and wetting agents to improve the separation of the aforementioned plastics. According to Wang et al. (2012), the floated products had waste PS purity and recovery rate of 90.12 percent and 97.45 percent, respectively, while the depressed products had waste ABS purity and recovery rate of 97.24 percent and 89.38 percent, respectively.

Plastic recycling in Rio de Janeiro:

Field research including visits to plastic recyclers, dealers, sorting plants for recycling cooperatives, selective garbage programs, and urban waste dumps reveals that the most common plastic materials dumped in Duque de Caxias and Rio de Janeiro City—which together have the highest plastic recycling capacity—are polyolefins and PET packaging plastic materials. Most of the recycled plastics have a low total value. According to Pacheco et al. (2012), selective waste separation is essential to producing high-quality plastic products at a reduced cost.

Solid waste management and plastic recycling in Qatar:

Plastic waste is the second most common type of MSW in Qatar. According to the life cycle assessment (LCA), recycling plastic waste is the most effective way to deal with municipal plastic waste. Nearly 2,000,000 tons of solid municipal waste are generated daily in Qatar for landfill and composting, which are regarded as the most suitable methods of waste disposal. This information includes the major constituents of the waste as well as the total amount of municipal solid waste generation, storage, collection, and disposal in Qatar. Lowering the Human Toxicity Potential (HTP) and Global Warming Potential (GWP) indicators might have a lessening effect on the environment (Al-Maaded et al., 2012).

Municipal solid waste characteristics and potential of plastic recovery in Malaysia:

Food waste, plastic, and paper make up the majority of the MSW at the Bakri Landfill in Malaysia. The sum of the aforementioned three elements makes up roughly 55% of the MSW. A daily total of about 35 tonnes of paper and soil are contaminating the MSW that contains the plastic waste. Elimination of the aforementioned impurities is crucial for recovering high-quality plastic. The city currently generates a lot of solid waste, so more attention is needed to help it recover (Kalanatarifard and Yang, 2012).

Recycling and recovery of post-consumer plastic solid waste in Europe:

The majority of plastics are not biodegradable; in almost a century, only 1-3 percent of their hydrocarbon content breaks down (Al-Salem et al., 2009). The high volume to weight ratio makes recycling plastics unappealing. The final cost of the recovered or recycled plastics depends on a number of factors, including collection, transportation, recycling, reprocessing, and the price of the virgin feedstock (Ambrose et al., (2002)). The only options left are mechanical and thermal recycling processes, the first of which is appropriate for homogeneous plastic waste, as landfilling plastic waste is no longer a practical option. When it comes to treating mixed plastic waste, pyrolysis is the best choice because it can yield feedstocks. Another method for burning waste plastics to produce heat energy is incineration; these plastics have a calorific value of between 30 and 40 MJ/kg. In addition to the different technical choices available, it is necessary to take into account several US laws and European directives pertaining to the management of plastic waste (Brems et al., 2012).

Preparation of nanocomposites:

It has been discovered that adding nano fillers, such as graphene, SiO2, nano clay, CaCO3, carbon nanotubes (CNTs), and mica, is an effective method for recycling polymers like PP, HDPE, and PVC mixes and composites. Compatibilizers are said to be more beneficial when used in blend and composite recycling because they improve the properties of recycled polymers and make them lighter, easier to process, and less expensive (Zare, 2013).

Recycling of bioplastics, their blends and bio composites:

A review has been conducted on the mechanical recycling of bioplastics, their blends, and thermoplastic biocomposites. The presence of shear stress, heat, oxygen, UV light, residual catalyst, and water during mechanical recycling causes the recycled products to degrade and alter their stability and functional qualities. It is challenging to obtain high-quality recycled and extruded products using the aforementioned method (Soroudi and Jakubowicz, 2013).

Recycling of waste from polymer materials:

Using the principles of mechanical and chemical recycling, polymeric wastes such as polyethylene, polypropylene, and polystyrene, as well as their blends and composites, have been processed. When recycling waste that has undergone intricate chemical treatments, mechanical recycling is preferred over chemical recycling. The qualities of the recycled products have been enhanced during mechanical recycling by the addition of a small amount of virgin polymers—either the same or different polymers to the waste in the presence of appropriate compatibilizers (Hamad et al., 2013).

Characterization of polyolefin packaging waste:

Pre-concentrate made from hand-picked mixed household waste from Romania. A method based on magnetic density separation (MDS) and hyperspectral imaging (HSI) has been reported to recover high-purity polyolefins from complex wastes at a low cost (Bakker et al., Serranti et al.,, 2009. 2011). In order to differentiate between the two polymer manufacturing processes (blow molding and injection molding) based on the physical characteristics of the flake, all necessary characterization and identification have been completed using hyperspectral imaging. It has been noted that density separation can be used to recycle polyolefins from packaging waste. Procedures based on hyperspectral imaging have also been used to establish quality control measures for recycled goods (Hu et al., 2013).

Technical specifications for mechanical recycling of agricultural plastic waste:

There is a considerable amount of dirt in some of the film categories' agricultural plastic waste. It has been discovered that it is difficult to recycle plastic wastes like the above film waste because the film curls during the recycling process and plugs the extruder filter. Instead, thin agricultural wastes must be flaked or pelletized. In order to valorize the waste stream, Briassoulis et al. (2013), have implemented LabelAgri-Waste collection, a special shipping format for APW throughout Europe.

Tribo-electrostatic separation for granular plastic waste recycling:

Many hazardous substances, including polycyclic aromatic hydrocarbons (PAHs), polychlorinated biphenyls (PCBs), polychlorinated dioxins (PCDs), and polychlorinated dibenzofurans (PCDFs), can be produced during the incineration of municipal solid waste that contains plastics. These substances have the potential to induce mutagenesis, carcinogenesis, and teratogenesis (Chung et al., 2010). The development and efficacy of various devices have been discussed, and tribo-electrostatic separation for granular plastic waste has been reported (Wu et al., 2013).

Wastes of electrical and electronic equipment:

Metals, glass, and plastics are discovered in e-waste. In contrast to recovering plastics through chemical, mechanical, and thermal recycling methods, recovering glass requires the use of glass-to-glass and glass-to-lead recycling technologies, as well as recovery procedures for copper, lead, and precious metals like silver, gold, platinum, and palladium (Kang and Schoenung, 2005). Electrostatic separation techniques can be used to recover chopped electric wires or printed-board circuits (Li et al., 2008). Aerodynamic forces make electrostatic separation of fine particles like powder and micrometric-size particles less effective even though electrostatic forces can control the motion of charged particles leading to electrostatic precipitation, electrostatic powder coating, and electrostatic separation. The high voltage level, the disks' rotational speed, the fluidization rate, the total mass of the fluidized bed, and the granular mixture's composition ratio are some of the variables that affect efficiency. The correlation between the particle system, the particle properties, and the adapted chargeability must be established through the application of analytical methods (Landauer and Foerst, 2018, Bouhamri et al., 2019). It has been documented that waste electrical and electronic equipment can be recycled to create a blend of polycarbonate and polyamide plastics. A tribo-aero-electrostatic separator is used for the initial stages of separation. Granular waste that has undergone tribo-aero-electrostatic separation is subsequently processed in two free-fall electrostatic separators. The upper and lower portions of the same installation are the tribo-aero-electrostatic separator and the free-fall electrostatic separator. A MATLAB image processing program has been used to confirm the products' purity. There has been a report of a very high purity separation (approximately 95% for both PC and PA) at a recovery rate exceeding 70% (Aksa et al., 2013).

Figure 2.8. The schematic diagram for the recycling of electronic products.

Source: Njoku A, Agbalenyo M, Laude J, Ajibola TF, Attah MA, Sarko SB. Environmental Injustice and Electronic Waste in Ghana: Challenges and Recommendations. International Journal of Environmental Research and Public Health. 2024; 21(1):25. https://doi.org/10.3390/ijerph21010025

Use of recycled plastics in wood plastic composites:

The performance of waste plastics collected from different sources varies because of the different environmental conditions and recycling/reprocessing cycles they are subjected to. Waste plastic can be a significant feedstock for the production of wood plastic composites, which could be a promising material. It has been discussed how waste and recycled plastics differ mechanically and physically (Najafi, 2013).

Regeneration and recycling of waste thermosetting plastics:

The disintegration of the cross-linked structure during mechanical and thermal processing altered the phenolic resins' molecular structure and chemical characteristics. This leads to the formation of numerous mechanically active groups and a decrease in crosslink density. Composites made of a blend of polypropylene and phenolic resin are reshaped. The results of mechanical testing indicate that the recycling rate is increased by phenolic resin. It is discovered that the samples have high tensile and bending strengths (Wu et al., 2014).

The recovery potential of plastic wastes obtained from landfill mining:

It has been discovered that the most practical method for processing plastic waste from landfill mining under standard cleaning procedures is to incinerate it or treat it as residue-derived fuels for energy recovery. 69.13 percent of the stored wastes in an old landfill in central China, which has been in storage for 24 years, contains plastic waste. Of these, white PE plastic bags account for 11.34 percent, colored PE plastic bags for 29.77 percent, and other plastic bags for 28.02 percent. The remaining 30.87 percent of the wastes are other plastics, such as PP, PVC, and PS. The plastic surfaces of the excavated plastic bags contain a lot of impurities, so mechanical or chemical methods utilizing hydrogenation, gasification, or pyrolysis are not appropriate for removing them all. Incineration is the best technique for recycling these plastic wastes so that energy can be extracted from them (Zhou et al., 2014).

Selective floatation of polycarbonate and polystyrene waste plastics:

Using the selective floatation method, waste polycarbonate (PC) plastic has been separated from a mixture of PC and polystyrene (PS). Ammonia treatment has changed the wettability of the PC surface, resulting in PC as the floated product and PS as the submerged product. It is discovered that the submerged product's

PC purity is up to 98.21 percent, while the overflowed product's PS purity is up to 99.53 percent. According to Wang et al.(2016), the aforementioned method is less expensive than alternative separation procedures for waste PC and PS mixture.

Implications of the presence of phthalates in plastics recycling:

When making plastics, "phthalates," or phthalic acid esters, are frequently combined with polyvinyl chloride (PVC) as plasticizers. It has been noted that plastic products disposed of in residual or source-segregated waste have the same phthalate contents. Phthalates may not be eliminated after plastics recycling, as evidenced by the negligible difference between household residual waste plastics (RWP) and source-segregated waste plastics (SSWP) of recycled household plastics (RHP). In these situations, there's a chance that phthalates will be present during the recycling process, which could lead to phthalate accumulation and spreading. Both the virgin and recycled industrial plastic waste have no discernible phthalate contents. The presence of noticeable phthalate content suggests that phthalates are added during the manufacturing process (labeling, gluing, etc.) or other items with a higher phthalate content have contaminated the samples of household waste plastics (e.g. vinyl (PVC) (Pivnenko et al., 2016).

Thermal and catalytic pyrolysis of plastic solid waste:

Pyrolysis has been used to treat plastic solid waste (PSW) in an oxygen-free environment, which has reduced emissions of carbon monoxide and dioxide and prevented the production of dioxins. In addition to resolving the landfilling issue, pyrolysis has produced a wide variety of feedstock or as energy and reduced reliance on oil and gas. Operating parameters such as temperature, pressure, and residence time are highly adjustable for both thermal and catalytic pyrolysis, allowing for the production of any desired product mixture (Al-Salem et al., 2017).

Recycling of plastic solid waste and future applications:

The identification and separation processes for plastic solid waste (PSW), such as froth flotation and magnetic density separation, have been thoroughly covered along with the various recycling methods for PSW. The impact of different types of reinforcements, such as sand, fiber, ash, wood husk, rice husk, metal powder, etc. , has also examined virgin and recycled HDPE, LDPE, and nylon (Singh et al., 2017).

Mechanical and chemical recycling of solid plastic waste:

X-ray detection, magnetic density separation, froth flotation, mechanical recycling, and separation are a few approaches. The immiscibility of plastic blends in SPW and thermomechanical or lifetime degradation have been covered. By breaking down plastics into smaller molecules, chemical recycling can create fuels or virgin polymers that function nearly exactly like virgin monomers. Several methods of recycling chemicals, including gasification, hydrocracking, pyrolysis, fluid catalytic cracking, and chemolysis. have been examined, and it was discovered that contaminated and heterogeneous SPW may benefit more from the use of the chemical recycling process. It has been discovered that waste contaminations have an impact on product quality, which directly affects the recycled materials' economics (Ragaert et al., 2017).

Plastic waste to liquid oil through catalytic pyrolysis using natural and synthetic zeolite catalysts:

Using both natural and synthetic zeolite catalysts, plastic wastes containing PE, PS, PP, and PET have been catalytically pyrolyzed to produce styrene, ethylbenzene, benzene, azulene, naphthalene, and toluene along with a few aliphatic hydrocarbon compounds. The pyrolysis products are characterized using FTIR and GCMS analyzers. It has been noted that following appropriate refining or blending with commercial fuels, pyrolysis products may be used as value-added products (Miandad et al., 2017).

Separation and recovery of glass, plastic and indium from spent liquid crystal displays (LCD) panels:

It has been discussed how to recover valuable fractions like plastic, glass, and indium during the disassembly of spent LCDs using physico-mechanical pretreatments. Both wet and dry processes are used for recovery; from the wet process, 20 percent, 15 percent, and 40 percent of the feeding panels' weight are recovered as plastic, glass, and indium concentrate, respectively; from the dry process, only two fractions are separated, yielding approximately 11 percent and 85 percent of the weight being recovered as plastic and glass/indium mixture, respectively (Ferella et al., 2017).

Plastic flexible films waste management:

Clean and homogeneous post-industrial and post-consumer plastic waste is recycled using closed-loop or open-loop mechanical processes, depending on the quality and applications of the recycled materials. The least amount of material degradation is shown by the closed-loop mechanical recycling process method. Open-loop mechanical recycling is used to recycle post-consumer wastes from the packaging and agriculture industries because there are routes for selective waste collection available. The presence of contaminants in the waste has a negative impact on the quality of recycled plastics, so washing the wastes before processing has to be upgraded. The lowest rate of recycling of household flexible packaging is caused by inadequate sorting technologies. Horodytska et al (2018), have reported on a restricted number of Life Cycle Assessments (LCA) related to the waste management type mentioned above.

Gasification of waste plastics:

Numerous pilot-scale gasification operations involving waste plastics have been carried out. Any commercial plastic valorization process's capacity to handle plastic wastes with high impurity contents determines how successful it will be. It has been discovered that the co-gasification of plastics with coal and biomass valorization offers greater process flexibility. According to Lopez et al.,, pyrolysis and in-line reforming processes can produce syngas and hydrogen without the use of tar. 2018).

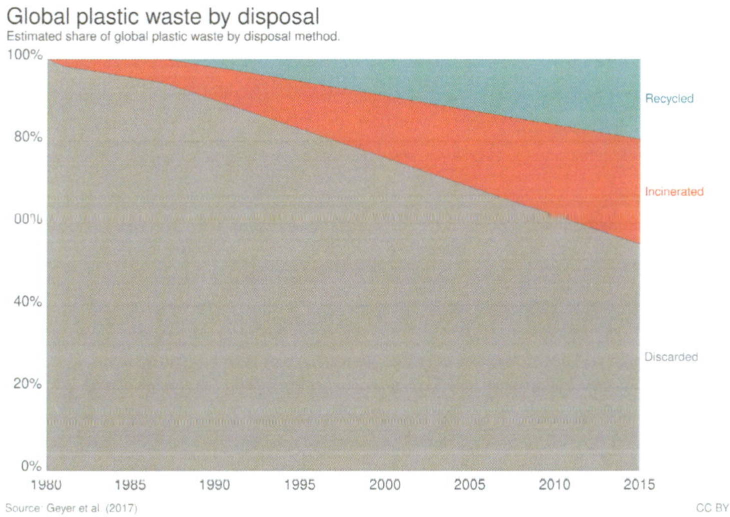

Figure 2.9. Disposal of waste plastics.

Source: By Our World In Data - https://ourworldindata.org/grapher/global-plastic-fate, CC BY 3.0, https://commons.wikimedia.org/w/index.php?curid=86934912

Recycling of polystyrene-based plastics:

The process of recycling plastics made of polystyrene has been examined, along with the different ways that these materials are separated out before recycling (Thakur et al., 2018).

Solvent-based separation and recycling of waste plastics:

Plastic recovery and separation using different solvent extraction techniques have been applied selectively. Degradation of the recovered monomers occurred when plastics were recovered using distillation techniques after dissolution. The hardest parts are figuring out the right solvent and purification methods. Since solid plastic waste is a blend of various plastic types, the properties and qualities of the recovered products differ greatly. In addition to the aforementioned, there are many drawbacks, such as the use of dangerous solvents and higher solvent costs, which restrict the use of the aforementioned process (Zhao et al., 2018).

Circular use of plastics-transformation of existing petrochemical clusters into thermochemical recycling plants:

The following steps have been taken to transform plastic waste (PW) into a thermochemical recycling plant that can recover 100% of plastics: (1) PW replaces feedstock partially (by 45% on a carbon basis); (2) PW replaces feedstock completely; (3) the process is electrified; and (4) 100% of carbon is recovered as monomers or permanent storage through oxy-combustion, carbon capture, and storage.

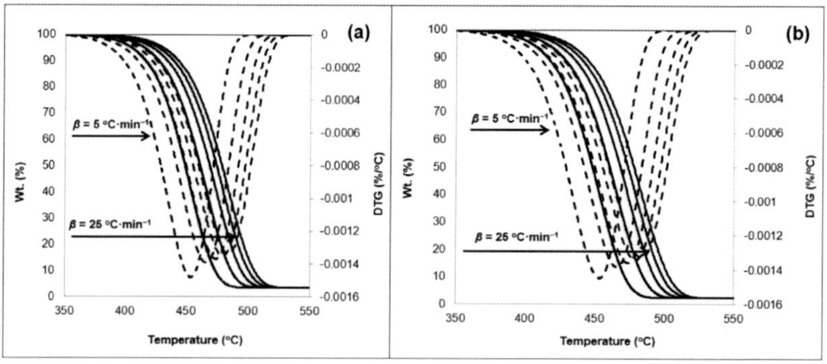

Figure 2.10. *TGA thermogram showing weight loss (—) and first derivative (- - -) of the 50/50 sample (**a**) pre-exposure and (**b**) exposed to 11 days in N_2 atmosphere with respect to heating rates (5 to 25 °C·min^{-1}) in sequential order.*

Source: Al-Salem SM, Behbehani MH, Karam HJ, Al-Rowaih SF, Asiri FM. On the Kinetics of Degradation Reaction Determined Post Accelerated Weathering of Polyolefin Plastic Waste Blends. International Journal of Environmental Research and Public Health. 2019; 16(3):395. https://doi.org/10.3390/ijerph16030395

The technical solution and process economics for closing the plastic material cycle have been studied using thermochemical recycling. According to Thurman et al. (2019), the aforementioned method can handle any kind of plastic waste, whether it is mixed or sorted, with nearly 100% carbon recovery.

Utilization and recycling of end-of-life plastics for the iron and steel industry:

In addition to energy savings, recycling waste plastics has been shown to reduce the carbon footprint by almost 30% when compared to other carbon sources in high-temperature material processing industries like the iron and steel sector, which accounts for 7% of CO2 emissions. Plastics also significantly lower emissions and add energy during the production of magnesia, a process with a high carbon footprint (Devasahayam et al., 2019).

Delineating the plastic waste status in the State of Qatar:

The State of Qatar is one of the nations that produces the most waste in the world; over the past few years, it has produced almost 1.4 kg of waste per person per day. Plastics make up roughly 13–14 percent of the MSW overall, of which some may be recovered. Using a SWOT analysis, the opportunities and problems for this kind of waste management in the state of Qatar have been examined (Hahladakis and Aljabri, 2019).

Dissolution and recovery of waste expanded polystyrene:

Expanded polystyrene has been recovered from waste expanded polystyrene (WEP) by dissolving it with commercial essential oils. Several instrumental techniques of analysis are employed to characterize the WEP and the recovered expanded polystyrene, including the oil utilized for the recovery. It has been stated that the recovery process is more affordable and environmentally benign (Gil-Jasso et al., 2019).

Micronized plastic waste recycling:

Recently, a mixture of granular pure virgin micronized white polyvinyl chloride (WPVC) and gray polyvinyl chloride (GPVC) particles with an average size of 100 μm and a small percentage of carbon were separated experimentally using triboelectrostatic separation. The effectiveness of the separation process has been noted, and it is dependent upon the disc's rotation speed, the rate of fluidization, the total mass of the fluidized bed, the ratio of the granular mixture's composition, and the applied potential difference (Bouhamri et al., 2019).

2.2 PLASTICS RECYCLING

Since the development of multiple pathways for the production of polymers from petrochemical sources, the plastics industry has experienced significant growth. When compared to many other material types, plastics offer significant advantages in terms of low weight, durability, and cost (Andrady and Neal 2009; Thompson et al., 2009a). All polymers, including thermoplastics and thermoset plastics, adhesives, and coatings, were estimated to be produced globally in 2007 at a rate of 260 million metric tonnes annually; synthetic fiber production was not included in this estimate (PlasticsEurope 2008b). This suggests a growth rate of roughly 9% historically.Approximately two-thirds of this production are thermoplastic resins, and their use is increasing at a rate of five percent per year. a. worldwide (Andrady 2003).

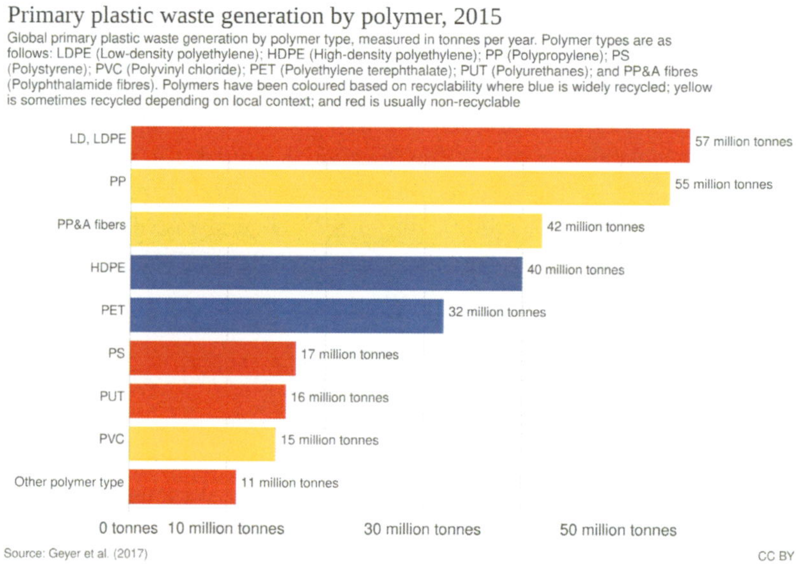

Figure 2.11: Recyclability.

Source: By Our World In Data - https://ourworldindata.org/grapher/plastic-waste-polymer, CC BY 3.0, https://commons.wikimedia.org/w/index.php?curid=86934997

Nowadays, petrochemicals made from fossil fuels like gas and oil make up practically all of the ingredients in plastics. From petrochemical feedstock, about 4% of the world's annual petroleum production is directly transformed into plastics (British Plastics Federation 2008). Since plastics must be manufactured, a comparable amount of additional fossil fuels must be consumed during the production process. But it's also possible to make the case that using lightweight plastics can cut down on the amount of fossil fuels used, for instance when replacing steel and other heavier conventional materials in transportation (Andrady and Neal 2009; Thompson et al., 2009b).

About fifty percent of plastics are used in single-use, disposable applications like packaging, agricultural films, and disposable consumer goods; twenty to twenty-five percent are used in long-term infrastructure like pipes, cable coatings, and structural materials; and the remaining portion is used in durable, intermediate-life consumer applications like furniture, appliances, cars, and so on. In the European Union (EU), 24.6 million tonnes of post-consumer plastic waste were generated in 2007 (PlasticsEurope 2008b). A breakdown of the UK's plastics consumption in 2000 and its contributions to waste generation demonstrate that packaging is the primary source of waste plastics, but it also shows that other sources, like end-of-life vehicles (ELV) and waste electronic and electrical equipment (WEEE), are growing in importance (Waste Watch 2003).

Table 2.2. Key advantages and disadvantages of plastics recycling technologies

Advantages	**Disadvantages**
Mechanical Recycling	
• Straightforward process and effective for most thermoplastics (**38,44,45**). • Least negative environmental impacts (**40,43**).	• Degrades recyclates' properties (**36,49,50,51,52,53**). • Limited recycling cycles (**54**). • Low tolerance to contaminants (**52,65**).
Solvent-Based Purification	
• Able to treat contaminated plastic waste using dissolution with precipitation (**86,87,88**). • Recovers high-quality polymers (**86,87,88**).	• Health and safety concerns arise from using strong and hazardous solvents (**86,88,93**). • Unit cost of solvents and its associated recovery processes can be costly (**53**).
Depolymerization	
• Recovers monomers (**95**). • Uses less energy than other chemical recycling processes (**102**).	• Low tolerance to contaminants (**76,91**). • Expensive and only applicable to certain polymers (**76,91**).
Pyrolysis	

Advantages	Disadvantages
• Straightforward and highly versatile process, where different treatment parameters yield different outputs (**110,111**).	• Not cost effective (**82**). • High energy requirements (**113**).

Source: Tan J, Jia S, Ramakrishna S. Accelerating Plastic Circularity: A Critical Assessment of the Pathways and Processes to Circular Plastics. Processes. 2023; 11(5):1457. https://doi.org/10.3390/pr11051457

Since plastics have only been produced in large quantities for the past 60 years or so, it is uncertain how long they will last in the environment. The bulk of polymers produced today will endure for at least decades, and most likely for centuries or even millennia, as the majority of plastics are not biodegradable (Andrady 1994). In fact, most plastics are incredibly durable. Given that rates of degradation depend on physical factors like temperature, oxygen content, and UV light exposure (Swift and Wiles 2004), even degradable plastics may last for a long time depending on local environmental factors. In contrast, biodegradable plastics require the presence of certain microorganisms. As a result, degradation rates differ significantly between terrestrial and marine environments, as well as landfills (Kyrikou & Briassoulis 2007). Even when a plastic object deteriorates due to weathering, the polymer itself might not always completely degrade in a significant amount of time; instead, it first breaks down into smaller pieces of plastic debris. Because of this, large amounts of end-of-life plastics are building up in landfills and as debris in the environment, which causes problems with waste management and environmental harm (see Barnes et al., 2009; Gregory 2009; Ohlmann et al., 2009; Ryan et al., 2009; Tenen et al., 2009; Thomas et al., 2009b).

Recycling can be considered a contemporary application of the concept of industrial ecology, as opposed to a natural ecosystem where wastes and products are the only things present (Frosch and Gallopoulos 1989; McDonough & Braungart 2002). Recycling is undoubtedly a waste-management strategy. One way to lessen the impact on the environment and the depletion of resources is through plastic recycling. Fundamentally, high recycling rates can enable a given level of product service with fewer material inputs than would otherwise be necessary, along with reductions in use, reuse, repair, and remanufacturing. Recycling can therefore result in increased eco-efficiency by reducing the amount of energy and material used per unit of output (WBCSD 2000). However, it should be noted that the system's ultimate sustainability will depend on its ability to maintain whatever residual level of material input, as well as energy inputs and the effects of outside influences on ecosystems.

2.2.1 Systems for Plastic Recycling

There are numerous ways to recycle plastic materials, and the ease of recycling varies depending on the type of polymer, the design of the packaging, and the product. For instance, recycling rigid containers made of a single polymer is easier and more cost-effective than recycling packages made of multiple layers and components.

PET, PE, and PP are thermoplastics with a high potential for mechanical recycling. Thermosetting polymers, like unsaturated polyester or epoxy resin, are incapable of being mechanically recycled; the only possible use for them is as filler materials after they have been ground into fine powder or reduced in size (Rebeiz and Craft 1995). This is so that they can't be remelted and reformed because thermoset plastics are manufactured with permanent cross-links. The practice of recycling cross-linked rubber from automobile tires to rubber crumb for subsequent manufacturing into other products is not new, but it is anticipated to increase due to the EU Directive on Landfill of Waste (1999/31/EC), which prohibits the disposal of tires and their waste in landfills.

Due to inherent molecular immiscibility and variations in macroscale processing requirements, the majority of plastic types are incompatible with one another, which poses a significant challenge to the production of recycled resins from plastic wastes. For instance, the evolution of hydrochloric acid gas from the PVC at a higher temperature needed to melt and reprocess PET will cause a small amount of PVC contamination present in a PET recycle stream to degrade the recycled PET resin. On the other hand, PET in a PVC recycling stream will solidify into lumps of crystalline PET that are not dispersed, greatly diminishing the value of the recycled material.

Therefore, adding recovered plastic to virgin polymer is frequently not technically possible without reducing at least some of the virgin plastic's quality attributes, like color, clarity, or mechanical qualities like impact strength. The majority of recycled resin applications involve either blending it with virgin resin (typically using polyolefin films for non-critical applications like garbage bags and irrigation or drainage pipes that are not pressure-rated) or using it in multi-layer applications where the recycled resin is layered between virgin resin surface layers.

The purity of the recovered plastic feed and the specifications needed to create the desired plastic product determine whether recycled plastic can replace virgin polymers. Because of this, current post-consumer waste recycling programs focus on the packages that are easiest to separate, like PET water and soft drink bottles and HDPE milk bottles, which can be positively identified and sorted out of a co-mingled waste stream. On the other hand, because multilayer/multi-component articles cause polymer-type contamination, there is little recycling of them. Therefore, collection, sorting, cleaning, size reduction and separation, and/or compatibilization to lessen contamination by incompatible polymers are some of the essential steps in post-consumer recycling.

(a) Collection

Both curbside collection and "bring-schemes" are options for collecting plastic waste. In the absence of either highly committed public behavior or deposit-refund schemes that impose a direct economic incentive to participate, bring-schemes typically yield low collection rates. As a result, curbside collection of recyclables is becoming more and more common alongside MSW. Most kerbside collections of co-mingled recyclables (paper/board, glass, aluminum, steel, and plastic containers) are made in order to optimize the cost-effectiveness of these programs. Although curbside collection programs have been very effective in recovering plastic bottle packaging from homes, only about 30 to 40 percent of post-consumer plastic bottles are recovered overall because a large portion of this type of packaging comes from food and drink that is consumed away from home. Therefore, if overall plastic packaging collection rates are to rise, it is critical to create efficient "on-the-go" and "office recycling" collection programs.

(b) Sorting

Co-mingled rigid recyclables are sorted using both manual and automatic techniques. Usually, automated pre-sorting is enough to produce a stream of plastics that is distinct from glass, metals, and paper (apart from when connected, e.g. such as closures and labels). Milk bottles made of clear PET and unpigmented HDPE are typically recognized and extracted from the stream. Nowadays, many plastic recycling facilities as well as operators of material recovery facilities use automatic container sorting. These systems typically sort the streams into clear and colored fractions using optical color recognition camera systems and analyze polymer types using Fourier-transform near-infrared (FT-NIR) spectroscopy. PET containers in the colors clear, light blue, dark blue, green, and other colors can be distinguished from one another using optical sorters. Sorting in series and employing multiple detectors can optimize sorting performance. Other technologies for sorting include X-ray detection, which is used to separate PVC containers, which are easily distinguished because they contain 59% chlorine by weight (Arvanitoyannis and Bosnea 2001; Fisher 2003). The majority of material recovery facilities and local authorities do not proactively gather used flexible packaging because there are currently flaws in the machinery that can easily separate flexibles. Trommels and density-based air-classification systems are widely used in plastic recycling facilities to remove small amounts of flexible materials, including labels and certain films. Nonetheless, advancements in this field and novel technologies—like air-classifiers, sophisticated hydrocyclones, and ballistic separators—will boost the recovery of flexible packaging after consumer use (Fisher 2003).

(c) Size reduction and cleaning

Usually, food residue, pulp fibers, and adhesives are eliminated from rigid plastics by grinding them into flakes and cleaning them. The newest wash plants use roughly half as much water—two to three m3 per tonne of material—as those from earlier generations. "Dry-cleaning," an innovative technology that uses friction to clean surfaces without the use of water, is one way to remove organics and surface contaminants from flakes.

(d) Further separation

Numerous separation techniques can be used after size reduction. Polyolefins (PP, HDPE, L/LLDPE) can be efficiently separated from PVC, PET, and PS in water using sink/float separation. While PS and PET can be separated by using different media, PVC and PET cannot be separated in this way because of the overlap in their density ranges. Low-density films can also be extracted from denser ground plastics using alternative separation methods like air elutriation, e.g. in taking off the PET flake labels (Chandra and Roy 2007).

Froth flotation is one technology that reduces PVC contaminants in PET flake (Drelich et al., FT-NIR or Raman emission spectroscopic detectors to facilitate flake ejection and employing varying electrostatic properties (Park et al., 1998; Marcos and Tenorio 2000)(). Since PVC turns black when heated, thermal kilns can be used to selectively degrade small amounts of PVC impurities for PET flake, allowing for color sorting.

There are many ways to sort flake, but conventional PET sorting systems are mainly limited to separating two types of materials from PET: (i) colored flakes from clear PET flakes and (ii) materials with different physical properties like density. Other contaminants like silicones and nylon can be eliminated with the use of novel techniques like laser sorting systems.

"Laser-sorting" distinguishes between different polymer types using emission spectroscopy. These systems, with their capacity to scan individual flakes and achieve up to 860 000 spectra s^{-1}, are expected to greatly enhance the ability to separate complex mixtures. One of their advantages is that, unlike traditional automatic systems, they can be used to sort various black plastics. The separation of automotive plastics and WEEE is expected to increase with the use of laser-sorting systems. These systems can also separate polymers based on their type or grade, and they can also separate HDPE from polyolefinic materials like PP. Still, this is a relatively new method that is only being utilized in a few European recycling centers.

(e) Current advances in plastic recycling

Over the past ten years, advances in recycling technologies have included more dependable detectors and advanced decision-making and recognition software. These advancements have increased the precision and efficiency of automated sorting; for instance, modern FT-NIR detectors have a fault-free operating lifespan of up to 8,000 hours.

Finding higher-value uses for recycled polymers in closed-loop processes—which can directly replace virgin polymer—has been another innovative endeavor. For instance, since 2005, the majority of PET sheets used in thermoforming in the UK have included 50–70% recycled PET (rPET) thanks to the use of A/B/A layer sheets, in which the inner layer (B) is made of rPET and the outer layer (A) is the virgin resin that has been approved for food contact. Because of the development of "super-clean" grades, food-grade rPET is now widely available in the market for direct food contact. These merely exhibit a minor reduction in transparency when compared to virgin PET, and they are utilized as a 30–50% substitute for virgin PET in numerous applications, as well as the entire content in certain bottles.

Table 2.3. *Substances of very high concern (SVHC) used as plastic additives; POPs (Persistent Organic Pollutants) and their regulation, according to (**63**), modified; sources for concentrations there. X = applicable; ppm = mg/kg; 1000 ppm = 0.1 wt.%*

Additive	Purpose	Plastic, Application	Conc. (wt.%)	REACH (58)	RoHS2 (59)	POP (60)
HBCD	Flame retardant	EPS, XPS in isolation material HIPS in EEE	0.7–2.5% 1–7%	X	–	Products: 100 ppm Waste: 1000 ppm
PBDEs	Flame retardant				Σ PBDEs: 1000 ppm	Waste ((**64**), Annex IV): 500 ppm; 350 ppm from 30.12.2025 on; 200 ppm from 30.12.2027 on

Additive	Purpose	Plastic, Application	Conc. (wt.%)	REACH (58)	RoHS2 (59)	POP (60)
TetraBDE (1)	Flame retardant	as c-PentaBDE (2) in PUR, former PC boards	0.5–5%	See PBDEs	See PBDEs	Substances (Annex I): 10 ppm per substance. Products (Annex I): 500 ppm for Σ PBDEs (4) (Annex I)
PentaBDE (1)	Flame retardant			See PBDEs	See PBDEs	
HexaBDE (1)	Flame retardant	as c-OctaBDE (3) in: ABS, HIPS, PBT, PA	12–18%	See PBDEs	See PBDEs	
HeptaBDE (1)	Flame retardant			See PBDEs	See PBDEs	
DecaBDE	Flame retardant	HIPS, PA, PO	5–16%	See PBDEs	See PBDEs	
PBBs	Flame retardant, plasticiser	ABS, foams, textiles, devices	10%		Σ PBBs: 1000 ppm	
DEHP	Plasticiser	PVC	30%	X		
BBP	Plasticiser	PVC	5–30%	X		
DBP	Plasticiser	PVC	1.5%	X		
DIBP	Plasticiser	PVC	like DBP	X		

(1) Production of these fabrics was discontinued decades ago, but they can still be found in old stocks and in recycled products. (2) Commercial (c) mixture mainly of isomers of pentaBDE and tetraBDE. (3) Commercial (c) mixture mainly of isomers of isomers of HeptaBDE and OctaBDE as well as a lower proportion of Nona- and HexaBDE. (4) By way of derogation, the manufacture, placing on the market and use of electrical and electronic equipment covered by Directive 2011/65/EC is permitted. Further exemptions apply to DecaBDE (certain aircraft and motor vehicles and their spare parts). Additives: BBP = benzyl butyl phthalate; DBP = di butyl phthalate; DecaBDE = decabromo diphenyl ether; DEHP = di ethyl hexyl phthalate; DIBP = di iso butyl phthalate; HBCD = hexa bromo cyclo dodecane; HeptaBDE = hepta-bromo-diphenyl ether; HexaBDE = hexa-bromo-diphenyl ether; PBBs = polybrominated biphenyls; PentaBDE = penta-bromo-diphenyl ether; TetraBDE = tetra-bromo-diphenyl ether. Plastics: ABS = acrylonitrile-butadiene-styrene copolymer; EPS = expanded polystyrene; HIPS = high impact polystyrene; PA = polyamide; PBT = polybutylene terephthalate; PO = polyolefins; PVC = polyvinyl chloride; XPS = extruded polystyrene.

Source: Lahl U, Zeschmar-Lahl B. Material Recycling of Plastics—A Challenge for Sustainability. Sustainability. 2024; 16(15):6630. https://doi.org/10.3390/su16156630

In addition to their bottle streams, several European nations, including Germany, Austria, Norway, Italy, and Spain, are already gathering rigid packagings, like trays, tubs, and pots, as well as a restricted quantity of post-consumer flexible packagings, like films and wrappers. New markets for the recyclables and advancements in sorting and washing technologies have made it feasible to recycle this non-bottle packaging. After conducting a preliminary study on mixed plastics recycling in the UK, the Waste Resource Action Programme (WRAP) is currently moving this to full-scale validation (WRAP 2008b). When one considers that there are estimated to be over a million tonnes of non-bottle plastic packaging produced annually in the UK (WRAP 2008a), compared to 525,000 tonnes of plastic bottle waste (WRAP 2007), the potential benefits of recycling mixed plastics are substantial in terms of resource efficiency, diversion from landfills, and emission savings.

2.2.2 Ecological Case for Recycling

Utilizing life-cycle analysis can help evaluate the possible advantages of recycling initiatives. Recycled plastics can directly reduce oil consumption and greenhouse gas emissions associated with the production of virgin polymer (minus emissions from recycling activities). Recycled plastics can be used to make goods that would otherwise have been made from new (virgin) polymers. However, requirements for polymer production will not be reduced if recycled plastics are used to create goods that were previously made of wood or concrete (Fletcher and Mackay 1996). Any such alternative material usage may have additional environmental costs or benefits, but these would need to be taken into account case-by-case and are unrelated to our discussion of the advantages of recycling. In this case, the main focus will be on recycling plastics to create goods that otherwise would have been made from virgin polymers.

Although feedstock (chemical) recycling technologies meet the material recovery principle in general, they are less energy-efficient and more expensive than mechanical recycling because the polymer must first be depolymerized and then re-polymerized. Because of the low cost of petrochemicals compared to the high process and plant costs of chemically recycling polymers, this has historically required very large subsidies.

Energy recovery from waste plastics (conversion to fuel, direct combustion for the production of electricity, use in blast furnaces and cement kilns, etc. because the waste plastic was manufactured using petrochemicals, can be used to reduce landfill volumes but does not lessen the demand for fossil fuels (Garforth et al., 2004). Their emissions are linked to health and environmental issues as well. Reducing the need for plastic production is one of the main advantages of recycling plastics. Even after accounting for the energy required for the collection, transportation, and reprocessing of the plastic, recycling has been demonstrated to save more energy than energy recovery (Morris 1996). For plastic recycling systems, life-cycle analyses have also been utilized to assess the overall environmental effects (Arena et al., 2003; Contini et al., 2005), and they discovered that mechanical recycling has more favorable environmental effects than landfilling and energy-recovering incineration.

Recycling PET bottles is thought to reduce greenhouse gas emissions by 1.5 tonnes of CO_2 equivalent per tonne of recycled PET (Department of Environment and Conservation (NSW) 2005), as well as net energy consumption and landfill space usage. A German life-cycle analysis (LCA) study (Patel et al., 2000) appears to have served as one of the bases for the estimate of an average net reduction of 1.45 tonnes of CO2-e per tonne of recycled plastic that has been suggested as a useful guideline to policy (ACRR 2004), which also discovered that the production of virgin polymers is replaced, accounting for the majority of the net energy and emission benefits. According to a recent life-cycle assessment (LCA) conducted especially for the production of PET bottles, using 100% recycled PET rather than 100% virgin PET would lower the full life-cycle emissions per bottle from 446 to 327 g CO2, or a 27% relative reduction in emissions (WRAP 2008e).

The least advantageous source of recycled polymer, mixed plastics, may nevertheless offer a net benefit of about 0.5 tonnes of CO2-e per tonne of recycled product (WRAP 2008c). The more effective method of recycling bottles as opposed to mixed plastics and the comparatively high emissions profile of virgin PET production account for the higher eco-efficiency of bottle recycling. As long as virgin polymer is substituted, the mixed plastics recycling scenario still has a positive net benefit—which was thought to be superior to the other options studied—of both landfills and energy recovery as solid refuse fuel.

2.2.3 Public Support for Recycling

The public's understanding of the importance of sustainable production and consumption is growing. This has prompted some manufacturers to create products with recycled content, other businesses to fill this consumer demand, and local authorities to organize recyclable collections. Studies on consumer preferences in marketing show that a sizable, if not overwhelming, percentage of people consider environmental values when making decisions about what to buy. Confirmation of recycled content and packaging suitability for recycling can be positive attributes for these customers, while overly optimistic claims about recyclability—where recyclability is potential rather than actual—can undermine consumer confidence. It has been observed that taking part in recycling programs is an environmentally friendly behavior that the general public engages in widely. In the UK, participation in recycling programs was 57% in a 2006 survey (WRAP 2008d), and in Australia, where curbside collection had been in place for longer, it was 80% (NEPC 2001).

Certain governments utilize regulations, like the EU Directive on packaging and packaging waste (94/62/EC), to promote post-consumer recycling. As a result, Germany established laws pertaining to extended producer responsibility, which gave rise to the die Grüne Punkt (Green Dot) program, which aims to recycle and recover packaging. Producer responsibility was implemented in the UK through a packaging recovery note generation and trading scheme, and more recently, a landfill levy to finance various waste reduction initiatives. Due to the aforementioned trends, recycled polymer's market value has significantly increased in recent years, indicating the viability of recycling.

Deposit-refund programs are another way to implement extended producer responsibility; these programs can apply to tires, batteries, and beverage containers, among other things. These programs, such as kerbside collection programs and container deposit programs in one Australian state that accepts PET soft drink bottles, can help increase collection rates. In this state, 74% of sales were collected in PET bottles, compared to 36% of sales in other states where kerbside collection was the only option. In comparison to other states, the percentage of bottles in litter was also lower (West 2007).

2.2.4 Economic Issues Relating To Recycling

The feasibility of recycling thermoplastics is influenced by two important economic factors. These are the costs of recycling in comparison to other acceptable forms of disposal and the price of recycled polymer compared to virgin polymer. When comparing variations in the quantity and quality of supply to virgin plastics, there are additional problems. Using recycled material may also be discouraged by a lack of knowledge regarding its quality, suitability for a given application, and availability.

Historically, landfills and incinerators have been the main methods used to dispose of waste. Based on the underlying geology and land-use patterns, landfill costs differ significantly between regions and can affect whether recycling is a viable alternative for disposal. For example, the hard volcanic bedrock beneath Japan makes the excavation required for landfills expensive, but in the Netherlands, the permeability from the sea makes it expensive. Exorbitant disposal costs provide a financial incentive for energy recovery or recycling.

Gathering discarded plastics from homes is more cost-effective in suburban areas where population density is high enough to realize economies of scale.

The best collection plan may differ depending on the area, the kind of housing (houses or big multi-apartment buildings), and the available sorting facilities. "Bring schemes," in which members of the public bring their own waste for recycling—for instance, when they visit a nearby town—are thought to be more economical in rural areas than curbside collection. In the UK, a lot of local governments and a few supermarkets run "bring banks" or even reverse vending machines. These latter techniques don't work well for collecting large amounts of post-consumer waste, but they can be a good source of relatively pure recyclables. It wasn't until the relatively recent introduction of curbside recycling that there was a noticeable increase in the collection of plastic bottle waste in the UK.

Since oil is the main raw material used to make plastic, the price of oil has an impact on the cost of virgin plastic. The price of virgin plastic sets the ceiling for recovered plastic prices because recovered plastic is usually of lower quality than virgin plastic. In the past few years, the price of oil has risen dramatically; from a range of about USD 25 per barrel to a price range between USD 50 and USD 150 since 2005. Recycling has thus become more financially appealing, even though rising oil prices also somewhat raise the cost of collection and reprocessing.

Technological developments in recycling can boost the economy in two main ways: first, by lowering the cost of recycling (through increases in productivity and efficiency); and second, by narrowing the difference in value between recycled and virgin resin. The latter is made possible by technologies that support closed-loop recycling by removing contamination and converting recovered plastic into food-grade polymer. rPET from clear bottles and, more recently, rHDPE from milk bottles have been demonstrated to be compatible with this technology (WRAP 2006; WRAP 2008b).

Therefore, recycling plastics without subsidies is becoming more and more feasible on a much larger geographic scale, as well as for post-consumer waste, whereas over ten years ago it was primarily only viable from post-industrial waste, or in locations where the cost of alternative forms of disposal was high.

2.2.5 Current Trends in Plastic Recycling

While the amount of mechanical recycling increased strongly at a rate of about 7% per year, plastic waste generation in Western Europe is growing at a rate of about 3% per year, roughly in line with long-term economic growth. Still, this only accounted for 14.8% of the total amount of plastic waste generated in 2003 (from all sources). When combined with feedstock recycling (1.7%), energy recovery (22.5%), and other factors, the total recovery rate from the 21.11 million tonnes of plastic waste produced in 2003 was about 39%. The trend of rising rates of energy recovery and mechanical recycling is still present, but so is the trend of rising waste generation.

2.2.6 Challenges and Opportunities for Improving Plastic Recycling

The next big issue facing the plastics recycling industry is how to recycle mixed plastic waste effectively. The benefit is that more plastic waste can be recycled by broadening the scope of post-consumer plastic packaging collection to include a greater range of materials and pack styles. There is a good chance that product design for recycling will help these kinds of recycling initiatives. According to a UK study, between 21 and 40 percent of the packaging in a typical shopping basket could not be recycled, even if it were collected (Local Government Association (UK) 2007). Therefore, a greater number of policies that encourage industry to use environmental design principles could have a significant effect on recycling performance by raising the amount of packaging that can be profitably collected and diverted from landfills (see Shaxson et al., 2009). The same reasoning holds true for durable consumer goods, where recycling and creating requirements for the use of recycled resins are essential steps in boosting recycling.

Since flexible packaging frequently causes issues during the collection and sorting stages, the majority of post-consumer collection schemes are designed for rigid packaging. Due to the different handling characteristics of rigid packaging, the majority of material recovery facilities today find it challenging to handle flexible plastic packaging. Moreover, investing in the required facilities for collection and sorting is less financially feasible due to the low weight-to-volume ratio of films and plastic bags. That being said, under the correct circumstances, this is feasible as plastic films are currently recycled from sources such as agricultural films and secondary packagings like pallet and box shrink-wrap. Strategies to increase the recycling of films and flexible packaging could involve investing in additional sorting and processing facilities at recovery facilities for handling mixed plastic wastes, or setting up separate collections. High-performance sorting of the input materials is required to guarantee that plastic types are separated to high levels of purity to successfully recycle mixed plastics; further development of end-markets is necessary for each polymer recyclate stream.

Reducing the variety of materials to a fraction of what is currently used could dramatically improve the efficacy of post-consumer packaging recycling. For instance, all rigid plastic packaging could be gathered and sorted to create recycled resins with the least amount of cross-contamination if rigid plastic containers, such as bottles, jars, and trays, were made entirely of PET,

HDPE, and PP and not clear PVC or PS, which are difficult to separate from mixed recyclables. Both the value of the recycled resins and the losses from rejected material would increase. To optimize recycling performance, labels and adhesive materials should also be chosen. By lowering waste fractions, energy, and water consumption, improvements in sorting and separation within recycling facilities offer additional potential for increased recycling volumes as well as improved eco-efficiency. Maximizing recycled resin volume and quality should be the main objective.

In conclusion, recycling is one tactic for managing plastic product waste at the end of its useful life. Both environmentally and economically, it makes more sense, and current trends show that the rate at which plastic waste is recovered and recycled has significantly increased. These trends are probably here to stay, but there are still some big obstacles to overcome, including technological ones as well as behavioral, economic, or social ones involving the collection of recyclable waste and the replacement of virgin materials.

Increased recovery rates of plastic waste and its diversion from landfills will be made possible by recycling a greater variety of post-consumer plastic packaging in addition to waste plastics from consumer goods and ELVs. Recycling waste plastics is a practical strategy to enhance the environmental performance of the polymer industry, especially when combined with initiatives to expand the use and specification of recycled grades as a substitute for virgin plastic.

REFERENCES

1. ACRR 2004*Good practices guide on waste plastics recycling* Brussels, Belgium: Association of Cities and Regions for Recycling
2. Aguado J., Serrano D. P., San Miguel G.2007European trends in the feedstock recycling of plastic wastes. *Global NEST J.* 9, 12–19
3. Aksa W., Medles K., Rezoug M., et al., Two stage electrostatic separator for the recycling of plastics from waste electrical and electronic equipment. *Journal of Electrostatics.* 2013;71:681–688.
4. Al-Maaded M., Madi N.K., Kahraman R., Hodzic A., Ozerkan N.G. An overview of solid waste management and plastic recycling in Qatar. *Journal of Polymers and the Environment.* 2012;20:186–194. doi: 10.1007/s10924-011-0332-2.
5. Al-Salem S.M., Antelava A., Constantinou A., Manos G., Dutta A. A review on thermal and catalytic pyrolysis of plastic solid waste (PSW) *Journal of Environmental Management.* 2017;197:177–198.
6. Al-Salem S.M., Lettieri P., Baeyens J. Recycling and recovery routes of plastic solid waste (PSW): A review. *Waste Management.* 2009;29:2625–2643.
7. Alter H. The recovery of plastics from waste with reference to froth flotation. *Resources, Conservation and Recycling.* 2005;43:119–132.
8. Ambrose C.A., Hooper R., Potter A.K., Singh M.M. Diversion from landfill: Quality products from valuable plastics. *Resources, Conservation and Recycling.* 2002;36:309–318.
9. Andrady A. L., Neal M. A.2009Applications and societal benefits of plastics. *Phil. Trans. R. Soc. B* 364, 1977–1984 (doi:10.1098/rstb.2008.0304)
10. Andrady A.1994Assessment of environmental biodegradation of synthetic polymers. *Polym. Rev.* 34, 25–76 (doi:10.1080/15321799408009632)
11. Andrady A.2003An environmental primer. In *Plastics and the environment* (ed. Andrady A.), pp. 3–76 Hoboken, NJ: Wiley Interscience
12. Anzano J., Casanova M.E., Bermudez M.S., Lasheras R.J. Rapid characterization of plastics using laser-induced plasma spectroscopy (LIPS) *Polymer Testing.* 2006;25:623–627.
13. APME 2004*An analysis of plastics consumption and recovery in Europe* Brussels, Belgium: Association of Plastic Manufacturers Europe
14. Arena U., Mastellone M., Perugini F.2003Life cycle assessment of a plastic packaging recycling system. *Int. J. Life Cycle Assess.* 8, 92–98 (doi:10.1007/BF02978432)

15. Arvanitoyannis I., Bosnea L.2001Recycling of polymeric materials used for food packaging: current status and perspectives. *Food Rev. Int.* 17, 291–346 (doi:10.1081/FRI-100104703)
16. ASTM Standard D5033 2000 *Standard guide to development of ASTM standards relating to recycling and use of recycled plastics* West Conshohocken, PA: ASTM International; (doi:10.1520/D5033-00)
17. Bakker E.J., Rem P.C., Fraunholcz N. Upgrading mixed polyolefin waste with magnetic density separation. *Waste Management.* 2009;29(5):1712–1717. doi: 10.1016/j.wasman.2008.11.006. (ISSN 0956-053X)
18. Barnes D. K. A., Galgani F., Thompson R. C., Barlaz M.2009Accumulation and fragmentation of plastic debris in global environments. *Phil. Trans. R. Soc. B* 364, 1985–1998 (doi:10.1098/rstb.2008.0205)
19. Barnes D.K.A., Galgani F., Richard C., Thompson R.C., Barlaz M. Plastic accumulation and fragmentation. *Philosophical Transactions of the Royal Society B.* 2009;364:1985–1998. doi: 10.1098/rstb.2008.0205.
20. Bendimerad S., Tilmatine A., Ziane M., Dascalescu L. Plastic wastes recovery using free-fall triboelectric separator. *International Journal of Environmental Studies.* 2009;66(5):529–538. doi: 10.1080/00207230902722838.
21. Bertolini G.E., Fontaine J. Value recovery from plastics waste by pyrolysis in molten salts. *Conservation & Recycling.* 1987;10(4):331–343.
22. Boss, A., Boström, J.O., Nilsson, P.H., *et al.*, 2011. New technology for recycling of plastics from cable waste. In: Proceedings of the Jicable' 11. Versailles.
23. Bouhamri N., Zelmat M.E., Tilmatine A. Micronized plastic waste recycling using two-disc tribo-electrostatic separation process. *Advanced Powder Technology.* 2019;30:625–631.
24. Brebu M., Bhaskar T., Muto A., Sakata Y. Alkaline hydrothermal treatment of brominated high impact polystyrene (HIPS-Br) for bromine and bromine-free plastic recovery. *Chemosphere.* 2006;64:1021–1025.
25. Brems A., Baeyens J., Dewil R. Recycling and recovery of post-consumer plastic solid waste in a European context. *Thermal Science.* 2012;16(3):669–685.
26. Briassoulis D., Hiskakis M., Babou E. Technical specifications for mechanical recycling of agricultural plastic waste. *Waste Management.* 2013;33:1516–1530.
27. British Plastics Federation. Oil consumption. 2008. See http://www.bpf.co.uk/Oil_Consumption.aspx. (20 October 2008)

28. Brock, J., 2020. Pandemic exposes cracks in oil majors′ bet on plastic. Reuters. Available at: https://www.reuters.com/article/us-health-coronavirus-plastic-analysis/pandemic-exposes-cracks-in-oil-majors-bet-on-plastic, (idUSKBN23B0UT).
29. Carvalho T., Durão F., Ferreira C. Separation of packaging plastics by froth flotation in a continuous pilot plant. *Waste Management.* 2010;30:2209–2215.
30. Chanda M., Roy S.2007*Plastics technology handbook* 4th Edn. Boca Raton, FL: CRC Press
31. Chauhan V., Varis J., Karki T. The potential of reusing technical plastics. *Procedia Manufacturing.* 2019;39:502–508.
32. Chung T.L., Liao C.J., Chang-Chien G.P. Distribution of polycyclic aromatic hydrocarbons and polychlorinated dibenzo-p-dioxins/dibenzofurans in ash from different units in a municipal solid waste incinerator. *Waste Management and Research.* 2010;28:789.
33. DEFRA 2007Waste strategy factsheets. See http://www.defra.gov.uk/environment/waste/strategy/factsheets/landfilltax.htm (26 November 2008)
34. Department of Environment and Conservation (NSW) 2005*Benefits of recycling* Australia: Parramatta
35. Department of Environment and Heritage (Australia) 2008. Plastic bags. See http://www.ephc.gov.au/ephc/plastic_bags.html (26 November 2008)
36. Department of Environment Heritage and Local Government (Ireland) 2007. Plastic bags. See http://www.environ.ie/en/Environment/Waste/PlasticBags (26 November 2008)
37. Devasahayam S., Raju B.G., Hussain M.C. Utilization and recycling of end of life plastics for sustainable and clean industrial processes including the iron and steel industry. *Materials Science for Energy Technologies.* 2019;2:634–646.
38. Drelich J., Payne J., Kim T., Miller J.1998Selective froth floatation of PVC from PVC/PET mixtures for the plastics recycling industry. *Polym. Eng. Sci.* 38, 1378 (doi:10.1002/pen.10308)
39. EEA 2008*Better management of municipal waste will reduce greenhouse gas emissions* Copenhagen, Denmark: European Environment Agency
40. Ferella F., Belardi G., Marsilii A., Michelis I.D., Vegliò F. Separation and recovery of glass, plastic and indium from spent LCD panels. *Waste Management.* 2017;60:569–581.
41. Fisher M.2003Plastics recycling. In *Plastics and the environment* (ed.

Andrady A.), pp. 563–627 Hoboken, NJ: Wiley Interscience

42. Fisher, M., Kingsbury, T., Headley, L., 2004. Sustainable electrical and electronic plastics recycling. In: Proceedings of the IEEE International Symposium on Electronics and the Environment, Conference Record, pp. 292–297. Scottsdale, AZ. doi: 10.1109/ISEE.2004.1299732.
43. Fletcher B., Mackay M.1996A model of plastics recycling: does recycling reduce the amount of waste? *Resour. Conserv. Recycling* 17, 141–151 (doi:10.1016/0921-3449(96)01068-3)
44. Frosch R., Gallopoulos N.1989Strategies for manufacturing. *Sci. Am.* 261, 144–152
45. Garforth A., Ali S., Hernandez-Martinez J., Akah A.2004Feedstock recycling of polymer wastes. *Curr. Opin. Solid State Mater. Sci.* 8, 419–425 (doi:10.1016/j.cossms.2005.04.003)
46. Geyer R., Blass V.D. The economics of cell phone reuse and recycling. *The International Journal of Advanced Manufacturing Technology.* 2010;47:515–525. doi: 10.1007/s00170-009-2228-z.
47. Gil-Jasso N.D., Segura-Gonzáleza M.A., Soriano-Gilesa G., et al., Dissolution and recovery of waste expanded polystyrene using alternative essential oils. *Fuel.* 2019;239:611–616.
48. Gilpin R., Wagel D., Solch J.2003Production, distribution, and fate of polycholorinated dibenzo-p-dioxins, dibenzofurans, and related organohalogens in the environment. In *Dioxins and health* (eds Schecter A., Gasiewicz T.), 2nd Edn Hoboken, NJ: John Wiley & Sons Inc
49. Gregory M. R. 2009Environmental implications of plastic debris in marine settings—entanglement, ingestion, smothering, hangers-on, hitch-hiking and alien invasions. *Phil. Trans. R. Soc. B* 364, 2013–2025 (doi:10.1098/rstb.2008.0265)
50. Gu F., Guo J., Zhang W., Summers P.A., Hall P. From waste plastics to industrial raw materials: a life cycle assessment of mechanical plastic recycling practice based on a real-world case study. *Science of the Total Environment.* 2017;601–602:1192–1207.
51. Hahladakis J.N., Aljabri H.M.S.J. Delineating the plastic waste status in the State of Qatar: Potential opportunities, recovery and recycling routes. *Science of the Total Environment.* 2019;653:294–299.
52. Hamad K., Kaseem M., Deri F. Recycling of waste from polymer materials: An overview of the recent works. *Polymer Degradation and Stability.* 2013;98:2801–2812.
53. Hopewell J., Dvorak R., Kosior E. Plastics recycling: Challenges

and opportunities. *Philosophical Transactions of the Royal Society B.* 2009;364:2115–2126.

54. Horodytska O., F.J. Valdés F.J., Fullana A. Plastic flexible films waste management – A state of art review. *Waste Management.* 2018;77:413–425.
55. Hou P., Xu Y., Taiebat M., et al., Life cycle assessment of end-of-life treatments for plastic film waste. *Journal of Cleaner Production.* 2018;201:1052–1060.
56. Hu B., Serranti S., Fraunholcz N., Maio F.D., Bonifazi G. Recycling-oriented characterization of polyolefin packaging waste. *Waste Management.* 2013;33:574–584.
57. Institute for Local Self-Reliance 2002Western Europe's experience with refillable beverage containers. See http://www.grrn.org/beverage/refillables/Europe.html (26 November 2008)
58. Kalanatarifard A., Yang G.S. Identification of the municipal solid waste characteristics and potential of plastic recovery at Bakri Landfill, Muar, Malaysia. *Journal of Sustainable Development.* 2012;5(7) (ISSN 1913-9063 E-ISSN 1913-9071)
59. Kaminsky W. Chemical recycling of mixed plastics by pyrolysis. *Advances in Polymer Technology.* 1995;14(4):337–344.
60. Kaminsky W., Schlesselmann B., Simon C.M. Thermal degradation of mixed plastic waste. *Polymer Degradation and Stability.* 1996;53:189–197.
61. Kamo, T., 2011. Overview of the waste plastic recycling system in Japan and future tasks. In: Proceedings of the 6th International Symposium on Feedstock Recycling of Polymeric Materials ISFR 2011, pp 15–18. Toledo.
62. Kang H.-Y., Schoenung J.M. Electronic waste recycling: A review of U.S. infrastructure and technology options. *Resources, Conservation and Recycling.* 2005;45:368–400.
63. Kumar S., Panda A.K., Singh R.K. A review on tertiary recycling of high-density polyethylene to fuel. *Resources. Conservation and Recycling.* 2011;55:893–910.
64. Kyrikou I., Briassoulis D.2007Biodegradation of agricultural plastic films: a critical review. *J. Polym. Environ.* 15, 125–150 (doi:10.1007/s10924-007-0053-8)
65. Landauer J., Foerst P. Triboelectric separation of a starch-protein mixture – Impact of electric field strength and flow rate. *Advanced Powder Technology.* 2018;29:117–123.

66. Lazarevic D., Aoustina E., Bucletb N., Brandt N. Plastic waste management in the context of a European recycling society: Comparing results and uncertainties in a life cycle perspective. *Resources. Conservation and Recycling.* 2010;55:246–259.
67. Li J., Lu H., Liu S., Xu Z. Optimizing the operating parameters of corona electrostatic separation for recycling waste scraped printed circuit boards by computer simulation of electric field. *Journal of Hazardous Materials.* 2008;153:269–275.
68. Lindahl, M., Winsnes, M., 2005. Recycling of cable plastics – A life cycle assessment of several different alternatives. In: Proceedings of the Fourth International Symposium on Environmentally Conscious Design and Invers Manufacturing, Eco Design 2005, pp. 539–546. IEEE.
69. Local Government Association (UK) 2007*War on waste: food packaging study* UK: Local Government Authority
70. López A., Marcoa I., Caballeroa B.M., et al., Catalytic pyrolysis of plastic wastes with two different types of catalysts: ZSM-5 zeolite and Red Mud. *Applied Catalysis B: Environmental.* 2011;104:211–219.
71. Lopez G., Lopez G., Artetxe M., et al., Recent advances in the gasification of waste plastics. A critical overview. *Renewable and Sustainable Energy Reviews.* 2018;82:576–596.
72. Manzoor, J., Sharma, M., Sofi, R.I., Dar, A., 2020. Plastic Waste Environmental and Human Health Impacts. doi: 10.4018/978-1-5225–9452-9.ch002. Available at: https://www.researchgate.net/publication/338749809.
73. Marques G., Tenorio J.2000Use of froth flotation to separate PVC/PET mixtures. *Waste Management* 20, 265–269 (doi:10.1016/S0956-053X(99)00333-5)
74. McDonough W., Braungart M.2002*Cradle to cradle: remaking the way we make things* New York, NY: North Point Press
75. Miandad R., Barakat M.A., Rehan M., et al., Plastic waste to liquid oil through catalytic pyrolysis using natural and synthetic zeolite catalysts. *Waste Management.* 2017;69:66–78.
76. Morris J.1996Recycling versus incineration: an energy conservation analysis. *J. Hazard. Mater.* 47, 277–293 (doi:10.1016/0304-3894(95)00116-6)
77. Mwanza B.G., Mbohwa C. Drivers to sustainable plastic solid waste recycling: A Review. *Procedia Manufacturing.* 2017;8:649–656.
78. Naik R.K., Naik M.M., D'Costa P.M., Shaikh F. Microplastics in ballast

water as an emerging source and vector for harmful chemicals, antibiotics, metals, bacterial pathogens and HAB species: A potential risk to the marine environment and human health. *Marine Pollution Bulletin.* 2019;149

79. Najafi S.K. Use of recycled plastics in wood plastic composites – A review. *Waste Management.* 2013;33:1898–1905.
80. NEPC 2001 Report to the NEPC on the implementation of the National Environment Protection (used packaging materials) measure for New South Wales. Adelaide, Australia: Environment Protection and Heritage Council
81. Oehlmann J., et al.,2009. A critical analysis of the biological impacts of plasticizers on wildlife. *Phil. Trans. R. Soc. B* 364, 2047–2062 (doi:10.1098/rstb.2008.0242)
82. Oku, A., 2005. Chemical conversion of plastic wastes to monomers. In: Müller-Hagedorn, M., Bockhorn, H. (Eds.), Feedstock Recycling of Plastics. Selected Papers presented at Proceedings of the Third International Symposium on Feedstock Recycling of Plastics & Other Innovative Plastics Recycling Techniques, September 25–29, 2005. Karlsruhe.
83. Pacheco E.B.A.V., Ronchettia L.M., Masanet E. An overview of plastic recycling in Rio de Janeiro. *Resources, Conservation and Recycling.* 2012;60:140–146.
84. Park C.-H., Jeon H.-S., Park K.2007PVC removal from mixed plastics by triboelectrostatic separation. *J. Hazard. Mater.* 144, 470–476 (doi:10.1016/j.jhazmat.2006.10.060)
85. Patel M., von Thienen N., Jochem E., Worrell E.2000Recycling of plastics in Germany. *Resour., Conserv. Recycling* 29, 65–90 (doi:10.1016/S0921-3449(99)00058-0)
86. Perugini F., Mastellone M., Arena U. 2005 A life cycle assessment of mechanical and feedstock recycling options for management of plastic packaging wastes. *Environ. Progr.* 24, 137–154 (doi:10.1002/ep.10078)
87. Pivnenko K., Eriksen M.K., Martín-Fernández J.A., Eriksson E., Astrup T.F. Recycling of plastic waste: Presence of phthalates in plastics from households and industry. *Waste Management.* 2016;54:44–52.
88. Plastics Europe, 2019. Plastics- the Facts 2019. An Analysis of European Plastics Production, Demand and Recovery for 2018. Plastics Europe Association of Plastics Manufacturers. Brussels. Available at: http://www.plasticseurope.org.
89. PlasticsEurope 2008aEco-profiles of the European Plastics Industry. Brussels, Belgium: PlasticsEurope; See http://lca.plasticseurope.org/index.htm (1 November 2008)

90. PlasticsEurope 2008bThe compelling facts about Plastics 2007: an analysis of plastics production, demand and recovery for 2007 in Europe. Brussels, Belgium: PlasticsEurope
91. Ragaert K., Delva L., Geem K.V. Mechanical and chemical recycling of solid plastic waste. *Waste Management.* 2017;69:24–58.
92. Rebeiz K., Craft A. 1995 Plastic waste management in construction: technological and institutional issues. *Resour., Conserv. Recycling* 15, 245–257 (doi:10.1016/0921-3449(95)00034-8)
93. Rebeiz K.S., Craft A.P. Plastic waste management in construction: Technological and institutional issues. *Resources, Conservation and Recycling.* 1995;15(1995):245–257.
94. Rigamonti L., Grossoa M., Møllerb J., et al., Environmental evaluation of plastic waste management scenarios. *Resources, Conservation and Recycling.* 2014;85:42–53.
95. Ryan P. G., Moore C. J., van Franeker J. A., Moloney C. L.2009Monitoring the abundance of plastic debris in the marine environment. *Phil. Trans. R. Soc. B* 364, 1999–2012 (doi:10.1098/rstb.2008.0207)
96. Serranti S., Gargiulo A., Bonifazi G. Characterization of post-consumer polyolefin wastes by hyperspectral imaging for quality control in recycling processes. *Waste Management.* 2011;31(11):2217–2227. doi: 10.1016/j.wasman.2011.06.007. (ISSN 0956-053X)
97. Shaxson L.2009Structuring policy problems for plastics, the environment and human health: reflections from the UK. *Phil. Trans. R. Soc. B* 364, 2141–2151 (doi:10.1098/rstb.2008.0283)
98. Shen L., Worrell E., Patel M.K. Open-loop recycling: A LCA case study of PET bottle-to-fibre recycling. *Resources, Conservation and Recycling.* 2010;55:34–52.
99. Siddique R., Khatib J., Kaur I. Use of recycled plastic in concrete: A review. *Waste Management.* 2008;28:1835–1852.
100. Singh N., Hui D., Singh R., et al., Recycling of plastic solid waste: A state of art review and future Applications. *Composites Part B.* 2017;115:409–422.
101. Sinha V., Patel M., Patel J. 2008 PET waste management by chemical recycling: a review. *J. Polym. Environ.* (doi:10.1007/s10924-008-0106-7)
102. Smith M., Love D.C., Rochman C.M., Neff R.A. Microplastics in seafood and the implications for human health. *Current Environmental Health Reports.* 2018;5:375–386. doi: 10.1007/s40572-018-0206-z.
103. Soetaert W., Vandamme E.2006The impact of industrial

biotechnology. *Biotechnol. J.* 1, 756–769 (doi:10.1002/biot.200600066)

104. Song J. H., Murphy R. J., Narayan R., Davies G. B. H.2009Biodegradable and compostable alternatives to conventional plastics. *Phil. Trans. R. Soc. B* 364, 2127–2139 (doi:10.1098/rstb.2008.0289)
105. Soroudi A., Jakubowicz I. Recycling of bioplastics, their blends and bio composites: A review. *European Polymer Journal.* 2013;49:2839–2858.
106. Swift G., Wiles D.2004Degradable polymers and plastics in landfill sites. *Encyclopedia Polym. Sci. Technol.* 9, 40–51
107. Tarantili P.A., Mitsakaki A.N., Petoussi M.A. Processing and properties of engineering plastics recycled from waste electrical and electronic equipment (WEEE) *Polymer Degradation and Stability.* 2010;95:405–410.
108. Teuten E. L., et al., 2009 Transport and release of chemicals from plastics to the environment and to wildlife. *Phil. Trans. R. Soc. B* 364, 2027–2045 (doi:10.1098/rstb.2008.0284)
109. Thakur S., Verma A., Sharma B., et al., Recent developments in recycling of polystyrene based plastics. *Current Opinion in Green and Sustainable Chemistry.* 2018;13:32–38.
110. Thompson R. C., Moore C. J., vom Saal F. S., Swan S. H.2009bPlastics, the environment and human health: current consensus and future trends. *Phil. Trans. R. Soc. B* 364, 2153–2166 (doi:10.1098/rstb.2009.0053)
111. Thompson R. C., Swan S. H., Moore C. J., vom Saal F. S.2009aOur plastic age. *Phil. Trans. R. Soc. B* 364, 1973–1976 (doi:10.1098/rstb.2009.0054)
112. Thompson R.C., Moore C.J., vom Saal F.S., Swan S.H. Plastics, the environment and human health: Current consensus and future trends. *Philosophical Transactions of the Royal Society of London. Series B, Biological Sciences.* 2009;364(1526):2153–2166. doi: 10.1098/rstb.2009.0053.
113. Thunman H., Vilches T.B., Seemann M., et al., Circular use of plastics-transformation of existing petrochemical clusters into thermochemical recycling plants with 100% plastics. *Materials and Technologies.* 2019;22
114. Uhde Inventa-Fischer 2007Flakes-to-resin (FTR)—recycling with FDA approval. See http://www.uhde-inventa-fischer.com/ (3 November 2008)
115. Wang C., Wang H., Gu G., et al., Ammonia modification for flotation separation of polycarbonate and polystyrene waste plastics. *Waste Management.* 2016;51:13–18.
116. Wang H., Xiao-lei C., Yang B., Chao G., Li Z. Application of dissolved

air flotation on separation of waste plastics ABS and PS. *Waste Management.* 2012;32:1297–1305.

117. Waring R.H., Harris R.M., Mitchell S.C. Plastic contamination of the food chain: A threat to human health? *Maturitas.* 2018;115:64–68.
118. Waste Watch 2003*Plastics in the UK economy* London, UK: Waste Watch & Recoup
119. WBCSD 2000Eco-efficiency: creating more value with less impact. See http://www.wbcsd.org/includes/getTarget.asp?type=d&id=ODkwMQ (1 November 2008)
120. Wemken N., Drage D.S., Abdallah M.A., Harrad S., Coggins M.A. Concentrations of brominated flame retardants in indoor air and dust from Ireland reveal elevated exposure to decabromodiphenyl ethane. *Environmental Science & Technology.* 2019;53(16):9826–9836. doi: 10.1021/acs.est.9b02059.
121. West D.2007*Container deposits: the common sense approach v.2.1* Sydney, Australia: The Boomerang Alliance
122. Wong S.L., Ngadi N., Abdullah T.A.T., Inuwa I.M. Current state and future prospects of plastic waste as source of fuel: a review. *Renewable and Sustainable Energy Reviews.* 2015;50:1167–1180.
123. Wongthatsanekorn W. A goal programming approach for plastic recycling system in Thailand. *International Journal of Industrial and Manufacturing Engineering.* 2009;3(1):23–28.
124. WRAP 2006*WRAP food grade HDPE recycling process: commercial feasibility study* London, UK: Waste & Resources Action Programme
125. WRAP 2007. In *Annual local authorities Plastics Collection Survey 2007* London, UK: Waste Reduction Action Plans
126. WRAP 2008a*Domestic mixed plastics packaging: waste management options* London, UK: Waste & Resources Action Programme
127. WRAP 2008bLarge-scale demonstration of viability of recycled PET (rPET) in retail packaging. http://www.wrap.org.uk/retail/case_studies_research/rpet_retail.html.
128. WRAP 2008c*LCA of management options for mixed waste plastics* London, UK: Waste & Resources Action Programme, MDP017
129. WRAP 2008d*Local authorities Plastics Collection Survey* London, UK: Waste & Resources Action Programme
130. WRAP 2008eStudy reveals carbon impact of bottling Australian wine in the UK in PET and glass bottles. See http://www.wrap.org.uk/wrap_corporate/news/study_reveals_carbon.html (19 October 2008)

131. Wright S.L., Thompson R.C., Galloway T.S. The physical impacts of microplastics on marine organisms: A review. *Environmental Pollution.* 2013;178:483–492.
132. Wu G., Li J., Xu Z. Triboelectrostatic separation for granular plastic waste recycling: A review. *Waste Management.* 2013;33:585–597.
133. Wu Z.W., Liu G.F., Song S.X., Pan S.B. Regeneration and recycling of waste thermosetting plastics based on mechanical thermal coupling fields. *International Journal of Precision Engineering and Manufacturing.* 2014;15(12):2639–2647.
134. Yates M.R., Barlow C.Y. Life cycle assessments of biodegradable, commercial biopolymers – A critical review. *Resources, Conservation and Recycling.* 2013;78:54–66.
135. Zare Y. Recent progress on preparation and properties of nanocomposites from recycled polymers: A review. *Waste Management.* 2013;33:598–604.
136. Zhao Y.B., Lv X.D., Ni H.G. Solvent-based separation and recycling of waste plastics: A review. *Chemosphere.* 2018;209:707–720.
137. Zhou C., Fang W., Xu W., Cao A., Wang R. Characteristics and the recovery potential of plastic wastes obtained from landfill mining. *Journal of Cleaner Production.* 2014;80:80–86.
138. Zhou Y., Guan J., Wang J., Wu W., Zhang C. Life cycle analysis on regenerating and energy recovery of plastic solid waste management in Shanghai. *Applied Mechanics and Materials.* 2015;768:426–433.

Chapter **3**

Recycling of the Engineering Plastics PC, HIPS, ABS and PA, Their Blends and Composites

Contents

INTRODUCTION

Without a doubt, plastics play a crucial role in daily life. Plastic components can be found in practically every application, from food containers to aerospace applications (Figure 3.1). Engineering polymers are currently gaining a significant market share among other plastic categories. The phrase "engineering plastics" describes polymeric materials that, when compared to standard commodity plastics, exhibit superior mechanical and thermal properties, such as increased toughness, strength, and heat resistance. Their use has also increased over conventional engineering materials like metals due to their superior physical characteristics and operating temperatures. Engineering plastics can be used in demanding and load-bearing applications, such as automotive, electric and electronic equipment, etc., because of their improved performance. Poly(acrylonitrile—butadiene—styrene) (ABS), nylon or polyamide (PA), high impact polystyrene (HIPS), polycarbonates (PC), and other materials are representative examples of engineering plastics (Shrivastava 2018; Crawford and Martin 2020).

Their increased consumption as a result of their benefits also increased the amount of waste they produced, particularly from electric and electronic equipment. The growing amount of plastic waste coupled with its non-biodegradable nature necessitated the creation of ecologically friendly methods for handling it. Thus, engineering plastics were collected, handled, and recycled when their useful lives were coming to an end. Finding and implementing environmentally friendly ways to dispose of them is challenging, though, as a variety of additives are added to polymers to enhance their properties in addition to the wide range of polymers that are used (Tarantili et al., 2010; Buekens and Yang 2014; Crawford and Martin 2020; Charitopoulou et al., 2020).

There are numerous difficulties in handling and recycling post-consumed plastics (Figure 3.1). Specifically, some of the problems that need to be resolved during recycling are the existence of various polymer types (possibly with unknown compositions), polymeric blends or composites, multilayered structures with occasionally non-polymeric materials, and the presence of additives.

Over the past ten years, various techniques have been used to handle engineering plastics, including mechanical recycling, chemical recycling, energy recovery, landfilling, and primary recycling. Because landfilling contributes to environmental pollution, it is undesirable. Recently, the Ragaert et al., (2017) group reviewed processes for recycling plastic waste mechanically and chemically.

Table 3.1. *History of plastic development*

Chronicle Year	Representative Event	References
15th century	Start using natural rubber	American Mayans
1839	Co-heating of rubber and sulfur into elastic and plastic polymers	Charles Goodyear Father of vulcanized rubber
1869	Created the first synthetic plastic "celluloid."	John Wesley Hyatt
1909	Invented the first artificial plastic- phenolic resin. Commonly known as bakelite	Leo Hendrik Baekeland
1920	The term polymer and polymer officially appeared	Hermann Staudinger
1926	The earliest synthetic plastic Polyvinyl Chloride	Waldo Semon
1930	Extracted polystyrene	Eduard Simon
1933	Synthesized polyethylene	Eric Fawcett and Reginald Gibson
1954	The advent of polypropylene	Karl Ziegler and Giulio Natta
1967	The Birth of PET Bottle—Polyethylene Terephthalate	Nathaniel Wyeth
2004	Coined the word microplastics	Richard Thompsom

2016	Microplastic pollution is officially listed as the second-largest issue in environmental and ecological science.	UNEA United Nations Environment Assembly
2023	Taiwan banned the use of polyvinyl chloride in food packaging starting in July.	Executive Yuan Environmental Protection Agency PVC

Source: Lin Y-D, Huang P-H, Chen Y-W, Hsieh C-W, Tain Y-L, Lee B-H, Hou C-Y, Shih M-K. Sources, Degradation, Ingestion and Effects of Microplastics on Humans: A Review. *Toxics*. 2023; 11(9):747. https://doi.org/10.3390/toxics11090747

Chemical recycling appears to be the best option among the various recycling and recovery techniques (Yang et al., 2013; Charitopoulou et al., 2020). Within the category of chemical recycling techniques, pyrolysis is a sustainable method that yields secondary valuable materials, recovers monomers, and allows for the use of the liquid pyrolysis fraction as fuel for energy production (Achilias et al., 2012). Research is concentrated on it because it is thought to be the most promising technique for recycling waste engineering plastics.

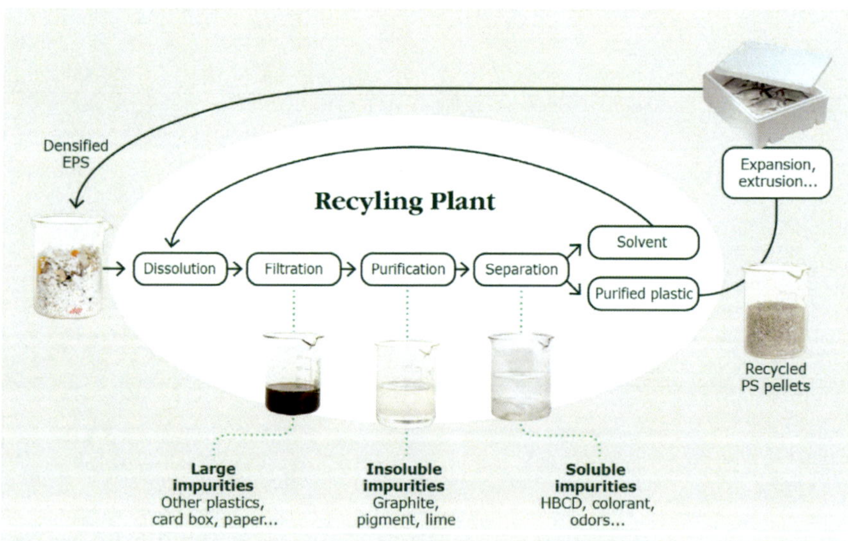

Figure 3.1. Challenges in post-consumed plastic recycling.

Source: *Pin J-M, Soltani I, Negrier K, Lee PC. Recyclability of Post-Consumer Polystyrene at Pilot Scale: Comparison of Mechanical and Solvent-Based Recycling Approaches. Polymers. 2023; 15(24):4714. https://doi.org/10.3390/polym15244714*

The statistics pertaining to the socio-economic practices of plastics recycling are another facet of the matter that requires our attention. Approximately 3,000 businesses in Europe deal with mechanical recycling, and another 100 businesses buy recycled plastics that have been mechanically processed (pellets and flakes). Merely 13% is directly incorporated into final goods, primarily inexpensive plant pots, grilles, and floor coverings. The European Union has an annual capacity of 5 million tons for mechanical recycling and only 50,000 tons for chemical recycling, according to the Institute for Prospective Technological Studies of the European Commission. The demand for recycled polymers cannot even be partially met by Europe's current consumption levels. The European market for recycled materials is undoubtedly immature at this point, requiring a more thorough and effective exchange of information between recycling companies and manufacturers of final goods. A significant rise in the need for plastics (particularly for nonwoven hygiene products, construction, agriculture, and food packaging) and an abundance of end-of-life streams in Europe should prove to be crucial advantages for the businesses. Recyclables can be optimized and intensified to reduce reliance on raw materials derived from oil (Ignatyev et al., 2014).

3.1 ENGINEERING PLASTICS: DEFINITION, PROPERTIES, AND TYPES

The most important component of the plastic industry is still, for the most part, plastic fabrication. It entails employing a variety of techniques to design and produce plastic products that satisfy end users' specifications using a range of plastic materials. However, some plastics that fall under the engineering plastics category have different plastic grades than regular plastic materials. Thus, the types, characteristics, and uses of engineering plastics are covered in this section.

3.1.1 What are Engineering Plastics?

A class of synthetic resins known as engineering plastics, or thermoplastic polymers, provide superior plastic properties and high performance over other common plastic materials.

More specifically, compared to commonly used commodity plastics, engineering plastics perform significantly better in structural applications due to their wide range of properties, particularly mechanical and/or thermal properties. They also resist significant mechanical stress, and environmental changes in terms of chemical and/or physical properties, and they remain stable across a broad temperature range.

Figure 3.2. Engineering plastic parts.

Source: *https://commons.wikimedia.org/wiki/File:Plastic_objects.jpg*

Additionally, part manufacturers have special access to engineering plastics in their raw form so they can use plastic engineering to create the desired end products.

3.1.2 Engineering Plastics vs Commodity Plastics: What's the Difference?

Plastic grades used in high-volume applications where technical performance is not critical, such as everyday household items, are classified as commodity plastics. Polypropylene, polyethylene, polyvinyl chloride, polyurethane, polystyrene, and other polymers are examples of commodity plastics. Commodity plastics are still widely used and accessible despite not being as advanced as engineering plastics. They are utilized for disposables, apparel, containers, packaging, and other things.

Engineering plastics like polycarbonate, polyamide, polyphenylene oxide, etc. are the result of ongoing research into better plastic materials and technological advancements.

Because of their excellent performance and combined strength, these plastic materials frequently gradually replace more conventional engineering materials like metals and ceramics for particular applications. Additionally, standard engineering plastics offer the following benefits in contrast to commodity plastics:

- Heat stability up to temperatures between 120 to 150°C.
- Excellent machinability and moldability into various shapes.
- Lightweight, higher strength, and minimal friction.
- Decent chemical resistance, resilience, stiffness, durability, and electrical properties.

For these reasons, engineering plastics are particularly useful in the manufacturing of mechanical parts for a variety of industries, including consumer goods, aerospace, automotive, medical, and electrical and electronics. However, engineering plastic materials are more expensive than commodity plastics, which makes them suitable for use in low-volume or smaller-product applications.

3.1.3 Properties of Engineering Plastics

Many plastic grades cut across multiple industries for a long time. There are currently more than 50 distinct types of plastic materials, and because of inherent qualities and design considerations, each type of plastic material is frequently restricted to use for particular applications. As a result, not every plastic grade is appropriate for use in technical and engineering settings.

Figure 3.3. Properties of engineering plastics.

Source: *By Phrontis - Own work, CC BY-SA 3.0, https://commons.wikimedia.org/w/index.php?curid=15436176*

Nevertheless, below are some general properties of engineering plastics:

- Good mechanical properties (abrasion resistance, impact strength)
- Optical Properties (translucent and transparent)
- Chemical Resistance

- Dimensional Stability
- Low Flammability
- Thermal Resistance (minimum and maximum operating temperature)
- Water repellent
- Sliding Properties
- Electrical Properties

3.1.4 A Brief Overview of Plastic Engineering

The creation of plastic components from appropriate plastic materials is the main goal of the field of plastic engineering. It involves a number of procedures, including these:

- Procurement of suitable plastic materials for projects
- Plastic product design based on specifications
- Manufacturing Plastic Parts
- Performance testing of fabricated plastic products

Figure 3.4. Plastic engineering.

Source: *By Maarten Gielen - I'm the author, CC BY-SA 4.0, https://commons.wikimedia.org/w/index.php?curid=6096649*

Above all, in order to meet the requirements for particular applications while minimizing operating costs, plastic engineers must address and quantify the critical properties of raw plastic materials. Therefore, during the manufacturing processes, plastic engineers face certain challenges due to the inherent characteristics of plastic materials.

For precisely and effectively machining plastic components made of engineering plastics, CNC plastic machining services offer practical answers. Because it is more accurate, efficient, and works with a variety of engineering plastic materials, this automated technology is useful in the production of engineering plastic products.

3.1.5 Common Types of Engineering Plastics

The uses of engineering plastics are numerous. As a result, selecting the appropriate material for your machining project can be challenging at times. Additionally, different engineering plastic kinds have different properties that determine how they should be used for different machining tasks and purposes. However, you can use plastic prototyping options to help you make better decisions. These represent a few of the most popular varieties of engineering plastics.

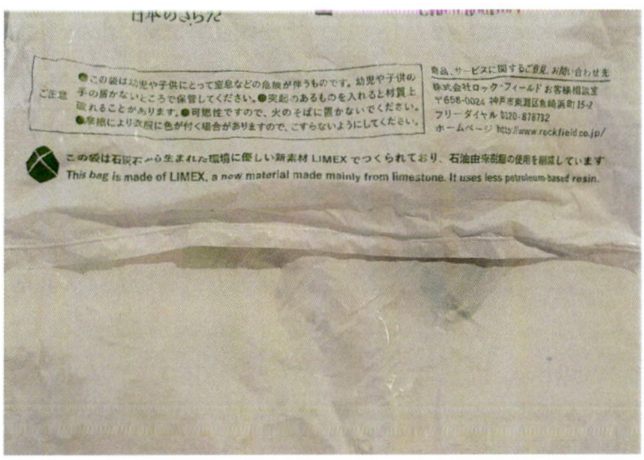

Figure 3.5. Plastic materials.

Source: *By Syced - Own work, CC0, https://commons.wikimedia.org/w/index.php?curid=116802895*

1. Polycarbonates (PC)

Carbonate groups are a component of PC plastics' chemical composition. This kind of engineering plastic is made up of thermoplastic polymers that combine impact strength, durability, stiffness, and toughness. They are simple to shape or machine into the correct shapes. Furthermore, in contrast to ordinary plastics, polycarbonates have good electrical, weathering, and thermal properties. Additionally, some PC plastic grades have optical transparency. Because of these combined qualities, it can be used to make optical discs, bumpers, and motorcycle helmets.

2. Acrylonitrile Butadiene Styrene (ABS)

As the name implies, acrylonitrile and styrene are polymerized in the presence of polybutadiene additive to form ABS, a copolymer. Because of its many qualities, including heat resistance, flame retardancy, chemical resistance, electrical insulation, easy coloring, impervious surface, impact strength, etc., ABS performs well. Furthermore, the proportionate composition of the monomers that comprise ABS can be altered to further improve its properties related to insulation and heat resistance. These characteristics, along with the inexpensive production cost, make it a good choice for 3D printing and CNC machining. Professionals frequently use ABS on automobile bumpers, wheel covers, and dashboard trims.

3. Polymethylmethacrylate (PMMA)

Another popular engineering plastic used for better performance is PMMA plastic, also affectionately known as acrylic, particularly in the building sector. It looks good and has excellent chemical resistance, temperature stability, optical transparency, and surface strength. In addition, PMMA compares favorably to aluminum due to its excellent wear resistance and lightweight nature. The icing on the cake would be appropriate processing, such as molding or machining, to particular plastic components like shields and taillights.

4. Polyphenylene Oxide (PPO)

Without PPO, the list of typical engineering plastic types is still lacking. High rigidity and hardness, impact strength, wear resistance, good dimensional stability, and electrical properties are all provided by this crystalline thermoplastic grade. Not to mention, modified PPO provides remarkable resistance to heat distortion from above 100°C to 195°C. Because of this, this engineering plastic material is used by the majority of manufacturers for instruments, electrical components, and automobile parts that need to have high heat resistance and stability.

5. Polyoxymethylene (POM)

This engineering thermoplastic has excellent hardness, stiffness, and resistance to wear. Likewise, POM plastic is easily processed into a wide range of high-performing components. POM is thus mostly utilized in gear wheels, fasteners, vehicle components, ball bearings, and skeleton-functioning products.

6. Polyethylene (PE)

Because of its affordability and versatility, polyethylene is one of the engineering plastics that is most frequently used. There are several varieties of it, including high-density polyethylene (HDPE) and low-density polyethylene (LDPE). While HDPE is more suited for products like bottles, pipes, and containers due to its rigidity, LDPE is more flexible and frequently used in packaging materials.

7. Polypropylene (PP)

Because of its superior chemical resistance, polypropylene is frequently used in applications where high temperatures or corrosive substances are present. It is utilized in a wide range of goods, including textiles, medical equipment, and automobile parts in addition to food containers.

8. Polyamide (PA)

Nylon, or polyamides, are a class of materials with a wide range of properties. They have good dimensional stability, outstanding wear resistance, and high strength. Applications for nylon include everything from textiles to mechanical parts.

9. Polyether Ether Ketone (PEEK)

High-performance engineering plastic PEEK has remarkable chemical and heat resistance. In applications involving extreme conditions, such as aerospace, medical, and electronics, it is frequently selected.

10. Polytetrafluoroethylene (PTFE)

Teflon®, a popular brand name for PTFE, is recognized for its chemical resistance and non-stick qualities. It is utilized in seals, gaskets, and cookware coatings, among other applications.

11. Polyethylene Terephthalate (PET)

Because of its clarity, mechanical qualities, and barrier resistance, PET is frequently used. It is utilized in textiles, food packaging, and beverage bottles.

3.1.6 Applications of Plastic Engineering Products

From an engineering perspective, engineering plastic components or products are classified into various groups according to the needs and specifications of their intended use. The following are some uses for products designed with plastic engineering:.

Figure 3.6. Types of engineering plastics.

Source: *By Peteruetz - Own work, CC BY-SA 4.0, https://commons.wikimedia.org/w/index.php?curid=61192008*

1. Mechanical Plastic Units

A few examples are gears, couplings, and cams. Important characteristics of these plastic parts include high tensile and impact strength as well as good stability over an extended period of time at high temperatures.

2. Chemical and Heat-Resistant Plastic Components

Water pump housings, fuel rails, valve covers, and seats are a few examples. These engineering plastic products need to be exceptionally resistant to high temperatures and corrosion-filled environments.

3. Electrical Plastic Parts

Excellent stability, tensile strength, and electrical resistance are required for these parts. Relays and connectors are two examples.

4. Low Friction Components

The lowest possible coefficient of friction is the most important prerequisite for using these parts. Wear-resistant surfaces, bearings, guides, and slides are a few examples.

3.1.7 Advantages of Engineering Plastic

Here are advantages of engineering plastic are given below:

1. Superior Mechanical Properties

The remarkable mechanical properties of engineering plastics are among their greatest benefits. These materials are appropriate for applications that require structural integrity because of their high strength, toughness, and impact resistance. Engineering plastics, as opposed to conventional plastics, are perfect for parts that are subjected to mechanical forces because they can bear large loads and dynamic stresses.

2. Lightweight Design

Because of their famed lightweight, engineering plastics are used in many products to reduce weight. This benefit is especially important in sectors like aerospace and automotive, where it's critical to minimize weight without sacrificing performance. Reducing emissions, increasing overall efficiency, and improving fuel efficiency are all possible outcomes of using engineering plastics instead of heavier materials like metals.

3. Chemical Resistance

Excellent resistance to a broad variety of chemicals, solvents, and corrosive substances is provided by many engineering plastics. They are appropriate for uses where exposure to abrasive environments is a concern because of this feature. Engineering plastics can withstand harsh chemical environments and still retain their structural integrity and functionality in a variety of applications, including chemical storage tanks and chemical processing industry components.

4. Thermal Stability

A number of engineering plastics have remarkable thermal stability, meaning they can maintain their characteristics at a variety of temperatures. Because of this benefit, these materials can be used in high-temperature applications like industrial machinery, electronic devices, and automotive engine components. Products are more reliable and last longer when they can operate consistently in different temperatures.

5. Electrical Insulation Properties

Applications requiring superior electrical insulation qualities frequently select engineering plastics. By blocking the flow of electrical current and lowering the possibility of short circuits or electrical failures, these materials can effectively isolate electrical components. This benefit is critical in industries where dependable insulation is essential, such as electronics and telecommunications.

6. Dimensional Stability

In many industries, it's imperative to maintain exact dimensions and tolerances, and engineering plastics shine in this regard. These materials ensure that products maintain their intended size and shape even under a variety of conditions because they exhibit low levels of shrinkage and excellent dimensional stability. This attribute holds significant value in sectors such as mold manufacturing and precision engineering.

7. Design Flexibility

Engineering plastics make it simple to mold into intricate designs that would be difficult to accomplish with other materials. This design freedom creates chances for creative product creation and effective production methods. The functionality and overall performance of the product are improved by the ability to modify individual components to meet specific needs.

8. Reduced Friction and Wear

Because of their low friction coefficients and natural lubrication, some engineering plastics can prolong component life and minimize wear. Applications involving sliding or rotating parts, like bearings, gears, and seals, benefit greatly from this feature. The application of engineering plastics can result in more dependable equipment and reduced maintenance expenses.

9. Environmental Considerations

Engineering plastics have benefits in terms of sustainability and recyclability in today's eco-aware society. Since many of these materials are recyclable, they lessen the impact that products have on the environment and encourage the use of circular economy principles. Furthermore, the low weight of engineering polymers helps to save energy when they are being used and transported.

3.1.8 Disadvantages of Engineering Plastic

It's critical to recognize that they have some drawbacks and restrictions as well. When choosing engineering plastics for particular applications, it's important to carefully consider their drawbacks, just like with any other material.

1. Lower Temperature Resistance

Certain engineering plastics may not be ideal for high-temperature applications, even with their remarkable thermal stability. While some engineering plastics are resistant to high temperatures, others may degrade, change in size, or lose mechanical properties in the presence of excessive heat. This may limit their applicability in situations where a constant resistance to high temperatures is needed.

2. UV Degradation

Sunlight-induced ultraviolet (UV) radiation can cause a lot of engineering plastics to break down. UV exposure over time may cause the material to become discolored, embrittled, and lose some of its mechanical properties. This restriction may affect outdoor applications where prolonged exposure to sunlight is inevitable, such as automotive parts and outdoor signage.

3. Cost

Engineering plastics can be more expensive to make and buy than regular plastics. Higher production costs are a result of the sophisticated formulations and specialized manufacturing techniques needed to achieve their enhanced properties. The choice to employ engineering plastics may be influenced by this cost consideration, especially in situations where financial constraints are tight.

4. Limited Availability of Grades

Even though there is a large range of engineering plastics available, not all grades might be easily obtained for every particular use. Some engineering plastics with highly specialized properties may be produced by a small number of manufacturers or have a restricted supply. The viability of employing these materials for some projects may be impacted by their restricted availability.

5. Processing Challenges

Processing and manufacturing of engineering plastics can be difficult. Certain materials may not be appropriate for certain production techniques due to their high melting temperatures or particular processing needs. For instance, adjustments to injection molding may be necessary to handle the elevated processing temperatures of specific engineering plastics, resulting in elevated processing expenses and intricacy.

6. Environmental Impact

Even though engineering plastics have advantages over other materials in terms of lightweight design and recycling, certain formulations may still be environmentally hazardous. Many additives used in processing or property enhancement may have environmental effects, and not all engineering plastics decompose or recycle readily. It's crucial to weigh the benefits against any possible environmental consequences.

7. Limited Transparency

Certain standard plastics and materials like glass offer greater transparency and optical clarity than other engineering plastics. This restriction may affect uses for optical lenses or displays, for example, where transparency and visual clarity are crucial.

8. Chemical Compatibility

Even though a lot of engineering plastics have good chemical resistance, not all chemicals or solvents will work with them. Certain engineering plastics can only be used in applications involving particular chemical environments due to the fact that some aggressive chemicals can cause swelling, embrittlement, or even chemical reactions with them.

9. Brittleness at Low Temperatures

Low temperatures can cause some engineering plastics to break easily. This may have an adverse effect on their toughness and impact resistance, making them less appropriate for use in environments with frequent temperature fluctuations or in cold climates.

3.2 PROPERTIES AND APPLICATIONS OF THE MOST REPRESENTATIVE ENGINEERING POLYMERS: ABS, PC, HIPS, PA

ABS is a widely used thermoplastic that is simple to mold. It is created by copolymerizing acrylonitrile and styrene with butadiene present (Figure 3.7). This gives it the toughness that is typical of butadiene rubber along with the strength and rigidity of the first two monomers. It is a hard, resilient plastic with good wear resistance, low water absorption, good dimensional stability, and a relatively high modulus. ABS is less expensive than acetals or nylon. ABS is susceptible to exposure to acids, alkalis, ketones, esters, and chlorinated solvents, though. By altering the relative amounts of the aforementioned monomers (styrene, acrylonitrile, and butadiene), its characteristics can be enhanced and changed. The circumstances in which the material was transformed into the finished product have an impact on the final properties as well. ABS materials age differently depending on how much polybutadiene is present and whether or not antioxidants are added. ABS finds many uses in the manufacturing of pipes, automobile parts, wheel covers, helmets, consumer electronics products, toys, sporting goods, and so on. It can also be utilized to create light, stiff, molded products. (Crawford 2003; Mc Keen 2019; Crawford and Martin 2020).

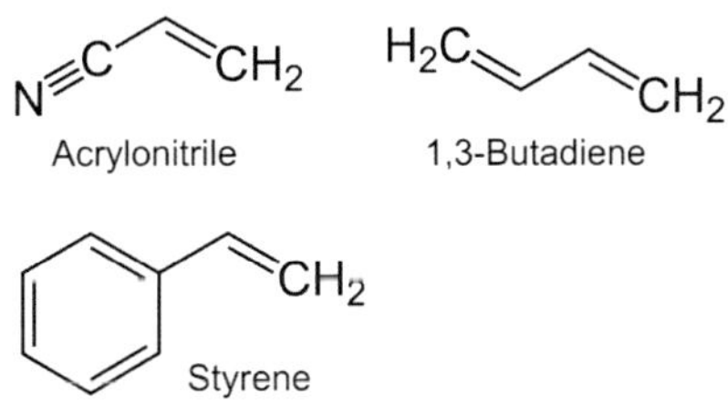

Figure 3.7. Monomers of ABS.

Source: *https://upload.wikimedia.org/wikipedia/commons/0/05/ABS_Monomers_V3.svg*

Many bisphenols have been studied in PCs; the most widely used ones are commercially derived from bisphenol-A (BPA). BPA-based PC (Figure 3.8) is an amorphous engineering polymer with a Tg ~50°C that has exceptional toughness, glass-like transparency and clarity, good heat resistance, good creep resistance, excellent electrical properties, and intrinsic flame retardancy. Its exceptional outdoor weathering resistance allows it to be used outdoors. However, hydrocarbon solvents and alkaline solutions can attack it chemically.

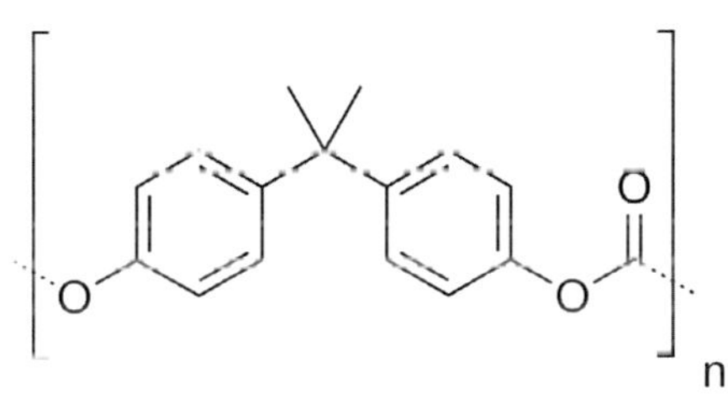

Figure 3.8. Structural units of PC.

Source: *By Ed (Edgar181) - Own work, Public Domain, https://commons.wikimedia.org/w/index.php?curid=52831270*

PCs have many uses in construction, data storage (such as CDs, DVDs, and Blu-ray discs), eyeglasses, camera parts, LCD/LED screens, mobile phone cases and covers, light covers, automotive headlamp lenses, and more because of their excellent qualities (Crawford 2003; Melton et al., 2011; Mc Keen 2019; Crawford and Martin 2020).

Another popular thermoplastic that comes in a variety of grades and impact strengths is polystyrene (PS), despite having a brittle nature by nature. The general-purpose PS (GP-PS) and high impact PS (HIPS) are two popular types of PS. The latter (Figure 3.9) is created when the PS matrix is filled with polybutadiene rubber, which greatly reduces the brittleness of the polymer and increases its toughness. High impact grades typically include 6–12% elastomer. To increase the polymer's resistance to impact and ability to deform before breaking, elastomers are added to the matrix. Polystyrene could also be offered in an enlarged version. The use of unmodified HIPS in external applications affects them more than GP-PS properties do. In particular, oxygen and solar radiation, particularly that which is near the UV end of the spectrum, cause the embrittlement and yellowing of materials. The butadiene elastomer portion is where most of these modifications occur. To get around these problems, materials are given additives like UV stabilizers to enhance their performance and stop the material from becoming brittle and yellowed from prolonged exposure to sunlight when unmodified HIPS is exposed to it. The aforementioned indicates that HIPS outperforms pure PS in terms of physical performance. Because of this, HIPS is one of the most sought-after polymers for use in a wide range of products, including toys, remote controls, electric lawn and garden equipment, small appliances, refrigeration accessories, and sterile products like labware and medical equipment (Massey 2003, 2005, 2007; Mc Keen 2019).

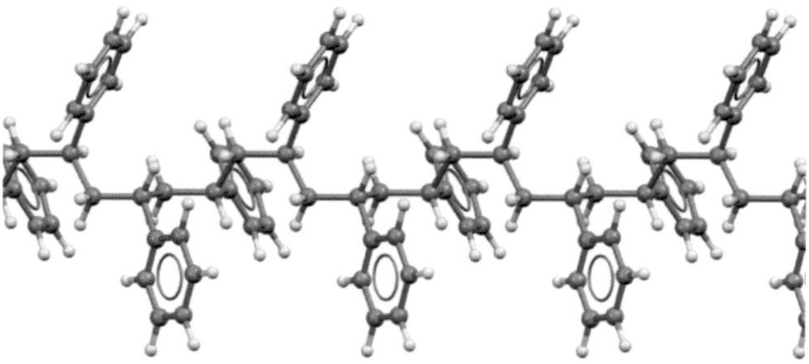

Figure 3.9. Structural units of polystyrene.

Source: *By Ben Mills - Own work, Public Domain, https://commons.wikimedia.org/w/index.php?curid=104148877*

The first engineering thermoplastics to be sold commercially were nylon or polyamides (PA) (Figure 3.10), which had superior performance to metals in certain situations and could be stiff and strong. Because of their high-water absorption, PA have high toughness in addition to high wear resistance, low coefficient of friction, excellent chemical resistance, and low dimensional stability (Crawford 2003). When designing a product, it is important to consider that PAs have a tendency to absorb moisture because this affects not only dimensional stability but also other properties. Most of the time, glass reinforcement is used to try to solve this issue (Crawford and Martin 2020). Nylon comes in a variety of forms, including nylon-6, nylon-11, nylon-12, and nylon-6. According to Mc Keen (2019), the numerical suffixes denote the quantity of carbon atoms found in the amine and acid molecular structures, respectively. Owing to their special qualities, PA can be utilized in electrical and electronic equipment, mechanical bearing and gear applications, automobiles, and fibers (Melton et al., 2011; Crawford and Martin 2020).

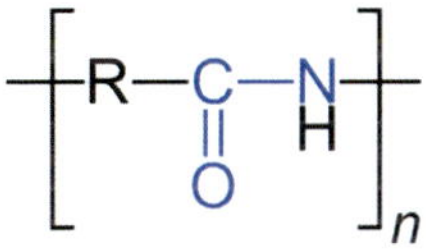

Figure 3.10. Structural units of aliphatic PA.

Source: *By Roland.chem - Own work, Public Domain, https://commons.wikimedia.org/w/index.php?curid=37642164*

3.3 PROPERTIES AND APPLICATIONS OF COMPOSITES AND BLENDS

A miscible polymeric blend, to start with a basic term, is made up of two or more homopolymers and looks to be homogenous all the way down to the macromolecular level, where the domain size is equivalent to the macromolecular dimension. Notably, overall performance requires fine-grained homogeneity, but some degree of micro-heterogeneity is typically expected; as a result, the blend maintains the unique characteristics of the individual polymer components. This implies that the characteristics governing each person may also have an impact on their blend. Cost dilution and property combination are two concepts that are of commercial interest in blend systems. Most engineering, industrial, household, electrical, automotive, and plastic parts made today are blends that end up in the trash after their useful lives (Alston et al., 2011). Conversely, the polymeric composite materials are produced when inorganic particles such as silicates, glass, clays, oxides, complexes, minerals, and carbon are incorporated into the polymer phase in different forms, volumes, or sizes, rendering them completely immiscible with the matrix. When a coupling agent is introduced, the physical forces of interaction between the two phases may become stronger. In terms of mechanical, thermal, and chemical reinforcement, as well as cost-effectiveness, the composite material proves to be highly advantageous. In general, the filler and the matrix retain many of their unique characteristics. As before, polymeric composite materials have enormous applications across a wide range of industries (Mittal 2015; Crawford 2020).

3.4 RECYCLING OF ENGINEERING PLASTICS, COMPOSITES AND BLENDS

Engineering plastics are thought to be the fastest-growing sector of the plastics industry. As a result of the products that are discarded from these applications, a significant amount of waste, particularly waste of electric and electronic equipment (WEEE), is produced annually. Furthermore, plastics can linger in the environment for a long time after their useful life has ended because they are not biodegradable and are resistant to environmental deterioration. Therefore, it is both vital and difficult to develop recycling methods and strategies for disposing of plastic waste in an environmentally friendly manner. Although those initiatives and practices date back to 1990, people in developed nations have only lately come to know about them and have largely embraced them.

Table 3.2. *Comparison of degradation methods*

Category	Degradation Method	Advantages	Disadvantages
Physical degradation	Thermal degradation Mechanical degradation	Ease of operation	Limited scope of application Toxic gas release Not easily degraded completely
		Takes less time	Limited scope of application Toxic gas release Not easily degraded completely
		Can recover energy	Limited scope of application Toxic gas release Not easily degraded completely

Category	Degradation Method	Advantages	Disadvantages
Chemical degradation	Photodegradation	More environmentally friendly	Incomplete degradation Time-consuming Affected by the environment
		High versatility	Incomplete degradation Time-consuming Affected by the environment
			Incomplete degradation Time-consuming Affected by the environment
Biodegradation	Microbial decomposition Metabolic Mechanisms in Organisms	More environmentally friendly	Potentially harmful to organisms in case of incomplete degradation
		Less extra pollution	Affected by the environment
		It can be used in a targeted manner	Influenced by microbiota

Source: *Lin Y-D, Huang P-H, Chen Y-W, Hsieh C-W, Tain Y-L, Lee B-H, Hou C-Y, Shih M-K. Sources, Degradation, Ingestion and Effects of Microplastics on Humans: A Review. Toxics. 2023; 11(9):747. https://doi.org/10.3390/toxics11090747*

The primary barriers to the recycling of engineering plastics are their diverse polymer composition and the numerous, frequently harmful additives (such as carbon fibers, flame retardants, colorants, fillers, reinforcing glass, and UV and thermal stabilizers) that are added to them. The current engineering plastics disposal methods used are landfilling, primary recycling (recycling without quality losses), mechanical/solvent-based or secondary recycling, chemical/feedstock or tertiary recycling, and energy recovery or quaternary recycling. These methods are in line with the larger framework for plastic recycling (Achilias et al., 2012; Wagner et al., 2019).

Large areas of soil become covered in waste and contaminated during landfilling as a result of contaminants leaching; additionally, since plastics are non-biodegradable materials, it will take many years for them to naturally degrade.

Research has concentrated on recycling techniques that can be used to eliminate plastic waste from landfills. These techniques include primary recycling, recycling without sacrificing quality, energy recovery (quaternary), mechanical or secondary recycling—downcycling into lower qualities—and chemical or tertiary recycling—recovery of chemical constituents.

- Re-extrusion, or primary recycling, involves reintroducing plastic scrap into the processing line's heating cycle to boost output. Its affordability and ease of use continue to make it a very popular method. Nevertheless, it is only applicable to clean, uncontaminated single-type waste.
- In mechanical recycling, plastic waste is reprocessed and modified using mechanical and physical methods to create plastic products that are either nearly identical to the original products or perform at a lower level. Sorting and separation are necessary for heterogeneous plastics before recycling, as mechanical recycling is only applicable to homogeneous plastics. In mechanical recycling, it is necessary to determine whether plastics contain brominated flame retardant (BFR) before applying BFR to prevent the formation of potentially harmful compounds like polybrominated dibenzo-p-dioxins/furans (PBDD/Fs). Its primary flaw is that the product's qualities deteriorate with each cycle; additionally, each polymer has a finite number of reprocessing cycles that it can withstand. The existence of mixed plastic waste, or polymer blends, presents another difficulty because different polymer types have varying melting points and processing temperatures. The processing temperature is typically set to the highest melting component in these situations. However, this could lead to overheating, potential degradation of the components with lower melting points, and consequently, a reduction in the final properties.
- Plastic wastes are transformed into lower-molecular-weight products, such as fuels, monomers, or secondary valuable products that can be used as refinery feedstock, through chemical or feedstock recycling. Chemical reactions involving reagents and solvents cause conversion to occur. Since it produces valuable products or monomers, as was previously mentioned, it is an environmentally friendly method.
- In the process of energy recovery, plastics are burned in a boiler or other industrial machinery to produce heat and electricity by utilizing their high energy value. However, incomplete incineration may result in the formation and release of hazardous materials into the atmosphere, including furans and dioxins, which could have an adverse effect on the environment.

In summary, plastics are broken down into smaller molecules (primarily liquids and gases) during chemical recycling, which can be utilized to create new, valuable products. For this reason, chemical recycling is regarded as both an environmentally and financially viable process. Chemical recycling also appears to be more advantageous than other current methods; this is evident when one considers that, with limited pretreatment, it is possible to treat both heterogeneous and contaminated polymers during chemical recycling. In addition, the process uses a very small amount of energy when compared to mechanical recycling or energy recovery.

3.5 CHEMICAL RECYCLING OF ENGINEERING PLASTICS: PC, HIPS, ABS

Engineering plastics are broken down into smaller molecules during chemical recycling so that they can be fed into the creation of new chemicals. Compared to primary recycling, mechanical recycling, and energy recovery, this method has more advantages (Figure 3.11). For example, it can be applied to heterogeneous and contaminated polymers with minimal pretreatment beforehand. Numerous engineering polymers appear to be beneficial for this kind of recycling; polyamides, such as nylon-6 and nylon-6,6, for example, can be effectively depolymerized to yield monomers that can be recycled or utilized as feedstock (Al-Salem et al., 2009).

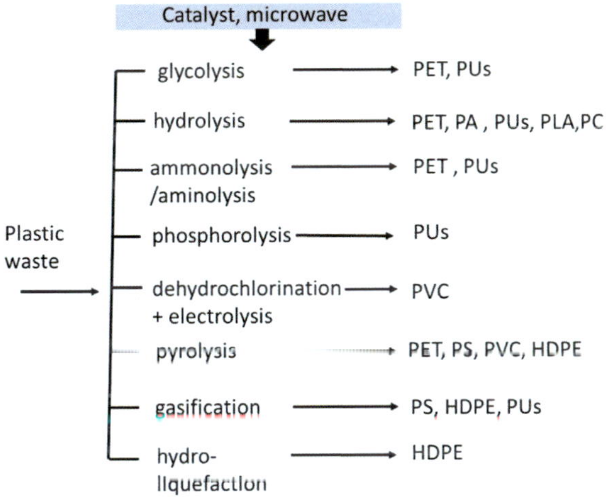

Figure 3.11. Chemical recycling vs other recycling techniques.

Source: *Damayanti D, Saputri DR, Marpaung DSS, Yusupandi F, Sanjaya A, Simbolon YM, Asmarani W, Ulfa M, Wu H-S. Current Prospects for Plastic Waste Treatment. Polymers. 2022; 14(15):3133. https://doi.org/10.3390/polym14153133*

The two primary pathways for chemical recycling are thermolysis and solvolysis . Solvolysis is the process of dissolving polymers in a solvent and treating them with or without initiators and catalysts. Additionally, it can be applied as a pretreatment before thermochemical procedures like pyrolysis. Polymers are heated during thermolysis in an inert atmosphere (such as the N_2 atmosphere) without the presence of air, oxygen, or vacuum. It is made up of several steps, such as hydrogenation, gasification, and pyrolysis (both catalytic and thermal) (Vinu et al., 2016; Sahin and Kirim 2018).

Table 3.3. *The advantages and drawbacks of various methods of recycling plastic waste*

Methods	Advantages	Limitations
Mechanical	- It is the easiest process for recycling metal matrix composites, and it is especially well-suited for fiber-reinforced polymers (FRP), where fiber breaking is accomplished through shredding.	- The decreased melt viscosity due to hydrolytic and thermal depolymerization.
	- The recycling facilities are simple and economical, and they use less energy and resources than chemical or physical recycling processes.	- The generation of cyclic and linear oligomers affects the printability and dyeability of the final product.
Chemical	- The greater rates of the monomer with a shorter reaction time.	- The expensive investment in developing technical infrastructure/ processes.
	- Higher potential for profitability through new materials application.	- The feasibility of an industrial scale has not yet been completely established.
	- The most cost-effective approaches for high-performance recycling composites.	- High temperature and much energy are needed.
Biological	- The procedure is easy to follow and is also environmentally friendly.	- The depolymerization is extremely slow for the high molecular weight of hydrophobic plastic polymers.

Methods	Advantages	Limitations
	- Cost-effective process.	

Source: *Damayanti D, Saputri DR, Marpaung DSS, Yusupandi F, Sanjaya A, Simbolon YM, Asmarani W, Ulfa M, Wu H-S. Current Prospects for Plastic Waste Treatment. Polymers. 2022; 14(15):3133. https://doi.org/10.3390/polym14153133*

Thermal cracking of polymers takes place during pyrolysis at high temperatures in an inert atmosphere. Numerous studies on pyrolysis have been conducted using a range of temperatures (from 300 to 900°C); it was generally discovered that the ideal temperatures for pyrolysis of plastic waste were between 500 and 550°C. In addition, various retention durations and heating rates ranging from 4 to 25°C/min for the pyrolysis process were investigated to optimize the procedure. Pyrolysis may occur in the presence of catalysts (catalytic pyrolysis) or in their absence (thermal pyrolysis). Engineering plastic waste is transformed into liquids, gases, and solid residues (chars) through the process of pyrolysis. Engineering plastics undergo pyrolysis via the following mechanisms:

- ***Depolymerization***: where polymers are consecutively broken down into their monomers from the end groups
- ***Random chain-scission***: where the chain of the polymer (e.g. PA) breaks down randomly and later leads to fragments with irregular lengths and different carbon numbers
- ***Chain-stripping***: side groups or reactive substitutes on the polymer chain are eliminated
- ***Cross-linking***: evolution of a chain network in the presence of heat, with cyclization or cross-linking reactions occur; generally, it appears at thermosetting polymers

Large hydrocarbon molecules are broken down into smaller molecular weight products during the hydrogenation or hydrocracking process. The entire process takes place in an atmosphere of hydrogen at moderate temperatures (between 150 and 400°C), high pressure (about 100 atm), and both (Sahin and Kirim 2018).

Indirect combustion, or partial oxidation, of polymers takes place during gasification at temperatures as high as 1600°C while oxygen is present. CO and H2 are the two principal products that are produced as a result. This mixture of gas can be utilized to power a gas engine or, in the end, to undergo the Fischer-Tropsch process, which turns it into hydrocarbon fuel. The ideal outcome is almost always to obtain petrochemicals or condensable liquids as the primary products. As pyrolysis is preferred over gasification, several steps are required in the gasification process to produce liquid products (Vinu et al., 2016; Sahin and Kirim 2018).

3.5.1 Pyrolysis of Engineering Plastics

In terms of recycling engineering plastic waste, pyrolysis appears to be one of the better options. Given that only 10% of the energy content of waste is used in the process of converting it into useful hydrocarbon products, one of its many benefits is the recovery of both material and energy from polymer waste. Furthermore, because pyrolysis occurs without oxygen, the production of harmful compounds like dioxins and furans may be inhibited if flame retardants are added to plastic waste (Liu et al., 2016).

The pyrolysis process, and particularly the quality and distribution of the products, is highly dependent on many variables, such as temperature, heating rate, residence time, operating pressure, and the presence or absence of catalysts. One of the most important factors in pyrolysis is temperature; at high temperatures (>500°C), the formation of gases and chars is encouraged, while at lower temperatures (between 300 and 500°C), the liquids are enhanced. The distribution of the products as well as the overall pyrolysis process depend critically on the existence or lack of catalysts. When compared to thermal pyrolysis, conventional (or non-catalytic) pyrolysis has a higher potential for producing liquids with higher quality and commercial value at lower temperatures and limes because it does not encourage the cracking reactions aided by the catalyst (Panda et al., 2010).

Conventional Pyrolysis of Engineering Plastics

Several engineering polymers have been reported to undergo non-catalytic pyrolysis in literature, with varying temperatures, residence times, and other conditions. The random chain-scission mechanism of HIPS degradation makes it one of the most researched polymers. According to Ojha and Vinu (2015), the primary result of pyrolysis is styrene (monomer), which is followed in lesser amounts by trimer, styrene dimer (like 1,2-diphenylpropane, 1,10-dipheny1-1,3-butadiene, etc.), and other aromatic compounds. Styrene monomers, dimers, and trimers are formed more frequently when slow pyrolysis conditions are used. Another study found that pyrolysis of HIPS and PC at 600°C produced products that were consistent with those of the prior study. Specifically, they found that pyrolysis of HIPS produced a small number of aromatic compounds with two rings (mostly styrene dimer) and single ring aromatic compounds (mostly toluene and styrene). Following PC's pyrolysis, phenolic compounds were found, including those with two rings (such as the monomer phenol, 4,40-bis(1-methylethylidene)) and those with one ring, such as phenol, 4-methylphenol and 4-ethylphenol, 4-isopropyl, and isopropenyl phenols).

Based on the investigation's findings, it's critical to emphasize that the type of polymer being pyrolyzed has a significant impact on the distribution of the products. For example, the total gas yield for PC was higher than the gas yield for HIPS. There were also observed variations in the liquid fraction, as was previously mentioned. In particular, phenols predominated in the liquid fraction obtained from the pyrolysis of PC, whereas aromatic hydrocarbons were abundant in the HIPS liquid fraction. Consequently, the type of polymers has a significant impact on the distribution of the products in addition to the previously discussed effect of pyrolysis temperature. Based on the aforementioned observation, it could be advantageous to pyrolyze engineering plastics following an initial separation of the constituent polymer types.

Similarly, pyrolysis of PA produces a variety of other products, including CO_2, unsaturated and cyclic hydrocarbons, and N-containing compounds like amines, amides, and nitriles. For example, the main products of pyrolysis of nylon-6 are ε-caprolactam, acyclic amides, CO_2, and nitriles; however, the pyrolysis of nylon-6,6 is more intricate. In the latter instance, the principal products obtained are N-containing compounds, CO_2, cyclopentanone, and numerous low molecular mass unsaturated and cyclic hydrocarbons (Bozi and Blazso 2009). Moreover, pyrolysis of ABS produces a number of different products. When Brebu et al., (2000) investigated the pyrolysis of ABS using various pyrolysis temperatures, they discovered that many products were produced, such as α-methylstyrene, toluene, ethylbenzene, acrylonitrile, and acetonitrile. The temperature at which the degradation takes place is crucial, as it has a substantial impact on the rate at which compounds evolve or transform, as well as the quantity and quality of the pyrolysis oils that are produced. Additionally, they discovered that the products of ABS degradation were significantly impacted by altering the atmosphere (such as the N_2 dynamic atmosphere) or the residence time.

The majority of the time, the high added-value chemicals (phenols, aromatic hydrocarbons, etc.) that are produced during pyrolysis are suitable for use as feedstock in the petrochemical industry. Therefore, pyrolysis can be regarded as a sustainable and environmentally friendly process based on the products that are produced. However, flame retardants—the most widely used kind of flame retardant—represent one representative category of additives found in polymers. These additives have the potential to produce a variety of hazardous brominated compounds in the liquid fraction that results from pyrolysis. This is not desirable because it prevents the liquid fraction from being recycled.

Table 3.4. The products distribution catalytic and co-pyrolysis of PP

Feed-stocks	Catalyst	Condition			Yield of Product, %			Calorific Values of Liquid Product, kJ/kg	Major Product
		Reactor	T, °C	t, min	Solid	Liquid	Gas		
PP	Spent FCC	Quartz Tube	510	60	NA	62	38	NA	Olefins, alkane
PP	Calcium bentonite clay	Batch	500	NA	0	88.5	~11.5	44,370	Alkene
PP + Lig-nocel-lulosic biomass; (1:2)	Spent FCC	Quartz Tube	510	60	12	52	36	NA	Aro-matics, olefins, alkanes, oxygen-ates
PP	Bentonite clay	Fixed bed	500	10	NA	90.5	NA	44,763	Aro-matics, alkanes, alkenes
PP	FCC	Stirred semi-batch	450	NA	3.6	92.3	4.1	NA	Olefins, paraffins, naph-thene, aromatics
PP	Fe-SBA-15	Batch	540	300	2–0.8	73–77	24–21	NA	CH_4, C_2H_6, C_3H_6, and C_4
PP	Spent FCC	Batch	300	NA	2.3	72.4	23.7	43,435	Paraffin, olefins, naph-thene, aromatic
Pp	10% dolomite	Batch	400–500	90	NA	85.2	NA	43,000–46,000	Alkanes, alkenes
PP	Spent FCC	Stirred semi-batch reactor	400	NA	2	85	13	NA	Olefin, paraffin, naph-thene, aromatic

Feed-stocks	Catalyst	Condition			Yield of Product, %			Calorific Values of Liquid Product, kJ/kg	Major Product
		Reactor	T, °C	t, min	Solid	Liquid	Gas		
PP	USY	Batch	450	45	1.2	82	16.8	NA	C_9, C_{12}, C_{15}, C_{18} and C_{21}
PP	Sulfated zirconium hydroxide	Batch	500	NA	<1	84.1	15	193.8	Paraffin, olefins
PP	Kaolin clay	Batch	450	30	23.67	67.5	8.85	46,470	Aromatics, olefins, amines, sulfide, hydroxyl

(**Source**: *Damayanti D, Saputri DR, Marpaung DSS, Yusupandi F, Sanjaya A, Simbolon YM, Asmarani W, Ulfa M, Wu H-S. Current Prospects for Plastic Waste Treatment. Polymers. 2022; 14(15):3133. https://doi.org/10.3390/polym14153133*

Pretreatment procedures, either prior to or during pyrolysis, have been extensively researched and are crucial in obtaining bromine-free products in these situations. Solvent extraction is one of the most often used pretreatments used before thermal pyrolysis for the identification and removal of (brominated) flame retardants. Typically, soxhlet extraction is used, which is a well-liked technique because of its low cost, even though it necessitates a lot of solvent and time. Over time, numerous sophisticated solvent extraction methods have been developed in addition to the conventional soxhlet extraction. Among these are microwave-assisted extraction (MAE), pressurized liquid extraction (PLE), ultrasonic assisted extraction (UAE), and supercritical fluid extraction (SFE), which require less solvent and less time (Vilaplana et al., 2009).

Catalytic Pyrolysis

In catalytic pyrolysis, a catalyst is added, and because of the process's lower temperatures and shorter reaction times, the product quality is higher and the rate at which plastic is converted into oils is higher. To improve product selectivity and optimize product distribution, catalysts are also utilized in pyrolysis. These catalysts produce yields that are comparable to those of conventional fuels like gasoline and diesel. Typically, catalysts are divided into two categories: heterogeneous and homogeneous. Heterogeneous catalysts are typically in solid form, whereas homogeneous catalysts are typically in a single phase (reactants, products, and catalysts are in the same phase), i.e., liquid solution. Aluminum chloride ($AlCl_3$), potassium hydroxide (KOH), sodium hydroxide (NaOH), and sodium methoxide (CH_3ONa) are the most commonly used homogeneous catalysts for plastic pyrolysis. Acidic catalysts such as hydrochloric, phosphoric, sulfuric, and organic sulphonic acids are frequently employed. Because base acid catalysts have less activity than homogeneous ones, the former are preferred.

Heterogeneous catalysts are primarily solids that are mixed with either liquid or gaseous constituents. Because they are simpler to separate from the liquid pyrolysis product and can be recycled and reused, heterogeneous catalysts are more widely used. The fact that heterogeneous catalysts are non-corrosive gives them a significant advantage over homogeneous ones. Important characteristics of the pyrolysis process are the surface area, porosity, functional group arrangement, and acidity. The following are the catalysts that are most commonly used: zinc oxide (ZnO); magnesium oxide (MgO); calcium carbonate ($CaCO_3$); calcium carbide (CaC_2); silicon dioxide (SiO_2); aluminum oxide (Al_2O_3); silicon dioxide alumina doped (SiO_2-Al_2O_3); zeolite Socony mobil-5 (ZSM-5); red mud; and fluid catalytic cracking (FCC). In petroleum refineries, FCC catalysts are extensively utilized for the conversion of heavy oils into gasoline and liquid petroleum. Zeolite crystals and a non-zeolite acid matrix composed of silica-alumina with a structural binder make up the three components of the FCC. Lewis acid sites, which serve as electron acceptors, and Brönsted acid sites, which contain ionizable hydrogen atoms, are features of silica-alumina catalysts. The pyrolysis of waste plastics produces liquid oils, but the acidity of these catalysts affects this process. There is less production of liquid oils when the acidity is very high. The most crucial characteristic of this kind of catalyst is the existence of open pores in zeolites. Elevated acidity zeolites exhibit greater activity during the cracking process, leading to a decrease in heavy fractions and an increase in light olefin production.

3.5.2 Other Thermo-Chemical Recycling Methods

Other thermo-chemical recycling methods used for plastics recycling include:

- ***Hydrocracking or hydropyrolysis***: The addition of H_2 is the primary distinction between this method and pyrolysis. The process operates at temperatures between 375 and 400°C and high hydrogen pressures of almost 70 atm.
- ***Gasification***: Through this process, nearly all organic feed is transformed into a gaseous mixture primarily made up of CO_2, CO, H_2, and CH_4. The term "syngas" refers to the mixture of hydrogen

and carbon monoxide. An oxidation agent is needed for the process, which is often a solution of steam and either air or pure oxygen. Catalysis can be used to further convert syngas into CH_3OH.

3.5.3 Treatment of Polymeric Blends or Composite Materials

The cleanliness of the plastics stream is a critical factor in the efficient recycling of plastics. This means that there are many difficulties in reprocessing mixed polymer waste or polymer composites. One way to start improving waste management is to remove any parts of items that are made of metal, glass, or paper. Only manufacturing waste (10% of total waste) is primarily composed of one material, making recycling easier because it isn't contaminated (or less purified) with other materials. Usually, these wastes are recycled again by reintroducing them into production lines.

Waste that is easily managed—that is, waste that is composed of mixtures of polymers on a macroscopic or microscopic scale—is transformed directly without first being sorted in order to minimize waste volume. Even though the two components of the blend have theoretically similar or close chemical properties, processing can drastically alter the composition (introducing moieties or unsaturations, for example) and structure (changing a linear polymer into a branched polymer), which can hinder the subsequent formation of a homogenous material. For instance, even with the tiniest addition of r-ABS, the tensile characteristics of ABS/r-ABS blends declined during the initial reprocessing (La Mantia and Scaffaro 2014). For this reason, the r-in-name labeling of polymers is mandated by current market purchasing trends. PMMA may be added to r-PC/r-ABS blends as a third component, which has been shown to have a significant impact on the morphology. This is another example of a blend (La Mantia and Scaffaro 2014).

However, the separation of constituents in microscopic mixtures or intimately connected (soldered, interpenetrated) units is challenging, if not impossible, and necessitates complex operations. The organic matrix composites eventually fit into that category. The waste from the automotive industry, which contains thermoplastics along with fibers, fillers, polymer mixes, and multilayer composite materials, is the most representative example (Sabi u 2018).

Waste is sorted based on the fundamental polymeric material. The procedures are as follows: (a) shredding (size reduction); (b) metal separation; (c) spraying (wiping waste); and (d) repeating certain steps as needed. The goal is to enhance the quality, productivity, and competitiveness of industrial products by achieving high levels of purity in feedstock and utilizing cutting-edge technological processes.

In that feedstock category, thermochemical recycling can also be separated into three primary groups:.

- ***Pyrolysis***: Organic materials undergo thermal degradation in this process, which occurs between 400 and 800°C. When there is no oxygen present, the material breaks down into gases (hydrocarbons, NH_3, and HCl) and leaves behind 25–40% of oils and other long-lasting residues.
- ***Hydrogenation***: The material is processed at 300–500°C in an H_2 atmosphere under pressures of 100–400 bar, yielding 65–90% oils, 10–20% gases, and up to 20 percent solid residues.
- ***Cracking***: The substance is heated to 400–600°C and pressured marginally above atmospheric to create waxy, oligomeric liquids that can be further catalyzed. The process itself is powered by the gases that are produced (Pinto and Rossi, 2003).

REFERENCES

1. Abnisa F, Wan Daud WMA (2014) A review on co-pyrolysis of biomass: an optional technique to obtain a high-grade pyrolysis oil. Energy Convers Manag 87:71–85. https://doi.org/10.1016/j.enconman.2014.07.007
2. Achilias D, Andriotis L, Koutsidis IA, Louka DA, Nianias NP, Siafaka P, Tsagkalias I, Tsintzou G (2012) Recent advances in the chemical recycling of polymers (PP, PS, LDPE, HDPE, PVC, PC, nylon, PMMA). In: Achilias D (ed) Material recycling—trends and perspectives. InTech Open, London. https://doi.org/10.5772/33457
3. Aguado J, Serrano DP, Escola JM (2006) Catalytic upgrading of plastic wastes. In: Scheirs J, Kaminsky W (ed) Feedstock recycling and pyrolysis of waste plastics: converting waste plastics into diesel and other fuels. Wiley, pp 73–110
4. Al-Salem SM, Antelava A, Constantinou A, Manos G, Dutta A (2017) A review on thermal and catalytic pyrolysis of plastic solid waste (PSW). J Environ Manag 197:177–198. https://doi.org/10.1016/j.jenvman.2017.03.084
5. Al-Salem SM, Lettieri P, Baeyens J (2009) Recycling and recovery routes of plastic solid waste (PSW): a review. Waste Manag 29:2625–2643. https://doi.org/10.1016/j.wasman.2009.06.004
6. Al-Salem SM, Lettieri P, Baeyens J (2010) The valorization of plastic solid waste (PSW) by primary to quaternary routes: from re-use to energy and chemicals. Prog Energy Combust Sci 36:103–129. https://doi.org/10.1016/j.pecs.2009.09.001
7. Alston S, Allan C, Arnold C, Bridget S (2011) Environmental impact of pyrolysis of mixed WEEE plastics part 1: experimental pyrolysis data. Environ Sci Technol 45(21):9380–9385
8. Antonakou EV, Kalogiannis KG, Stephanidis SD, Triantafyllidis KS, Lappas AA, Achilias DS (2014) Pyrolysis and catalytic pyrolysis as a recycling method of waste CDs originating from polycarbonate and HIPS. Waste Manag 34(12):2487–2493
9. Anuar Sharuddin SD, Abnisa F, DaudWMA W, Aroua MK (2016) A review on pyrolysis of plastic wastes. Energy Convers Manag 115:308–326. https://doi.org/10.1016/j.enconman.2016.02.037
10. Areeprasert C, Khaobang C (2018) Pyrolysis and catalytic reforming of ABS/PC and PCB using biochar and e-waste char as alternative green catalysts for oil and metal recovery. Fuel Process Technol 182:26–36
11. Bhaskar T, Hall WJ, Mitan NMM, Muto A, Williams PT, Sakata Y (2007)

Controlled pyrolysis of polyethylene/polypropylene/polystyrene mixed plastics with high impact polystyrene containing flame retardant: effect of decabromo diphenylethane (DDE). Polym Degrad Stab 92:211–221

12. Bozi J, Blazso M (2009) Catalytic modification of pyrolysis products of nitrogen-containing polymers over Y zeolites. Green Chem 11:1638–1645
13. Brebu M, Azhar Uddin M., Muto A, Sakata Y (2000) Composition of nitrogen-containing compounds in oil obtained from acrylonitrile-butadiene-styrene thermal degradation. Energy Fuels 14:920–928
14. Brebu M, Bhaskar T, Murai K, Muto A, Sakata Y, Uddin MDA (2004) Thermal degradation of PE and PS mixed with ABS-Br and debromination of pyrolysis oil by Fe- and Ca-based catalysts. Polym Degrad Stab 84:459–467. https://doi.org/10.1016/j.polymdegradstab.2004.02.003
15. Buekens A, Huang H (1998) Catalytic plastics cracking for recovery of gasoline-range hydrocarbons from municipal plastic wastes. Resour Conserv Recycl 23(3):163–181
16. Buekens A, Yang J (2014) Recycling of WEEE plastics: a review. J Mater Cycl Waste Manag 16:415–434. https://doi.org/10.1007/s10163-014-0241-2
17. Castaño P, Elordi G, Olazar M, Aguayo AT, Pawelec B, Bilbao J (2011) Insights into the coke deposited on HZSM-5, Hβ and HY zeolites during the cracking of polyethylene. Appl Catal B Environ 104:91–100
18. Charitopoulou MA, Kalogiannis KG, Lappas AA, Achilias DS (2020) Novel trends in the thermo-chemical recycling of plastics from WEEE containing brominated flame retardants. In: Environmental science and pollution research. Springer. https://doi.org/10.1007/s11356-020-09932-5
19. Chen D, Yin L, Wang H, He P (2014) Pyrolysis technologies for municipal solid waste: a review. Waste Manag 34:2466–2486
20. Covaci A, Voorspoels S, de Boer J (2003) Determination of brominated flame retardants, with emphasis on polybrominated diphenyl ethers (PBDEs) in environmental and human samples—a review. Environ Int 29:735–756. https://doi.org/10.1016/S0160-4120(03)00114-4
21. Crawford RJ (2003) Plastics Engineering. In: Meyers RA (ed) Encyclopedia of physical science and technology, 3rd Edn. Academic Press, Elsevier, USA, pp 457–474
22. Crawford RJ, Martin PJ (2020) Plastics engineering, (4th Edn). Butterworth-Heinemann, Elsevier
23. Foy GF (1969) Engineering plastics and their commercial development. Adv Chem Ser (American Chemical Society, Minneapolis)

24. Hudson R (1994) Commodity plastics as engineering materials. Rapra Technology Ltd, UK
25. Ignatyev IA, Thielemans W, Vander Beke B (2014) Recycling of polymers: a review. Chemsuschem 7:1579–1593. https://doi.org/10.1002/cssc.201300898
26. Kahlen S, Wallner GM, Lang RW (2010) Aging behavior of polymeric solar absorber materials—part 1: engineering plastics. Sol Energy 84:1567–1576
27. Khaliq A, Rhamdhani MA, Brooks G, Masood S (2014) Metal extraction processes for electronic waste and existing industrial routes: a review and Australian perspective. Resources 3:152–179. https://doi.org/10.3390/resources3010152
28. La Mantia FP, Scaffaro R, Recycling polymer. In: Blends LA, Utracki CA Wilkie (eds). Polymer Blends Handbook. https://doi.org/10.1007/978-94-007-6064-6_23
29. Liu W-J, Tian K, Jiang H, Yu H-Q (2016) Lab-scale thermal analysis of electronic waste plastics. J Hazard Mater 310:217–225
30. Ma C, Sun L, Jin L, Zhou C, Xiang J, Hu S, Su S (2015) Effect of polypropylene on the pyrolysis of flame retarded high impact polystyrene. Fuel Process Technol 135:150–156. https://doi.org/10.1016/j.fuproc.2014.12.011
31. Ma C, Yan Q, Yu J, Chen T, Wang D, Liu S, Bikane K, Sun L (2019) The behavior of heteroatom compounds during the pyrolysis of waste computer casing plastic under various heating conditions. J Clean Prod 219:461–470
32. Ma C, Yu J, Chen T, Yan Q, Song Z, Wang B, Sun L (2018) Influence of Fe based ZSM-5 catalysts on the vapor intermediates from the pyrolysis of brominated acrylonitrile-butadiene-styrene copolymer (Br-ABS). Fuel 230:390–396
33. Ma C, Yu J, Wang B, Song Z, Xiang J, Hu S, Su S, Sun L (2016) Chemical recycling of brominated flame retarded plastics from e-waste for clean fuels production: a review. Renew Sust Energ Rev 61:433–450
34. Ma C, Yu J, Yan Q, Song Z, Wang K, Wang B, Sun L (2017) Pyrolysis-catalytic upgrading of brominated high impact polystyrene over Fe and Ni modified catalysts: influence of HZSM-5 and MCM-41 catalysts. Polym Degrad Stab 146:1–12
35. Maisel F, Chancerel P, Dimitrova G, Emmerich J, Nissen NF, Schneider-Ramelow M (2020) Preparing WEEE plastics for recycling—how optimal

particle sizes in preprocessing can improve the separation efficiency of high quality plastics. Resour Conserv Recycl 154:104619. https://doi.org/10.1016/j.resconrec.2019.104619

36. Marra A, Cesaro A, Belgiorno V (2018) Separation efficiency of valuable and critical metals in WEEE mechanical treatments. J Clean Prod Accept Manuscr 186:490–498. https://doi.org/10.1016/j.jclepro.2018.03.112
37. Massey LK (2003) High impact polystyrene (HIPS), permeability properties of plastics and elastomers, 2nd Edn . In: A guide to packaging and Barrier materials, Plastics Design Library, Andrew W, Elsevier, pp 315–316
38. Massey LK (2005) High impact polystyrene, the effects of UV light and weather on plastics and elastomers, 2nd Edn . Plastics Design Library, Andrew W, Elsevier, pp 231–238
39. Massey LK (2007) High impact polystyrene (HIPS), the effect of sterilization methods on plastics and elastomers, 2nd end. In: The definitive user's guide and databook, Plastics Design Library, Andrew W, Elsevier, pp 179–185
40. Mc Keen LW (2019) The effect of UV light and weather on plastics and elastomers, 4th Edn . Plastics Design Library, Andrew W, Elsevier
41. Melton GH, Peters EN, Arisman RK (2011) Engineering thermoplastics. In: Kutz M (ed) Applied plastics engineering handbook. Elsevier, Andrew W, pp 7–21
42. Miandad R, Barakat MA, Aburiazaiza AS, Rehan M, Nizami AS (2016) Catalytic pyrolysis of plastic waste: a review. Process Saf Environ Prot 102:822–838. https://doi.org/10.1016/j.psep.2016.06.022
43. Mittal V (2015) Advances in synthesis and properties of engineering polymers. In: Mittal V (ed) Manufacturing of nanocomposites with engineering plastics. Woodhead Publishing, Elsevier, pp 1–13
44. Nnorom IC, Osibanjo O (2008) Sound management of brominated flame retarded (BFR) plastics from electronic wastes: state of the art and options in Nigeria. Resour Conserv Recycl 52:1362–1372. https://doi.org/10.1016/j.resconrec.2008.08.001
45. Ojha DK, Vinu RJ (2015) Resource recovery via catalytic fast pyrolysis of polystyrene using zeolites. Anal Appl Pyrolysis 113:349–359
46. Panda AK, Singh RK, Mishra DK (2010) Thermolysis of waste plastics to liquid fuel: a suitable method for plastic waste management and manufacture of value added products—a world prospective. Renew Sust Energ Rev 14:233–248. https://doi.org/10.1016/j.rser.2009.07.005

47. Pardos Marketing (2020) https://www.pardos-marketing.com/hot04.htm. Accessed 27 Sep 2020
48. Parker D, Bussink J, van de Grampel HT, Wheatley GW, Dorf E-U, Ostlinning E, Reinking K, Schubert F, Junger O, Wagener R (2012) Polymers, high-temperature, Ullmann's Encyclopedia of industrial chemistry. Wiley-VCH
49. Peters EN, Arisman RK (2000) engineering thermoplastics. In: Craver CD, Carraher Jr CE (ed) Applied polymer science: 21st century, Elsevier, pp 177–196
50. Pinto KNC, Rossi JL (2003) Recycling of polymeric matrix composites. Mater Sci Forum 416–418(1):444–448. https://doi.org/10.4028/www.scientific.net/MSF.416-418.444
51. Ragaert K, Delva L, Van Geem K (2017) Mechanical and chemical recycling of solid plastic waste. Waste Manag 69:24–58
52. Rahimi A, García JM (2017) Chemical recycling of waste plastics for new materials production. Nat Rev Chem 1(46):1–11. https://doi.org/10.1038/s41570-017-0046
53. Sabău E. (2018) Recycling of polymeric composite materials. In: Udroiu R and Bere P (ed) Product lifecycle management—terminology and applications. InTech Open. https://doi.org/10.5772/intechopen.81281
54. Sahin O, Kirim Y (2018) Material recycling. In: Dincer I (ed) Comprehensive energy systems. Elsevier, pp 1018–1042
55. Shrivastava A (2018) Introduction to plastics engineering. Elsevier, Plastics Design Library
56. Tarantili PA, Mitsakaki AN, Petoussi MA (2010) Processing and properties of engineering plastics recycled from waste electrical and electronic equipment (WEEE). Polym Degrad Stab 95:405–410
57. Velghe I, Carleer R, Yperman J, Schreurs S (2011) Study of the pyrolysis of municipal solid waste for the production of valuable products. J Anal Appl Pyrol 92:366–375
58. Vilaplana F, Ribes-Greus A, Karlsson S (2009) Microwave-assisted extraction for qualitative and quantitative determination of brominated flame retardants in styrenic plastic fractions from waste electrical and electronic equipment (WEEE). Talanta 78:33–39. https://doi.org/10.1016/j.talanta.2008.10.038
59. Vinu R, Ojha DK, Nair V (2016) Polymer pyrolysis for resource recovery. In: Reedijk J (ed) Reference module in chemistry, molecular sciences and chemical engineering. Elsevier Inc., Waltham. https://doi.org/10.1016/B978-0-12-409547-2.11641-5

60. Vouvoudi EC, Rousi AT, Achilias DS (2017) Thermal degradation characteristics and products obtained after pyrolysis of specific polymers found in waste electrical and electronic equipment. Front Environ Sci Eng 11(5):9
61. Wagner F, Peeters JR, De Keyzer J, Janssens K, Duflou JR, Dewulf W (2019) Towards a more circular economy for WEEE plastics—part a: development of innovative recycling strategies. Waste Manag 100:269–277
62. Wu H, Shen Y, Ma D, An Q, Harada N, Yoshikawa K (2015) Fuel oil production from two-stage pyrolysis-catalytic reforming of brominated high impact polystyrene using zeolite and iron oxide loaded zeolite catalysts. Open J Ecol 5:136–146
63. Yang R (2020) Polymer degradation and stability. In: Narain R (ed) Polymer science and nanotechnology fundamentals and applications. Elsevier, pp 125–148
64. Yang X, Sun L, Xiang J, Hu S, Su S (2013) Pyrolysis and dehalogenation of plastics from waste electrical and electronic equipment (WEEE): a review. Waste Manag 33:462–473. https://doi.org/10.1016/j.wasman.2012.07.025
65. Zhao L, Chen DZ, Wang ZH, Ma XB, Zhou G (2011) Pyrolysis of waste plastics and whole combustible components separated from municipal solid wastes: comparison of products and emissions. In: Proceedings of the thirteen international waste management and landfill symposium, Sardinia, pp 117–118

Chapter 4

Recycling of Waste Rubbers into Value-Added Products

Contents

INTRODUCTION

Nowadays, polymeric materials are essential to nearly every industry, including sports, construction, automotive, and packaging, because they are inexpensive, durable, and easy to manufacture. Rubber has become very popular among the many varieties of polymeric materials because of its elastomeric qualities, durability, and ease of manufacture. Rubber has the ability to undergo reversible deformation due to its elastomeric properties, which are also greatly influenced by rubber's composition. The elastomeric properties and, subsequently, the types of applications of rubber have been determined by the chemical and physical properties of the materials based on rubber.

Apparently, there are 25 kinds of rubber materials present including natural rubber, silicone, styrene-butadiene rubber (SBR), fluorocarbon rubber, ethylenepropylene-diene monomer (EPDM) rubber, nitrile-butadiene rubber, polyurethane and so on. These rubbers are widely used in various types of end products such as seals, gaskets, sports goods, diaphragms and tubes etc. (Fazli and Rodrigue 2020).

Over 65% of the rubber industry is made up of the tire industry, which is one of the most dominant sectors of rubber-based goods (Marín-Genescà et al., 2020). Car tires are typically made up of 42–48 percent rubber, 24–28 percent carbon black, 12–16 percent metal, 4-6 percent textiles, and roughly 9–12 percent additives (Rajan et al., 2006; Sienkiewicz et al., 1997; Akiba and Hashim. 2012). However, the application of rubber in various automotive industries is also determined by the quality of the rubber-based materials used in different tire types. This, in turn, depends on the vulcanization process and the kinds of stabilizers, anti-oxidants, and anti-ozonants used (Kocsis Karger et al., 2013; Ramarad et al., 2015; Ikeda et al., 2018). Tire longevity is increased by adding different types of fillers, antioxidants, and curative agents, which strengthen the tire's resistance to temperature-based, photochemical, and biodegradation degradation. While the tire and automotive industries benefit greatly from high-quality rubbers, their non-biodegradable nature poses a serious and concerning risk to these products' long-term sustainability (Fukumori et al., 2002). Because there are so many rubber-based products and scrap tires, managing waste rubbers has become a difficult problem.

Global estimates indicate that at the end of its life cycle, only 3–15% of rubber waste is recycled, whereas 5–23% of waste rubber is repurposed for the creation of new products with enhanced properties. About 25–60% of waste rubber was used for energy recovery, and the remaining 20–30% was either stockpiled or sent to the worst possible place—a landfill. The graphic depiction of the tire waste stacks' current state also highlights the concerning state of affairs. While precise data on the management of rubber waste is hard to come by, the state of affairs today indicates that a number of initiatives have been taken to address the waste rubber issue. However, the characteristics and nature of the rubber-based products also determined the final use of the rubber waste, i.e., the rubber's microstructural makeup determined the final use of the rubber-based products.

Figure 4.1. Waste tire rubber at landfill area.

Source: *https://encrypted-tbn0.gstatic.com/images?q=tbn:ANd9GcRxKU5xeBj9N b6eE0v8LwWXFN7r8FYfkiY_aA&s*

4.1 THE VALUE OF DIFFERENT RECYCLING TECHNOLOGIES FOR WASTE RUBBER TIRES

Rubber waste can come from a variety of sources, but in most cases, used or discarded tires with vulcanized rubber are the biggest source—they can account for up to 50% of all rubber waste. Additional significant sources include rubber waste from the conveyor belt and automotive industries, as well as municipal solid wastes like old clothes and shoe soles. These sources indicate that there are various types of waste rubber, including natural rubber (NR) and synthetic rubbers like silicone rubber, ethylene-propylene-diene monomer (EPDM), styrene-butadiene rubber (SBR), and nitrile-butadiene rubber (NBR). Rubber products and materials present a significant waste stream, and finding a suitable and sustainable method for reusing waste rubber has long been a global challenge. Traditionally, waste rubber (i.e. end-of-life (EOL) tires were disposed of in landfills, where they can create a dangerous ecosystem because they contain potentially harmful materials like antioxidants and antiozonants, which are used in the tire manufacturing process. BHA and BHT stand for butylated hydroxyanisole and butylated hydroxytoluene, respectively (Armada et al., 2022) in addition to 6PPD, N′-(1,3-dimethylbutyl)-N′-phenyl-p-phenylenediamine (Hu. , 2022)), in addition to heavy metals that may seep into the groundwater. Furthermore, mosquitoes may use these abandoned tires as a breeding ground, endangering both human health and the environment (Yee, 2008; González et al., in 2020).

Large landfill sites have emerged and grown over the centuries due to a lack of regulations and appropriate recycling options. As of right now, the EU discards 356 million tires annually, while the global total is one billion tons (Cummins, 2019). Tire stockpiles created an additional environmental hazard when they ignited, posing a risk of spreading uncontrollably and consuming millions of tires. This happened frequently in the US during the 1980s and 90s, and more recently in Seseña, Spain in 2016 and at the largest tire dump in the world in Sulaibiya, Kuwait in 2020. Rubber tires, on the other hand, do not ignite easily because rubber does not conduct heat well; therefore, ignition requires prolonged periods at very high temperatures. In fact, arsonists or extremely poor management at tire collection and destruction sites were the causes of a number of these accidents. These fires usually rage for many days or even months. In Heyope, Wales, a fire that started in 1989 and spread to 10 million tires still raged 15 years later due to the tires' dense packing, which made it difficult for firefighters to put out. All of these fires released dangerous greenhouse gases into the atmosphere, such as carbon dioxide, carbon monoxide, and sulfur oxides, and contaminated residential drinking water wells and soils with carbon black. They also seriously damaged the local ecosystems.

The United Nations Environmental Programme (UNEP) acknowledges that waste tires are becoming a growing issue for national and local governments to ensure a sustainable management of solid waste due to the growing issues related to end-of-life tires. Up to and including more than 50% of EOL tires were either illegally disposed of or dumped in landfills until about 20 years ago. Laws began to address that, for example. The EU landfill directive 1999/31/EC forbade the stockpiling of entire tires in landfills in 2003, and in 2006, tires were shred. However, these sizable landfill sites still exist in part due to the lack of appropriate reuse options for waste rubber. Globally, there are thought to be 4 billion of these tires in stockpiles and landfills right now.

Rubber waste is a problem for recycling and reuse because, unlike thermoplastic materials, it is a thermosetting polymeric material that cannot be remolded by simply heating it. This is in addition to the waste's massive quantity. During the vulcanization process, sulfur is used to crosslink rubber polymers, producing products that are difficult to break down and extremely strong chemical bonds. To create new rubber products, it has been suggested to use a variety of techniques to break these bonds and enable rubber to be revulcanized by adding fresh sulfur. Figure 4.2 provides an overview of these technologies and shows how they could lead to possible end products. However, the majority of these techniques also break the carbon bonds in addition to the sulfur crosslinks. Low-quality rubber products are usually the result of this, e.g., playground rubber mats or roofing materials, where the rubber's original physical properties are either completely or partially destroyed. Selective removal of sulfur crosslinks is necessary to enable the reuse of rubber for high-value products, like rubber tires, but this is an extremely difficult procedure.

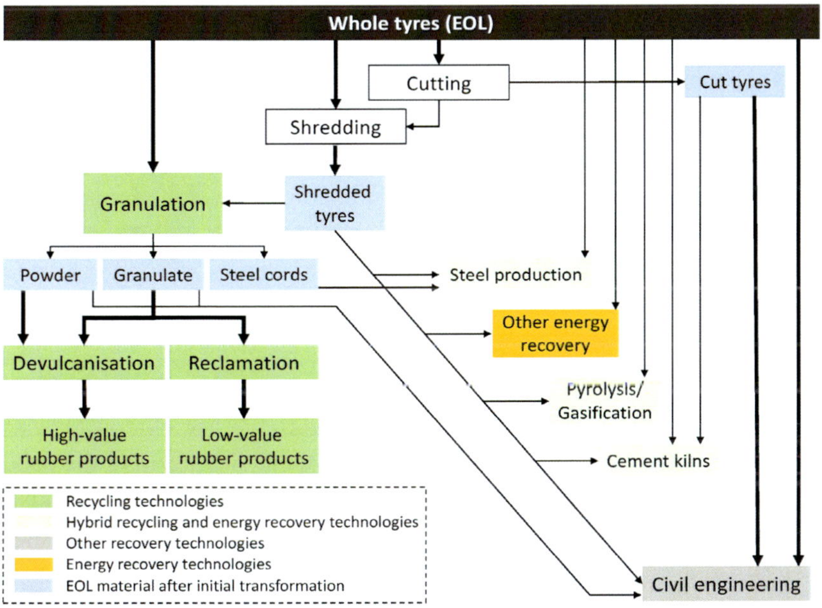

Figure 4.2. Different recovery routes of end-of-life rubber tire.

Source: *Goevert D (2024) The value of different recycling technologies for waste rubber tires in the circular economy—A review. Front. Sustain. 4:1282805. doi: 10.3389/frsus.2023.1282805*

4.1.1 Microstructural Composition

Elastomers

NR is a common elastomer component used in the production of rubber. The ability of NR with high MW and long chain branches to crystallize quickly under stretching results in high tensile strength and resistance to tear growth. In the tire manufacturing process, NR is typically blended with other synthetic rubbers, including butadiene rubber (BR), hydrogenated nitrile-butadiene rubber (HNBR), SBR, NBR, and EPDM, to enhance its tensile strength and resistance to tear growth.

Fillers

To increase the rubber's strength, various fillers have been used, including clay, carbon black (CB), and precipitated silica. Strong polymer-filler interactions and the creation of a flexible filler network are how this is accomplished. Because stiffening fillers (like silica and CB) increase the shear strength and chain entanglements between the polymer chains, they improve the tensile and tear strength, hardness, and rupture modulus of rubber. Clay-based fillers, such as montmorillonite, synthetic mica, and saponite, are utilized in the rubber industry because they exhibit superior mechanical properties in comparison to CB. For instance, Okada documented the advantageous impact of 10 vol. percentage of organoclay in NBR to attain tensile strengths comparable to 40 vol% CB. However, the fillers' size, shape, and molecular makeup may have an impact on the rubber microstructure.

Other Additives

Rubber compounds can be made more durable and their crosslinking reaction accelerated by using a variety of materials. Zinc oxide, for example, has been utilized as an activator in the vulcanization process. To increase the rubber's processability, paraffinic oils, naphthenic oil, treated distillate aromatic extract (TDAE), and mild extract solvate (MES) are used. However, the kind and quantity of filler added are highly dependent on the rubber matrix that is being utilized.

4.1.2 Rubbers Types

Rubbers can be divided into a variety of groups, such as natural and synthetic, saturated and unsaturated, etc. However, there are general rubbers and special rubbers based on the application and properties needed. While special rubbers have unique qualities like elastic properties, aging resistance, chemical resistance, swelling resistance in non-polar oils, and thermal stability/fire resistance, general rubbers are relatively inexpensive materials that are produced and consumed in large quantities.

Natural Rubber (NR)

In the rubber industry, one of the most widely used types of rubber is natural rubber (NR). It essentially came from the vegetable Hevea brasiliensis. With the chemical formula cis-1, 4-polyisoprene, which is essentially formed by the isoprene monomer unit, natural rubber is classified as a type of biopolymer. The high elastic properties of NR ($T_g \sim -70$ °C) make it suitable for a variety of applications, including sports goods, tires, gloves, and erasers. Although NR has always been vulcanized to create a highly rigid 3D network of rubbers, allowing NR-based products to be used for soft to hard applications, NR's highly flexible structural properties also allow it to be easily elongated reversibly in high deformation. Fillers and other stabilizers were added because direct application of NR is challenging due to its extremely low stability in the presence of oxygen and ozone.

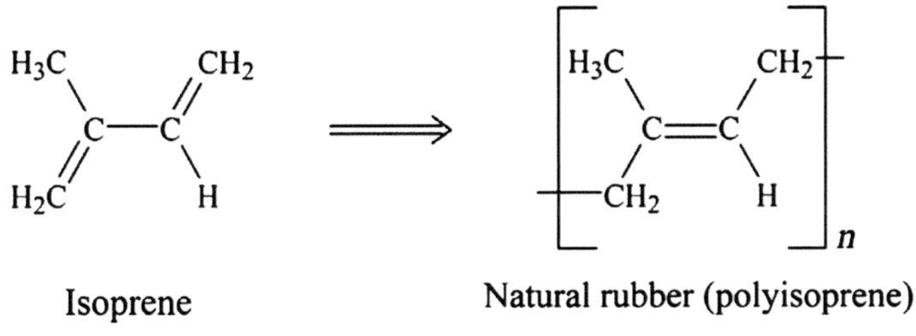

Figure 4.3. Chemical structure of isoprene and natural rubber (NR) (polyisoprene).

Source: *Fazli A, Rodrigue D. Waste Rubber Recycling: A Review on the Evolution and Properties of Thermoplastic Elastomers. Materials (Basel). 2020 Feb 8;13(3):782. doi: 10.3390/ma13030782. PMID: 32046356; PMCID: PMC7040846.*

Styrene-Butadiene Rubber

SBR is primarily made up of copolymers of butadiene and styrene. The two main methods used to prepare the SBR were anionic copolymerization in the presence of different types of solvent systems and free radical copolymerization in the presence of emulsions.

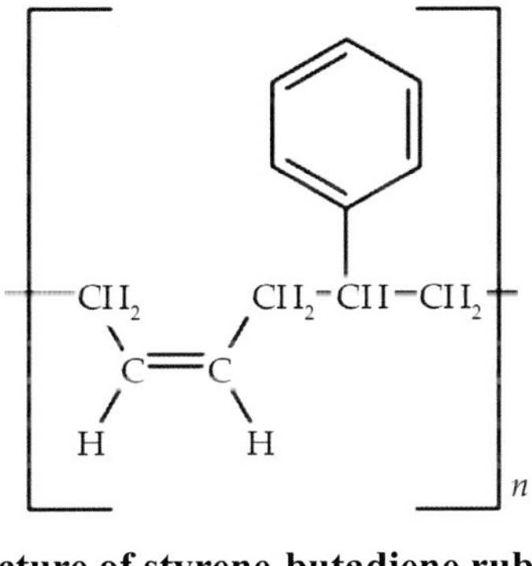

Figure 4.4. Chemical structure of styrene-butadiene rubber (SBR).

Source: *Fazli A, Rodrigue D. Waste Rubber Recycling: A Review on the Evolution and Properties of Thermoplastic Elastomers. Materials (Basel). 2020 Feb 8;13(3):782. doi: 10.3390/ma13030782. PMID: 32046356; PMCID: PMC7040846.*

The amount of styrene and the structure of the polymeric chains are the primary determinants of the properties of the SBR. SBR typically vulcanizes when exposed to different sulfur agents. Due to the low mechanical strength of SBR, different reinforcing fillers were added to the SBR matrix. One of the rubbers most frequently used for a variety of automotive applications is SBR. SBR has demonstrated its great potential for the tire industry due to its high resistance to crack formation, abrasion strength, and thermal stability.

Nitrile-Butadiene Rubber

The acrylonitrile and butadiene units that make up the chemical foundation of NBR are essentially produced by their radical copolymerization at low temperatures, between 5 and 30 °C. The preparation technique, the branching molecules, and the temperature during the copolymerization process all affect the structure of NBR. Furthermore, the amount of acrylonitrile in the NBR essentially determines its glass transition temperature (Tg). Due to its low rate of crystallization under stress, NBR typically exhibits low tensile strength, but it also demonstrates a high level of resistance to non-polar solvents, oils, and different types of motor fuel. The high resistance power of NBR towards different types of oils and fuels makes it a popular material for oil transport equipment, sealing tubes, and other similar appliances that withstand oil.

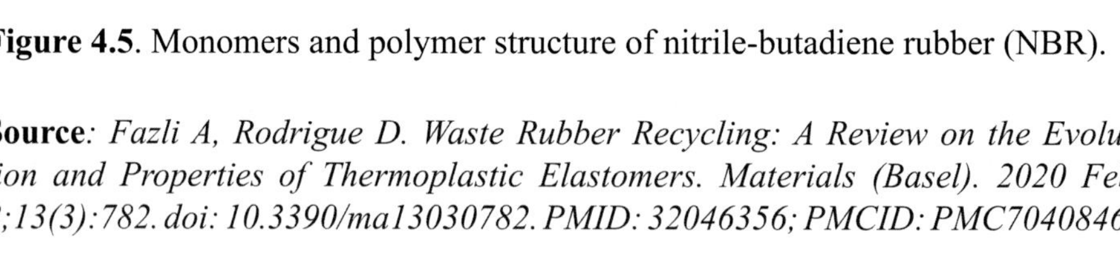

Figure 4.5. Monomers and polymer structure of nitrile-butadiene rubber (NBR).

Source: *Fazli A, Rodrigue D. Waste Rubber Recycling: A Review on the Evolution and Properties of Thermoplastic Elastomers. Materials (Basel). 2020 Feb 8;13(3):782. doi: 10.3390/ma13030782. PMID: 32046356; PMCID: PMC7040846.*

Ethylene-Propylene-Diene Monomer

A terpolymer of ethylene, propylene, and a non-conjugated diene with residual unsaturation in the side chain is known as ethylene-propylene-diene monomer or EPDM. In comparison to unsaturated rubbers (NR or SBR), this synthetic rubber with a non-polar backbone exhibits superior resistance to heat, light, and ozone. 5-Ethylidene-2-norborene (ENB) as a diene is one of the most significant grades of EPDM (Figure 4.6). The amount of ethylene and propylene determines the properties of EPDM. The outstanding resistance of the vulcanized EPDM to oxygen, ozone, and atmospheric aging up to 150 °C are its most notable characteristics. EPDM rubbers are widely used as sealing materials and can be cured using sulfur or peroxide systems. A few kinetic numerical models on the accelerated sulfur vulcanization of EPDM are available. Despite peroxidic curing, sulfur vulcanization of EPDM exhibits complex reactions induced by sulfur during crosslinking.

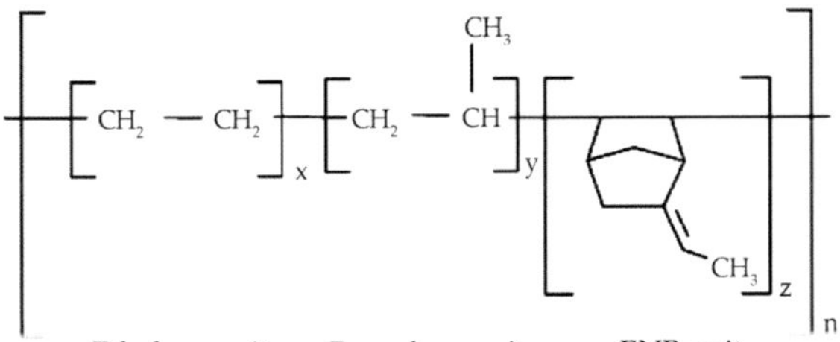

Figure 4.6. Chemical structure of ethylene-propylene-diene monomer (EPDM) containing 5-cthylidcnc-2-norborcnc (ENB) as a dicnc.

Source: *Fazli A, Rodrigue D. Waste Rubber Recycling: A Review on the Evolution and Properties of Thermoplastic Elastomers. Materials (Basel). 2020 Feb 8;13(3):782. doi: 10.3390/ma13030782. PMID: 32046356; PMCID: PMC7040846.*

Polyurethane (PU)

Another popular rubber that is being investigated in a number of the automotive and electrical industries is polyurethane.

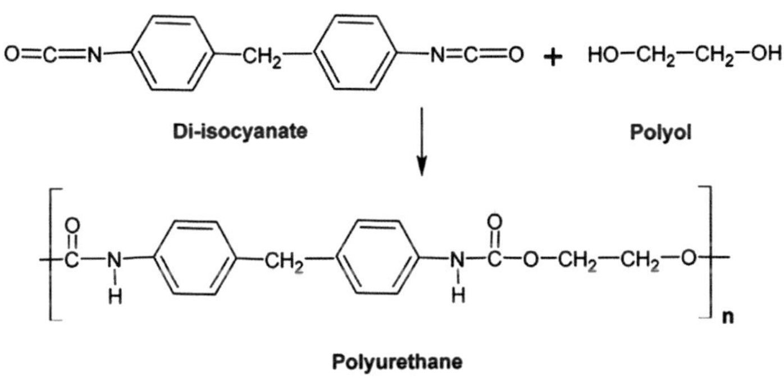

Figure 4.7. Schematic representation of polyurethane (PU) and its monomers.

Source: *Fazli A, Rodrigue D. Waste Rubber Recycling: A Review on the Evolution and Properties of Thermoplastic Elastomers. Materials (Basel). 2020 Feb 8;13(3):782. doi: 10.3390/ma13030782. PMID: 32046356; PMCID: PMC7040846.*

The process of polyaddition of polyols and diisocynates is typically used to create PU. Different kinds of PU with different properties can be obtained depending on the types of monomers used in the synthesis due to their excellent abrasion resistance, low friction coefficient, high resistive power against oxygen and ozone, and high tensile strength. PU is therefore frequently utilized in the automotive, flexible connection, and electric line industries.

Silicone Rubber

Silicone rubber, sometimes referred to as siloxanes, polyorganosiloxanes, or polysiloxanes, is made by either ring-opening polymerization of cyclotetrasiloxane, which is catalyzed by strong acids or bases, or multilevel hydrolysis and subsequent condensation of dimethyldichlorosilane in an acidic medium. Rather than carbon and hydrogen atoms, the polymer backbone is composed of a chain of silicon and oxygen atoms. Highly flexible silicone rubbers exhibit excellent stability across a broad temperature range (−70 °C to 250 °C). As seen in Figure 4.8, a typical polysiloxane is composed of letters that identify four main groups. Silicone rubber is primarily used in the production of tubing for the transportation of ozone because it is resistant to oxygen and ozone aging. Last but not least, silicon rubbers are extremely hydrophobic, sticky, and biocompatible, which makes them the perfect material for biocompatible devices like medical implants.

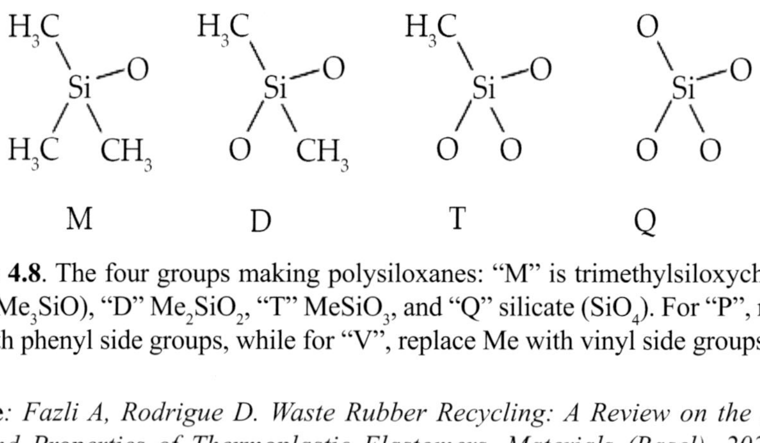

Figure 4.8. The four groups making polysiloxanes: “M” is trimethylsiloxychlorosilanes (Me_3SiO), “D” Me_2SiO_2, “T” $MeSiO_3$, and “Q” silicate (SiO_4). For “P”, replace Me with phenyl side groups, while for “V”, replace Me with vinyl side groups.

Source: *Fazli A, Rodrigue D. Waste Rubber Recycling: A Review on the Evolution and Properties of Thermoplastic Elastomers. Materials (Basel). 2020 Feb 8;13(3):782. doi: 10.3390/ma13030782. PMID: 32046356; PMCID: PMC7040846.*

4.2 VARIOUS RECYCLING TECHNIQUES

The recycling technology emerged in the early 1980s as a result of efforts by several European countries to recycle carbonic polymers, such as plastics and rubber. (Kim and Pal, 2010; statistics from Epro, 2018). According to Schonmayr (2017), recycling is the process of recovering materials, whether or not there has been a significant change in their chemical structure. There are four primary categories of recycling technologies, all of which center around the methodical processing and qualitative creation of products with added value.

4.2.1 Primary Recycling

Primary recycling, often referred to as re-extrusion or closed loop technique, is the recycling of single polymers that possess properties similar to the original material. Typically, the contaminations are cleaned up and mechanically removed during the process. Primary recycling, which is sometimes thought of as a rudimentary or traditional method of recycling polymers, can occasionally yield better materials than virgin ones (Sadat-Shojai and Bakhshandeh, 2011). Nonetheless, because of its ease of use and high conversion rate to virgin products, the process is highly well-liked in industries. Comparative molding via injection and mechanical recycling are included in the process, though it may vary depending on the need to produce high-quality goods (Barlow 2008).

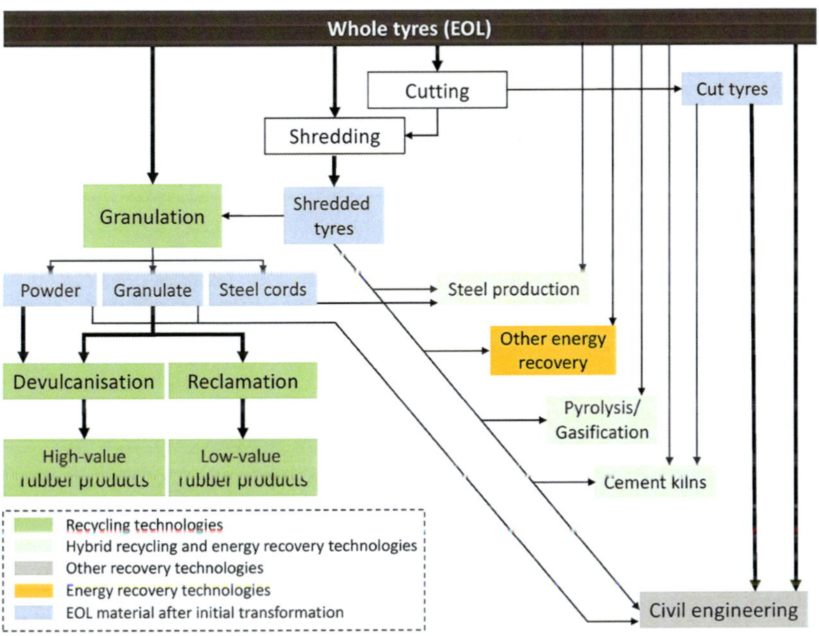

Figure 4.9. Different kinds of rubber recycling technologies.

Source: Goevert D (2024) The value of different recycling technologies for waste rubber tires in the circular economy—A review. Front. Sustain. 4:1282805. doi: 10.3389/frsus.2023.1282805

4.2.2 Secondary Recycling

The secondary recycling technique refers to the process of turning raw materials into less demanding products through mechanical recycling (Kumar et al., 2011). Cutting, grinding, mechanical separation, and floating separation techniques are used in the processing of secondary recycling to separate the polymeric flakes (Aznar et al., 2006; SubsTech 2006), After processing the rubbers or polymers to create granulates, washing and drying are necessary steps to make the necessary purifications. Occasionally, the final products may be sold after additives or pigments of some kind are added. On the other hand, before the final product was produced, the singular polymeric materials were processed through extrusion to form pellets (Kumar et al., Al Salem et al., 2011). The secondary recycling method involves several steps, such as blow molding, injection molding, screw extrusion, etc.

4.2.3 Screw Extrusion

The extrusion method can be used to process rubber polymers, plastic waste, leather, and agricultural waste, including food waste (Mikulionok and Radchenko 2012). Nonetheless, depending on the type of precursor, there are presently single and double screw extruders on the market, especially for recycling (Hopewell et al., 2009). There are extruders with adjustable sizes, shapes, and operating methods (Frame 1994). By heating the polymeric form to its melting point, the process can transform it into its new structure. In short, the precursor material is put into a barrel container, heated to its melting point, and then allowed to pass through a die with specific dimensions to form the final product according to the die's shape. The instrument is primarily made up of parts like a motor, die head, screw, hopper, and cylindrical barrel. A hopper is used to load materials under gravity; typically, the hopper is positioned in relation to heaters to maximize various temperature settings. Screws are used to process the material into dies while it is heated. Certain parameters, like screw speed and barrel length, are crucial for both proper processing and high-quality output. Screw speeds that are too high or too low can have a variety of effects on the recycling process, such as over melting the polymer or preventing it from fully forming (Sombatsompop and Panapoy 2000). The polymer's temperature rises steadily due to several heating zones.

4.2.4 Tertiary Recycling

Energy sustainability is the main component of the tertiary recycling process. All of the precursors used to make the processing materials are renewed during the process. The processes of gasification, chemolysis, pyrolysis, and cracking are used in tertiary recycling techniques. The fundamental idea behind these procedures was the de-polymerization of a polymer into its monomers. Basically, the two main methods that are available are chemical recycling and thermal recycling. Both heating and solvents can be used to accomplish the depolymerization process (solvolysis and thermolysis, respectively); the latter process is known as pyrolysis when it is carried out without the presence of air. While a similar degradation process in the presence of glycogens or ethylene glycols is known as glycolysis, it is known as methanolysis when it occurs in the catalytic presence of methanol (Kumar et al., 2011).

4.2.5 Quaternary Recycling

The most reliable method for recycling polymeric products back into their monomeric structural forms is quaternary recycling. Quaternary recycling also referred to as the energy recovery process, involves burning and other comparable techniques (Scott 2000). Process projects involve the reduction of waste materials in volume (Kumar et al., 2011). Additionally, the crude oil produced using this method is processed numerous times to produce an oil with a very high calorific value that is on par with that of conventional fossil fuels i.e., diesel and petroleum (Dirks 1996). Therefore, the discussion above draws a very clear line and explains why waste landfill sites shouldn't be used as a solution to environmental rubber polymer pollution (Singh et al., 2017). Actually, using the proper scientific recycling techniques to resolve the issue is necessary.

4.3 THERMOPLASTIC ELASTOMERS

To create TPE compounds, thermoplastic resins are frequently melted together with leftover rubber powder. TPE is a thermodynamically incompatible system made up of an elastomeric component as the soft fraction and a non-elastomeric material as the hard segment. The properties of glassy/semi-crystalline thermoplastics combined with soft elastomers, as well as the processability of thermoplastics, are advantageous to TPE compounds. TPE compounds can be made by extrusion, which involves shearing and heating the polymer melt until the hard domains separate, then cooling and solidifying the mixture. Thermoplastic polyurethane (TPU), thermoplastic vulcanizate (TPV), thermoplastic olefin (TPO), copolyester (COPE), styrene block copolymer (SBC), and polyether block amide (PEBA) are the different types of thermoplastic polymers.

4.3.1 TPE Structure

TPE compounds can be obtained by three different structures and morphologies as:

- Block copolymers consisting of elastic and non-elastic blocks;
- Rubber/thermoplastic blends;
- Dynamically vulcanized rubber/thermoplastic blends.

Block Copolymers

Block copolymer-based TPE is made up of multiple block copolymers whose ends can crystallize and link to one another to form a cross-linked network. The amorphous phase with characteristics similar to rubber makes up the majority of block copolymers. This category has seen the use of a number of copolymers, including TPU, SBC, PEBA, and COPE. A TPE copolymer's schematic has the rubbery blocks and stiff crystalline segments are represented as a continuous domain of soft rubbery chains. The soft rubber domains control TPE deformation because the hard blocks are crystalline and never distort. The material can be processed similarly to other thermoplastic polymers once it reaches the melt temperature, at which point the copolymer chains begin to flow.

Table 4.1. *Properties of HDPE*

Chemical formula	$(C_2H_4)_n$
Density (g/cm^3)	0.960
Specific Heat Capacity (J/KgK)	1.33
Thermal conductivity (W/mK, 23°C)	0.45
Mw (g/mol)	150,000
Mn (g/mol)	14,600
PDI (Mw/Mn)	10.3
O_2 permeability (cm^{3}μm/m^{2}h atm)	166–3041
CO_2 permeability (cm^{3}μm/m^{2}h atm)	9979–18,215
Water absorption (%, after 24 h)	<0.4
Resistance to ultraviolet	Good
Tg (°C)	102
Tc (°C)	116

Chemical formula	**$(C_2H_4)_n$**
ΔTc (J/g)	172
ΔTm (J/g)	215
Tm (°C)	134
Xc (%)	65
Storage modulus at 25°C (MPa)	1200
Storage modulus at 80°C (MPa)	500
Loss modulus at peak (°C)	80
Tensile strength (MPa)	15–20
Young's modulus (MPa)	1095
Flexural strength (MPa)	682
Elongation at break (%)	53
Flexural modulus (MPa)	904
Impact strength (kJ/m^2)	31
Hardness (Shore-A)	94
Dielectric constant (1 MHz)	2.3
Dielectric strength (kV/cm)	2.2

Source: *Olam, M. (2023). Mechanical and Thermal Properties of HDPE/PET Microplastics, Applications, and Impact on Environment and Life. IntechOpen. doi: 10.5772/intechopen.110390*

Rubber/Thermoplastic Blends Typical

TPE compounds are made by internally mixing (batch) or continuously extruding an elastomer and thermoplastic to create a direct melt blend. A popular kind of thermoplastic polyethylene (TPE) is called TPO, which is made by melting and blending rubber and polyolefins like HDPE, PP, LDPE, and linear low-density polyethylene (LLDPE). The thermoplastic phase is continuous, but during high shear processing, the rubber phase's size and shape may change due to coalescence or rupture, changing the TPO's morphology. Because the thermoplastic and dispersed rubber phases are not cross-linked, TPO is inexpensive and simple to make. TPO is widely used in the transportation industry, particularly for interior fascia and car exteriors.

TPE materials are predominantly produced using heat-resistant rubbers like EPDM, with NR being incorporated into TPE manufacturing, particularly following the emergence of dynamic vulcanization through phenolic curatives. In the realm of TPE production, compounds containing NR as the elastomer element combined with thermoplastics are referred to as TPNR. Typically, TPNR compounds are blended through internal mixers or co-rotating twin screw extruders. A variety of thermoplastics, such as polystyrene (PS), polyamide 6 (PA6), ethylene-vinyl acetate (EVA), and poly(methyl methacrylate) (PMMA), are commonly utilized in the production of TPNR. Moreover, various polyolefins, including PP, LDPE, and HDPE, have extensively been employed for the creation of TPNR. For instance, the fusion of NR and HDPE leads to a blend that leverages the outstanding processing characteristics of HDPE and the elastic properties of NR, resulting in TPNR suitable for automobile components. However, HDPE and NR, being nonpolar substances with significantly disparate melt viscosity and molecular weight, exhibit poor interfacial adhesion. Both compatibilizers and processing oils have been demonstrated to improve interactions between the two phases. Compatibilizers boost interphase compatibility, while processing oils contribute to their softening effect (plasticization), enhanced processability (lubrication), and elastic recovery.

Thermoplastic Vulcanizates

TPV materials are developed through the fusion of elastomers and thermoplastics at high temperatures, accompanied by intense mixing and either dynamic vulcanization or an in-situ crosslinking procedure. The dynamic vulcanization technique causes crosslinking within the elastomer element that is dispersed throughout the continuous thermoplastic medium, even when its proportion exceeds 50%. The size of the dispersed rubber particles, ideally being 1 micrometer, significantly influences the physical attributes of TPV.

The creation of TPV compounds is costly and intricate due to the need for advanced processing, as the dispersed rubber segment must undergo cross-linking during blending. The substantial quantity of rubber (over 50% in weight) with a high degree of cross-linking results in superior elasticity, with rubber serving as the primary phase. The evenly distributed rubber phase is crucial for achieving the desired mechanical properties of TPV. Conversely, a continuous plastic phase is crucial for facilitating appropriate processing. Ultimately, the transformation of the rubber phase from a continuous phase (in the initial blend) to a dispersed phase (in the TPV) plays a significant role in the manufacturing of TPV compounds. A significant amount of rubber (50–80 weight percent) is melt blended with the thermoplastic (20–50 weight percent), as Figure 4.11 illustrates percent) under shear stress and high temperatures. After adding the curing agents and other additives to the premixed blends, dynamic vulcanization is carried out under the same processing conditions in order to crosslink the rubber phase. Phase inversion results from rubber crosslinking and breaking up happening at the same time. The uniform dispersion of rubber particles in the thermoplastic matrix then necessitates vigorous mixing. Compatibility is necessary to produce thermoplastic laminates (TPV) with good mechanical strength and overall properties because the vulcanized rubber domain and thermoplastic matrix exhibit poor interfacial adhesion. Compatibilizers reduce the surface tension of the TPV components, which enhances interfacial adhesion.

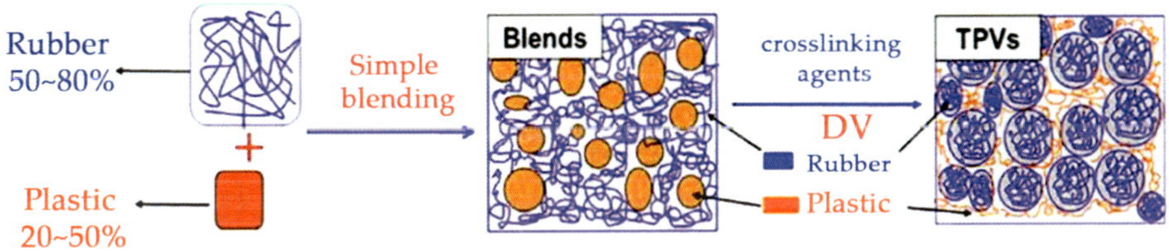

Figure 4.10. Processing steps to produce thermoplastic vulcanizates (TPV) compounds.

Source: Fazli A, Rodrigue D. Waste Rubber Recycling: A Review on the Evolution and Properties of Thermoplastic Elastomers. Materials (Basel). 2020 Feb 8;13(3):782. doi: 10.3390/ma13030782. PMID: 32046356; PMCID: PMC7040846.

4.4 "TRUE" DEVULCANISATION

The concurrent degradation of the rubber polymer and the resulting poor mechanical properties of the reclaimed rubber limit the usefulness of some of the above-mentioned technologies for the circular economy, the global rubber market, and its applications, even though some of them demonstrate desulfurization rates close to 100% and thus claim to demonstrate complete devulcanization. For a broad comparison of the benefits and drawbacks of the various methods for managing rubber waste and their primary uses. Therefore, the emphasis must change from recycling to using rubber waste as a valuable material for high-value products in order to solve the waste rubber issue. As previously indicated, to reuse waste rubber crumb for these purposes, the C-C bonds must remain intact while the S-S and C-S crosslinks between the rubber polymer chains must be broken. Reclamation cannot accomplish this; only "true" devulcanization can. To date, biotechnological techniques employing particular microorganisms and desulfurization enzymes have shown the greatest potential for achieving devulcanization without breaking down the rubber polymer. A number of critical factors are required for a devulcanization to be both successful and economically viable.

4.4.1 Rubber Granulation Technique and Particle Size

In order to provide a fine crumb or powder with a large surface area, waste rubber materials must first be reduced in size because these organisms cannot penetrate the rubber matrix. Tire scraps are usually first shredded into larger pieces, roughly 20 × 20 mm in size, and then ground into a fine crumb, usually less than 2 mm. In general, a smaller crumb means a larger surface area and more sulfur crosslinks for the desulfurizing microorganisms to break down. Mesh sizes typically tested for microbial desulfurization range from 12 (1,680 μm) to 70 mesh (210 μm). Rubber waste materials are typically ground using ambient milling, which can be done either wet or dry. Wet milling is typically chosen to cool the rubber during the milling process since ambient milling produces heat that may cause some deterioration of the rubber surface. As an alternative, rubber crumb can be created without heat degrading the rubber by using cryogenic grinding, a method first introduced in the 1960s.

After shredded tires are cooled in liquid nitrogen to a temperature of -196%, the brittle rubber becomes cold and is processed through a hammer mill, resulting in rubber particles of different sizes that are sieved to remove steel and remaining fibers. While cryogenic ground rubber has a smooth surface, ground rubber that has been milled usually results in rubber crumb with a rough surface. Water jet milling is a third technique that was created more recently. Here, high-pressure water jet systems (e.g. tire treads into fine rubber powder at pressures of up to 600 MPa without the need for chemicals or heat production. Hrdlička et al., The study conducted in 2022 examined the impact of three distinct methods and particle size on the characteristics of SBR, NR, and BR. The results indicated that water jet milling held the greatest potential, achieving the best tensile strength (18.4 MPa) and elongation at break (459 percent) when using extremely fine water jet ground rubber crumb, ranging from 0 to 100 μm. Undoubtedly, choosing the best milling technique and rubber particle size for a subsequent microbial treatment depends in large part on economics.

4.4.2 Selecting Desulfurizing Microorganisms

It is possible for certain species of both Bacteria and Archaea to break specific sulfur bonds on the surface of rubber crumbs. Previous studies have primarily focused on sulfur-oxidizing species, specifically those belonging to the genus Thiobacillus and the order Sulfolobales that are thermoacidophiles. For instance, during a 30-day cultivation of Thiobacillus ferreoxidans, a continuous rise in sulfate in the medium was utilized as a sign that the sulfur on the surface of GTR had been oxidized (Li et al., 2011). The same authors later verified the success of the devulcanization process by looking at swelling behavior, cure characteristics, and crosslink density. They found that the desulfurized GTR had improved mechanical properties when compared to the untreated controls (Li et al., 2012a). This was explained by the desulfurized GTR and NR's improved interfacial interaction. Then, Sphingomonas sp. was used in a study of a similar nature. , additionally showing a decrease in the amount of sulfur in GTR after 20 days of cultivation, along with better mechanical qualities and increased swelling values (Li et al., 2012b).

Another interesting phylum containing many members of gram-positive bacteria with demonstrated desulfurization capabilities are Actinobacteria. Various species of the genus Gordonia (ex Tsukamura) have been studied more recently in particular. Microbial treatment of vulcanized rubber for 20 days using *G. amicalis* and dibenzothiophene as substrate resulted in a reduction of surface sulfur content of up to 22.9% but also in a reduction of carbon (by 9.1%). Rubber underwent desulfurization and degradation as a result of the breaking of both S–S and C=C bonds, which converted into C=O bonds, according to FTIR analysis. Benefits from synergy were noted for a combination of *Gordonia* spp. including Sphingomonas sp. resulting in a 32.4% reduction in GTR surface sulfur (Cui et al., 2016). Following microbial treatment, mechanical characteristics and the interface coherence between desulfurized GTR and SBR matrix improved as well. These improvements were greater when a mixed consortium rather than a single bacteria was employed. A similarly high reduction in surface sulfur was also observed for *Gordonia desulfuricans* by Christofi et al., (2010) ranging between 23 and 35%. Bacterial persistence of Gordonia on rubber crumb, even after vulcanized GTR sterilization, was also shown for *G. desulfuricans* DSM 44462T. Desulfurized GTR vulcanisates demonstrated superior mechanical and rheological properties than untreated GTR vulcanisates in a different study conducted by the same authors with the same strain, and they were even comparable with an NR reference. Ghavipanjeh et al., examined eleven distinct strains with devulcanization characteristics, four of which were Gordonia strains. (2018) in a thorough investigation to assess 11 possible factors influencing the biological devulcanization of GTR. They discovered that the two most crucial variables in biological rubber devulcanization were the quantity and mesh size of GTR. Surprisingly, larger 45 mesh particle sizes produced better results than smaller 60–80 mesh particle sizes. Finer particle sizes are generally expected to result in better desulfurization because they give the bacteria a larger surface area to attach to and work on. However, by using finer particles, the authors also noticed mechanical and physical phenomena like shear tension and immersing ability, which may have an adverse effect on the rubber particles' ability to mix. This demonstrated that especially when using finer rubber particles, effective mixing can be a significant challenge during the microbial devulcanization of GTR.

4.4.3 Key Desulfurization Enzymes and Pathways

It is widely accepted that the same enzymes that catalyze the sulfur removal from fossil fuels are also involved in the devulcanization of rubber. Our knowledge of biodesulfurization has significantly increased over the past 20 years, mostly as a result of research on the genes involved in the breakdown of the model compound dibenzothiophene (DBT). Many DBT-desulfurizing Rhodococci and other isolates—especially *Rhodococcus* sp.—have been described and their desulfurization potential thoroughly investigated. The first isolated strain known to break down DBT is IGTS8. There are two main DBT desulfurization pathways known: the "Kodama" pathway and the "4S" pathway. The C–S bonds are cleaved and the sulfur portion is oxidized one after the other, but the C–C bonds remain intact, only in the 4S pathway. The step-wise reactions that convert DBT to DBT-5-oxide (DBTO), DBT-5, 5′-oxide (DBTO2), 2′-hydroxybiphenyl-2-sulfinate (HPBSi), and 2′-hydroxybiphenyl (HBP) without the degradation of either aromatic ring have been identified as being catalyzed by three enzymes: DszC, DszA, and DszB (Denome et al., 1993, 1994). Flavin-dependent monooxygenases DszC and DszA need FMNH2 or FADH2. Because of this, an extra enzyme called flavin reductase (DszD) is needed to catalyze the reduction of flavins by NAD(P)H, including FMN and FAD, to create reduced flavins. The final step of the pathway wherein HPBSi is desulfinated to produce HBP is then catalyzed by the desulfinase DszB.

The other main sulfur component of fossil fuels, benzothiophene (BTH), appears to be intractable to the majority of DBT-desulfurizing bacteria, despite having structurally similar structures. One of the rare microbes that has been found to selectively cleave sulfur from BTH without rupturing carbon-carbon bonds is *G. desulfuricans*. Other *Rhodococcus* sp. BTH-desulfurizing bacteria have been reported. *Paenibacillus* sp., T09. *Sinorhizobium* sp., A11-2. KT55, strain C-6 of *Gordonia terrae*, *Gordonia alkalivorans*, and *Rhodococcus* sp. strain SA11. The Dsz enzymes that break down DBT in the 4S pathway are thought to be related to the enzymes involved in the breakdown of BTH. Nevertheless, there is still some uncertainty regarding the analogues of DszA, DszB, and DszC in the biodesulfurization of BTH. Though G terrae has been shown to harbor the first operon for BTH biodesulfurization, Bds (Wang et al.,

2013), a complete characterization of the relevant biodesulfurization enzymes and their control remains elusive. To increase process efficiency, it is also necessary to clarify the mechanisms of feedback inhibition of sulfate, the most likely sulfurous end product of desulfurization, for strains that are relevant to the industrial sector. For instance, it was noted that sulfate represses the dsz promoter in *R erythropolis* IGTS8, and multiple investigations documented efforts to overcome sulfate repression for *Rhodococcus* sp. T09. Additionally, dszAB genes have previously been successfully expressed in *E. coli*, with the recombinant strain exhibiting no sulfate repression and growing 4–5 times faster.

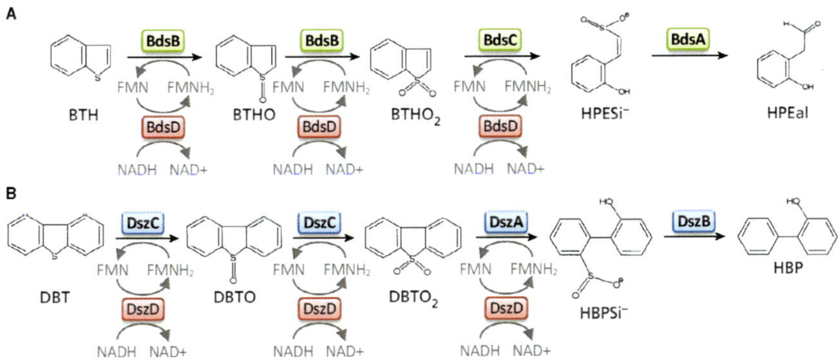

Figure 4.11. Comparison of the proposed "4S" BTH desulfurization pathway for *Gordonia desulfuricans* (A) with the DBT desulfurization pathway of *Rhodococcus* sp. strain IGTS8 (B). BTH, benzothiophene; BTHO, benzothiophene S-oxide; $BTHO_2$, benzothiophene S,S dioxide; $HPESi^-$, (Z)-2-(2H-hydroxyphenyl)-ethen 1-sulphinate; HBPEal, 2-(2H-hydroxyphenyl)-ethan 1-al.

Source: *Goevert D (2024) The value of different recycling technologies for waste rubber tires in the circular economy—A review. Front. Sustain. 4:1282805. doi: 10.3389/frsus.2023.1282805*

As these technologies are still being developed, life cycle data for rubber that has been microbially devulcanized are, to the best of the author's knowledge, not yet available. Therefore, it is currently not possible to fully analyze the potential economic and environmental benefits of devulcanized rubber products. However, considerable energy and cost savings can be anticipated given the generally mild conditions used for the majority of biotechnological processes. Since devulcanized rubber, as opposed to reclaimed rubber, can also be used for high-value rubber products like tires and selectively remove sulfur crosslinks without rupturing the carbon backbone of the rubber polymer, economic viability should be possible.

4.5 APPLICATIONS OF THE WTR RECYCLING

4.5.1 Waste Rubber as an Alternative Source for Traditional Fuels and Chemicals

Waste rubber, particularly waste tires, are pyrolyzed at temperatures between 45 and 250 degrees Celsius to produce a complex mixture of crude oils that may include heteroatoms or functional groups that are polar, aliphatic, aromatic, or both. Notably, compounds like benzene, toluene, xylene, limonene, and styrene have been reported and identified as content from the pyrolysis of tire waste (Williams 2013).

It can be challenging to separate the oil obtained from the pyrolysis of tire waste because it typically contains a variety of aromatic hydrocarbons and their derivatives combined with heteroatoms. GC-MS and FT-IR techniques have been widely used by researchers as a toolkit for the identification and analysis of compounds present in waste of this kind. While GC-MS can visualize the nature and type of compounds present, over half of the compounds are still anonymous (Miandad et al.,). In contrast, FT-IR can only determine the functional moieties of any materials. 2018).

Because it contains sulfur and may produce sulfur dioxide, which can be corrosive, the oil obtained from the pyrolysis of waste rubber or tires cannot be used directly as fuel in combustion processes. According to a recent study, hydrogen peroxide with a 25 vol percent concentration and formic acid with a pH of 4 . 0 are effective at removing sulfur from crude oil (Bunthid et al., 2010). They have reported that absorption across the surface of carbonic material has removed 70% of the sulfur content.

The obtained oil was first divided into three categories: low oil, which had a boiling point between 45 and 205 °C, middle oil, which had a boiling point between 205 and 300 °C, and heavy oil, which had a boiling point greater than 300 °C. Low-octane oils have been found to have a higher octane number; therefore, they may be utilized in the combustion process following hydro refining and reforming. The other fuels, however, can either be added straight to petroleum refineries or their quality can be raised by employing various organic catalysts. One research group performed tire pyrolysis and compared the resultant fuel to diesel; they came to the conclusion that these fuels could replace diesel in smaller engines (Islam and Nahian 2016). In light of this, pyrolysis of rubber or waste tires may be a very good alternative for a futuristic approach to the problem of waste rubber, one that can quickly assist and recover its cost value through the creation of valuable fuels. In addition, the fuel that is produced can be a valuable resource for the production of chemicals, thereby increasing the value of rubber pyrolysis (Williams, 2013).

4.5.2 Pyro-Gas

Along with fuels and char, pyrolysis also produces some combustible gases as byproducts. While the waste tire obtained gases typically consist of a mixture of several olefins, carbon oxides, hydrogen, and small amounts of sulfur and nitrogen compounds, the heating value of pyrolytic gas is approximately 2900 J/g. Out of all of them, hydrogen functions as the most effective and superior substitute for conventional fossil fuels, primarily because of its high energy density, carbon-free nature, and nearly zero greenhouse gas emissions. Recent studies have looked into the possibility of producing hydrogen in large quantities from used tires, which is a great alternative to fossil fuels. A team of scientists has recently reported creating carbon nanotubes and hydrogen gas from used tires through the use of Ni/Al_2O_3 as a catalyst for two-step pyrolysis in a fixed bed reactor. As a result, the large-scale industrial production of hydrogen gas using used tire rubber as a helpful precursor that can address various waste tire issues.

4.5.3 Ground Tire Rubber Recycling for Insulator Applications

Rubber recycling can occasionally be challenging due to the materials' intricate chemical and physical characteristics. For example, the crosslink structures of the rubber's vulcanized form make recycling them difficult; therefore, extra care must be taken when recycling rubber in this form. In order to improve their chemical and physical qualities, waste tires have recently drawn attention for use as filler material in polymeric matrixes. In addition to lowering the footprint of the rubber waste, this method of using the waste rubbers as ground tire rubbers (GTRs) also drastically lowers the cost of the finished goods. During this process, waste rubber materials are first shredded, or ground, into extremely small particles, usually smaller than a micrometer. These particles are then combined with various polymers, virgin rubbers, sports goods, automotive components, and other applications. Thermoplastic blends with GTR in particular demonstrated extremely sustainable and economical preparation techniques. Compatibility problems make it extremely difficult to mix these GTRs with polymers consistently. It has been observed that when the amount of GTR mixing in polymeric blends increases by more than 20 percent, compatibility problems arise. A team of researchers recently reported on their demonstration of the application of the GTR for the insulator in accordance with the Spanish UNE standards (Genesca et al., 2020) . This group combined the 200 μm-sized GTR with polyvinyl chloride (PVC), ethylene vinylacetate (EVA) copolymer, polypropylene (PP), high-density polyethylene (HDPE), polyamide 6 (PA 6), acrylonitrile butadiene-styrene (ABS), and polystyrene (PS) matrixes. For each chosen polymer, five specimens were made with varying amounts of GTR (0, 5, 10, 20, 40, 50, and 70%) and mixed for four minutes at different temperatures in a Brabender mixing machine. In the end, a hot plate press operating at different temperatures and pressures up to 100 KN was used to create films from the polymer/GTR blends. The entire test was conducted under ASTM-D0638-V standards. Following blend preparation, conductivity, electrical resistivity, and loss factor (Tg δ) were measured to examine the electrical properties of the polymer/GTR blends. The investigation's findings demonstrated that combining GTR with a chosen polymer demonstrates good insulating qualities in accordance with UNE and IEC standards. Further, results demonstrated by this group showed that the insulating properties of the polymer/GTR blend decreased significantly with a high content of GTR > 10%. For low content of GTR i.e. (5–10%), electrical insulating properties showed sufficient value of the electrical insulation for the use of the different kinds of electrical insulation applications.

4.5.4 Nanocarbons from Tyre Waste

Numerous studies have been carried out on the upcycling of waste tire rubbers (WTRs) into products with added value. Because of the invincible properties and applications of the synthesized nanocarbons, which are among the variety of value-added products from WTRs, researchers have shown a great deal of interest in this process. In this context, researchers have refocused on the pyrolysis approach, wherein the production of nanocarbons was carried out in the presence of multiple catalytic systems. The WTRs are pyrolyzed at 400–600 °C in an oxygen-free medium to produce oil, gas, and carbon char. The oil produced can be used in a variety of automotive applications and other related industries, and the gas produced can be used to power the pyrolysis plants. Another value-added product is carbon char, which was primarily utilized as a source of carbon black, solid fuel, or occasionally upgraded in the form of activated carbon. Yet, not much research has been done on the re-establishment of the carbon value from WTRs, despite the enormous applications of the carbon derived from WTRs. In the past 20 years, researchers have shown a great deal of interest in the nano-allotropic forms of carbon, such as carbon nanotubes, graphene nanosheets, nanocarbons, and other similar forms. The majority of material scientists working today are involved in the field of carbon nanomaterials because of these materials' exceptional mechanical and electrical properties, high surface area, high transparency, unique 1D, 2D, and 3D dimensions, porosity, and many other high-value added properties. These days, a wide range of applications include the development of extremely effective supercapacitors, batteries, solar cells, fuel cells, and many other areas. The pyrolysis method was used in 2004 to synthesize carbon nanoparticles. However, another study demonstrated the use of microwave plasma catalytic chemical vapor deposition (CVD) to synthesize carbon nanoparticles, using iron (Fe) and nickel (Ni) as a bimetallic catalytic system (Yu et al., 2002). There have also been reports of an electrolysis process for creating carbon nanoparticles, which involved utilizing carbon electrodes in molten alkali halide salts (Hsu et al., 1996). Numerous techniques have been employed to synthesize diverse nanocarbons from various sources; however, the pyrolysis approach remains one of the most effective methods for producing large quantities of nanocarbons from carbon-based precursor materials. The synthesis of nanocarbon from WTRs using a slow pyrolysis technique at 900 °C has been reported by several research groups. It has recently been shown that carbon nanoparticles (CNPs) can be produced sustainably by quickly pyrolyzing WTRs. The WTRs underwent rapid pyrolysis within the temperature range of 1550 °C for a duration of 5 s to 20 min. According to analysis of FESEM images, the CNPs thus produced had diameters of roughly 30–40 nm.

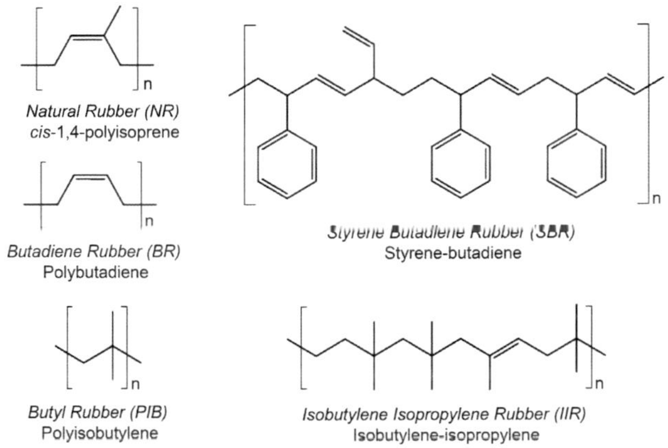

Figure 4.12. *Chemical structure of rubbers used in tire composition.*

Source: *Fazli A, Rodrigue D. Recycling Waste Tires into Ground Tire Rubber (GTR)/Rubber Compounds: A Review. Journal of Composites Science. 2020; 4(3):103. https://doi.org/10.3390/jcs4030103*

Since the rapid process in this work reduces the production of harmful gases, it has been reported to be a sustainable process. The process's sole disadvantage is that it requires a high temperature, which uses a lot of energy. Nevertheless, because it shows how to use an environmentally friendly and time-efficient method, it can be modified to produce CNPs on a large scale from WTRs. A group of researchers also presented their work on nanocarbons, reporting on the synthesis of CNTs from used tires through the use of CVD techniques. This study illustrated how to grow CNTs using a catalytic system based on cobalt. Although the average length and diameter of the CNTs that are thus obtained from scrap tires have not been mentioned, the SEM and TEM images of the CVD processed scrap tires over the catalytic system indicate that the obtained product primarily exists in the form of CNTs.

4.5.5 Waste Rubbers to Supercapacitors

Renovated recycling of used tires or rubbers into valuable products first emerged as a partial answer to the world's energy crisis. To obtain activated carbon (AC), a team of researchers employed crushed and chopped waste tires as a precursor, using potassium carbonate (K_2CO_3) as the material activator (Bello et al., 2018). For the activation, five samples with varying material weight ratios were prepared. Additionally, K_2CO_3 powder was mechanically mixed with each sample prior to being processed into a tube furnace for pyrolysis. The material was processed in a horizontal tube furnace at 800 °C for one hour while nitrogen was passed through to create an inert atmosphere. The material that was obtained underwent additional washings until its pH was entirely neutral. Activated carbon exhibited a superior mesoporous structural framework with a surface area ranging from 147 to 385 $m^2\ g^{-1}$. Additionally, AC was used as the active material to create an electrode for a supercapacitor device, which demonstrated a great 50 mAh g^{-1} discharge capacity at 0.25 Ag^{-1}. To obtain specific capacitance from the device, a redox electrolyte (1 M K3Fe(CN)6–1 M HNO_3) was used as the electrolytic medium. The device displayed a corresponding specific capacitance of 140 F g^{-1} at 0.25 A g^{-1}, and the energy efficiency was maintained at approximately 70% for 1000 cycles. An additional team of scientists extracted carbon onion from tire waste and combined it with sulfur-doped polyaniline (PANI) nanotubes adorned with a self-made hierarchical $Ni(OH)_2$ sponge for use in supercapacitors (Bhaumik et al., 2020).

Mesoporous $Ni(OH)_2$ over PANI nanotubes and carbon onions in a planned architecture created a remarkable composite with great electrochemical potential for supercapacitors. In short, anhydrous N-methyl-2-pyrrolidone (NMP) serves as the solvent during the creation of the slurry, which is made by homogeneously combining tire-derived carbon material, PANI, $Ni(OH)_2$, carbon black, and the binder polyvinylidene fluoride (PVDF) in a weight ratio of 80:15:5. In order to fabricate electrodes, they have additionally coated the resulting slurry with nickel foam in an amount of 3 to 7 mg and baked it for an entire night at 60 °C. They have utilized porous Whatman filter paper as the separator and 3 M KOH as the electrolyte. The device that was fabricated demonstrated exceptional specific capacitance of 622 F g^{-1} at 2 A g^{-1}, in addition to an exceptional retention capacity of 97% over 10,000 cycles. Huge energy and power densities of 70 Wh kg^{-1} and 136 kW kg^{-1}, respectively, were produced by the apparatus. The exceptional cyclic stability could be caused by the PANI nanotubes and nano sponge $Ni(OH)_2$ binding cooperatively on the surface of the carbon onion, stabilizing the volume changes during cycling. The work presents carbon onions made from waste tires as a great value-added product for supercapacitor applications. Likewise, the production of porous and hierarchical carbon from leftover rubber for the use as supercapacitor electrodes and methane storage.

They employed the activation followed by pyrolysis, using KOH as the activator. Under an inert atmosphere, waste rubber scrap was heated to 1000 °C for one hour at a rate of 5 °C per minute. The materials that were obtained were ground and combined with KOH at a 4:1 ratio. Additionally, the materials were allowed to be activated by KOH for 2.5 hours at various temperatures in an inert atmosphere in the pyrolysis furnace at 650, 750, 850, and 950 °C. Ultimately, the carbonic material was cleaned with a 1 M HCl solution and then double-distilled water until the pH of the removed liquid reached neutral to remove metallic impurities. The structure and morphology of the resulting activated carbon are tailored by the temperature variation, making it an excellent input for high-performance energy and methane storage. N-methyl 2-pyrrolidone (NMP) was used as the solvent to prepare a slurry containing carbon black, activated carbon, and PVDF in a ratio of 85:10:5. This allowed for the fabrication of the electrode device. The material slurry was prepared as described above and then cast onto 1cm^2 of nickel foam. For the purpose of creating electrodes, a mass loading of 3-5 mg/cm^2 was permitted. After that, the electrodes were oven-dried for the next 12 hours at 105 °C under 10 MPa. The device that was fabricated demonstrated an exceptional specific capacitance of 167.7 F/g at 0.5 A/g. Additionally, the sample with the highest surface area and microporous structure demonstrated the highest storage capacity, measuring 110.39 F/g. With a 40-fold increase in charge/discharge current density, the hierarchical and porous activated carbon-based electrode that was produced demonstrated a respectable rate capacitance retention of 86 percent (Li et al., 2020). Researchers Hubsuang et al 2017 treated rubber wood waste with H_3PO_4 and NaOH, and then used the resultant material as the electrode material for supercapacitors). In a nutshell, rubber waste was squarely chopped. To remove moisture, samples were dried in an oven set to 105 °C for a full day. After that, 30 grams of the sample were soaked in 300 cm^3 of H_3PO_4 or NaOH solution.

Additionally, pyrolysis was carried out for four hours at 800 °C and a heating rate of 5 °C under a 600 cm^3/min nitrogen flow. The produced porous carbonic material was thoroughly cleaned with double-distilled water until the pH reached 7, and then it was baked for the next twenty-four hours at 105 °C. Presently, the apparatus was constructed utilizing two symmetric carbon electrodes, graphite rods, and 1.0 M sulfuric acid as the current collector, electrolyte, and working electrodes, respectively. A representative surface area of 2.25 cm^2 was used for the mass loading of 0.18–0.22 g. The developed carbon electrodes were separated by a polyester sheet, which also serves as an intermediary bridge. Following performance testing, the CV plots were created at scan rates of 1, 5, 10, 20, 50, and 100 mV/s over the potential window of 0.0–1.0V. They revealed an excellent volumetric capacitance of 104 F/cm^3 and a specific capacitance of 129 F/g. This talk leads to the conclusion that the mechanics of capacitance storage in supercapacitors greatly depends on the volume and average size of the pores in addition to their specific surface area. Because it causes the layers of carbon to separate, the chemical activation of carbon typically increases surface area while simultaneously decreasing the amount of pollutants present. On the other hand, there are extremely few reports on the use of activated carbon derived from rubber as fuel cells, supercapacitors, and battery electrodes. Tire waste-derived carbon is a highly viable option for energy storage applications due to its high surface area, excellent porosity, large electrical conductivity, and delocalized π electrons.

4.5.6 Waste Tyres Recycling for Wastewater Treatment

The exceptional efficacy of activated carbon and other nanocarbons in eliminating pollutants from wastewater has garnered considerable interest. Due to activated carbon's high surface area and porosity, there has been a lot of interest in its potential for water purification. Rubbers have a high carbon content, which makes it possible to produce activated carbon in large quantities. Researchers have utilized waste tires to create an activated carbon framework and utilized it to absorb methylene blue dye. They have found that the activated carbon obtained from tires has a higher absorption rate than activated carbon that is sold commercially. Furthermore, according to research from another group, the waste rubber material can be used to produce activated carbon qualitatively by physical vapor deposition (PVD) using steam and carbon dioxide in addition to chemical activation using H_2SO_4, KOH, and H3PO4. This activated carbon can then be used for a variety of applications, particularly air treatments and industrial wastewater (Shah et al., 2006). The use of waste rubber tire-derived activated carbon has also been reported to remove pesticides like methoxychlor, methyl parathion, and atrazine. This study demonstrated that waste rubber tire-activated carbon functioned as a more affordable sorbent than commercially available carbon and demonstrated superior pesticide removal efficiency compared to other comparable commercially available products (Gupta et al., 2011). It has also been reported that this group effectively removed chromium (III) from wastewater by using activated carbon made from used rubber tires (Gupta et al., 2013). Thus, it demonstrated a practical strategy for the creation of affordable pathways for the elimination of heavy metal ions of the chromium (III) type from wastewater, as well as a means of addressing waste rubber disposal techniques.

4.5.7 Cementing Applications

The abundance of waste tire-based rubber (WTR) and its excessive use in human society have made it a serious ecological concern for the entire world. However, there are usually several ways to dispose of WTRs in order to manage the global situation, primarily through burning and landfilling. However, according to Weiguo et al., 2013, such a practice has detrimental effects on ecological growth, socioeconomic outcomes, and the universal health system). WTR and similar materials, however, can be applied in various construction contexts. They can also be used to replace natural aggregates in concrete mixtures, which will not only turn waste rubber into useful products but also serve as a relatively environmentally friendly solution to the severe environmental crisis. However, various waste rubber-derived materials have different characteristics, such as varying degrees of flexibility, random stiffness, and different particle sizes and shapes which have an impact on the characteristics of the materials they are incorporated into. Excellent qualities and good utilization potential of WTR as additional aggregates for the cementing concrete mixture have been reported by a number of researchers. However, the abrasion resistance of the cementing concrete mixture, along with thermal conductivity and chemical resistance, are the fundamental properties that are impacted by the addition of such WTR and based materials (Benazzouk et al., 2006).

WTRs and their analogs, including porous carbon, microrubber, rubber derived carbon nanomaterials (CNMs), etc. are capable of taking the place of fine aggregates (FA) in cement mixtures. The incorporating agents in concrete mixtures containing WTR and their equivalents exhibit an asymmetric structural network with porous morphology. Because WTR-based concrete has lighter structural networks that can trap the cementing mixture, the electron microscopy (SEM) image showed the empty space inside the structural network and concluded that the density of the composite decreased as the amount of WTR-based material increased. Additionally, it has been reported that the developed composite material exhibits good sound absorptivity because of the excellent porous nature of the waste rubber-based material. Additionally, rubber-based composites exhibit a wide range of mechanical strengths; for example, a composite containing 5% WTR demonstrated approximately 85% compressive and 96% flexural strength when compared to virgin concrete. The plastic failure at higher distortion resistance shown in the WTR-based composite is caused by overloading the material beyond its ultimate tensile strength. However, to improve the qualities of cementing composite, additional chemicals can be added as a binder in addition to rubber, or rubber-based CNMs can be used instead of just WTR.

On the other hand, applying limestone and silica fume coatings to the surface of concrete can improve the way the mixture interacts with WTR-based materials during mixing. This can also increase the overall strength by reducing the materials' absorption through capillaries. Because of its good durability, WTR-based concrete mixtures can be used to create shielding plasters. These unique qualities are primarily caused by the WTR-based concrete mixture's superior hydrophobic qualities over conventional ones; nevertheless, their effectiveness can be raised by adjusting the proportion of WTR-based materials. Researchers looking into the penetration of chloride ions in WTR-based concretes found that when WTR and its equivalents were used as suspension in concrete composites in place of 12.5 percent fine sand, there was a 56 percent increase in the penetration of chloride ions. Similar to this, a small addition of WTR-based materials can stop the chemical reactions between silica and alkali metals. Furthermore, the addition of WTR and its equivalents to the concrete mixture can help prevent the damage that occurs during drying and shrinkage. However, in some physical properties, like shrinkage properties, the addition of WTRs with less stiffness can be crucial in reducing the number of shrinkage-related cracks as well as the internal constraints (Khyaliya et al., 2017). While adding WTRs to cement reduces the material's elasticity modulus and shrinkage stress, which can also regulate shrinkage cracks to some extent, substituting 20% of the fine aggregates (FA) with WTRs can increase the material's resistance to shrinkage cracking. Compressive strength is one of the mechanical characteristics of WTR-based concretes that can be adjusted by adding rubber or other derived materials to the cementing concrete mixture.

Cementing concrete containing WTR with 5–50 percent fine aggregates has been found to reduce compression strength by about 4–70 percent; the sizes of the fine aggregate particles can range from 0.075 to 6 mm. Nevertheless, the mechanical and structural characteristics of the cementing concrete, as well as the percentage replacement level of WTR, determine whether the compression strength can be completely reduced by incorporating WTR-based cementing concretes. Because WTR functions as a space that controls and reduces material density, the decrease in compressive strength of WTR-based cementing concrete while increasing WTR content is primarily caused by a lower adhesion rate between WTR and the cementing concrete. Researchers were provided with microscopic observations, which verified the existence of voids and cracks in rubber-added concrete. These findings suggest a very weak bond between the rubber and cement. Additional research revealed that grounded WTRs can be used as fillers in concrete mixtures to partially replace natural sand. In one study, the impact of crumbed rubber was examined in concrete-based pavements, where the rubber was combined with the concrete to create rubber sand. The study demonstrated that even after the abrupt occurrence of failure cracks, concrete-based pavement with crumbed WTRs at the weight percent of 20 and 40 percent could hold themselves strongly without breaking into pieces.

The compressive and tensile stresses that result from the surface modeling of WTR particles close to the cementing concrete may also have some additional effects. These stresses are caused by the softness of the WTR particles, which first appears near the joint between the WTR particles and the cementing concrete.

4.6 CHALLENGES AND REALISTIC SOLUTIONS

Even though a number of techniques have been identified for recycling waste rubber, there is still a significant information and technological gap, which accounts for the small percentage of waste rubber that has been recycled. Real-time waste rubber recycling and full-proof concepts derived from laboratory-scale research lag behind due to the unestablished proper characterization of rubber waste. Situations sound particularly problematic when recycling techniques are chosen incorrectly. For instance, the pyrolysis of waste rubber produces carbon, gas, and oil, which are subsequently used in many different applications. However, the production of value-added products may suffer without adequate understanding of the chemical properties of the rubber-based materials, resulting in a loss of both time and money. Another drawback stems from the growth of large-scale, well-established pyrolysis plants, where it is now difficult to extract and dispose of hazardous gases. Furthermore, to develop some cutting-edge technologies for ecological and financial sustainability, further in-depth research on the upcycling of waste rubber is still needed, and it needs to be given careful consideration. Therefore, to remove all feasible recycling technologies at the ground level, an appropriate solution must be extracted. For a sustainable future, policymakers, government officials, and researchers need to have created a single window channel where the real-time solution can be implemented. Meeting by creating a proper industrial symbiosis between the industries and users can be one of the solutions. Numerous recycling methods, including WTR-derived nanocarbons for energy harvesting and cementing applications, have been developed, as was covered in earlier sections. These two fields are the most emerging fields of the twenty-first century, hence recycling methods may be driven in these sectors, where a possible industrial symbiosis and cost-benefit analysis can be met according to the conservation of "EEE" i.e. Energy, Ecology and Economy.

REFERENCES

1. Aggarwal, S., Karimi, I. A., and Reinaldi Ivan, G. (2013). In silico modeling and evaluation of *Gordonia alkanivorans* for biodesulfurization. Mol. Biosyst. 9, 2530.
2. Akiba M, Hashim AS (1997) Vulcanization and crosslinking in elastomers. Prog Polym Sci 22(3):475–521
3. Akindoyo JO, Beg M, Ghazali S, Islam M, Jeyaratnam N, Yuvaraj A (2020) Polyurethane types, synthesis and applications: a review. RSC Adv 6(115):114453–114482
4. Al-Salem SM, Lettieri P, Baeyens J (2009) Recycling and recovery routes of plastic solid waste (PSW): a review. Waste Manag 29(10):2625–43
5. Alves, L., Melo, M., Mendonça, D., Simões, F., Matos, J., Tenreiro, R., et al., (2007). Sequencing, cloning and expression of the dsz genes required for dibenzothiophene sulfone desulfurization from Gordonia alkanivorans strain 1B. Enzyme Microb. Technol. 40, 1598–1603.
6. Alves, L., Salgueiro, R., Rodrigues, C., Mesquita, E., Matos, J., Gírio, F. M., et al., (2005). Desulfurization of dibenzothiophene, benzothiophene, and other thiophene analogs by a newly isolated bacterium, Gordonia alkanivorans strain 1B. Appl. Biochem. Biotechnol. 120, 199–208.
7. Armada, D., Llompart, M., Celeiro, M., Garcia-Castro, P., Ratola, N., Dagnac, T. J., et al., (2022). Global evaluation of the chemical hazard of recycled tire crumb rubber employed on worldwide synthetic turf football pitches. Sci. Total Environ. 812, 152542.
8. Asaro, L., Gratton, M., Seghar, S., and Aït Hocine, N. (2018). Recycling of rubber wastes by devulcanization. Resour. Conserv. Recycl. 133, 250–262.
9. Aznar MP, Caballero MA, Sancho JA, Francs E (2006) Plastic waste elimination by cogasification with coal and biomass in fluidized bed with air in pilot plant. Fuel Process Technol 87(5):409–420
10. Barlow C (2008) Intelligent recycling (presentation). Institute for Manufacturing, University of Cambridge, Department of Engineering
11. Bassuoni MT, Nehdi ML (2007) Resistance of self-consolidating concrete to sulfuric acid attack with consecutive pH reduction. Cem Concr Res 37(7):1070–1084
12. Behnood A, Karimi MM, Cheraghian G (2020) Coupled effects of warm mix asphalt (WMA) additives and rheological modifiers on the properties of asphalt binders. Clean Eng Technol 1:100028
13. Behnood A, Olek J (2017) Stress-dependent behavior and rutting

resistance of modified asphalt binders: an MSCR approach. Constr Build Mater 157:635–646

14. Bhatnagar A, Hogland W, Marques M, Sillanpää M (2013) An overview of the modification methods of activated carbon for its water treatment applications. Chem Eng J 219:499–511
15. Bockstal, L., Berchem, T., Schmetz, Q., and Richel, A. (2019). Devulcanisation and reclaiming of tires and rubber by physical and chemical processes: a review. J. Clean. Prod. 236, 117574.
16. Campuzano, F., Martínez, J. D., Agudelo Santamaría, A. F., Sarathy, S. M., and Roberts, W. L. (2023). Pursuing the end-of-life tire circularity: an outlook toward the production of secondary raw materials from tire pyrolysis oil. Energy Fuels 37, 8836–8866.
17. Cañavate J, Carrillo F, Casas P, Colom X, Sunol JJ (2010) The use of waxes and wetting additives to improve compatibility between hdpe and ground tyre rubber. J Compos Mater 44(10):1233–1245
18. Castañón, A. M., Sanmiquel, L., Bascompta, M., Vega, Y., De La Fuente, A., Contreras, V., et al., (2021). Used tires as fuel in clinker production: economic and environmental implications. Sustainability 13, 10455.
19. Chou HT, Lee HJ, Lee CY, Tai NH, Chang HY (2014) Highly durable anodes of microbial fuel cells using a reduced graphene oxide/carbon nanotube-coated scaffold. Biores Technol 169:532–536
20. Christofi, N., Morton, J. G., and Bond, D. E. (2010). Rubber Treatment Method. US Patent No. 7,737,191B2.
21. Cuesta A, Dhamelincourt P, Laureyns J, Martinez-Alonso A Tascon JM, (1998) Comparative performance of X-ray diffraction and Raman microprobe techniques for the study of carbon materials. J Mater Chem 8(12):2875–2879
22. Cui K, Maruyama S (2019) Multifunctional graphene and carbon nanotube films for planar heterojunction solar cells. Prog Energy Combust Sci 70:1–21
23. Cui, X., Zhao, S., and Wang, B. (2016). Microbial desulfurization for ground tire rubber by mixed consortium-Sphingomonas sp., Gordonia sp. Polym. Degrad. Stab. 128, 165–171.
24. Cummins, A. (2019). Global ELT Management - A Global State of Knowledge on Regulation, Management Systems, Impacts of Recovery and Technologies. Geneva: World Bus. Counc. Sustain. Dev.
25. Delegan, Y., Kocharovskaya, Y., Frantsuzova, E., Streletskii, R., and Vetrova, A. (2021). Characterization and genomic analysis of Gordonia

alkanivorans 135, a promising dibenzothiophene-degrading strain. Biotechnol. Rep. 29, e00591.

26. Dimpe KM, Ngila JC, Nomngongo PN (2017) Application of waste tyre-based activated carbon for the removal of heavy metals in wastewater. Cogent Engineering 4(1):1330912
27. Fang Y., Zhan M., Wang Y. The status of recycling of waste rubber. Mater. Des. 2001;22:123–128.
28. Fazli A, Rodrigue D (2020) Waste rubber recycling: a review on the evolution and properties of thermoplastic elastomers. Materials 13(3):782
29. Fraile-Garcia E, Ferreiro-Cabello J, Defez B, Peris-Fajanes G (2016) Acoustic behavior of hollow blocks and bricks made of concrete doped with waste-tire rubber. Materials 9(12):962
30. Gilbert, S. C., Morton, J., Buchanan, S., Oldfield, C., and McRoberts, A. (1998). Isolation of a unique benzothiophene-desulphurizing bacterium, Gordona sp. strain 213E (NCIMB 40816), and characterization of the desulphurization pathway. Microbiology 144, 2545–2553.
31. Gonen T (2018) Freezing-thawing and impact resistance of concretes containing waste crumb rubbers. Constr Build Mater 177:436–442
32. Gupta VK, Gupta B, Rastogi A, Agarwal S, Nayak A (2011) Pesticides removal from waste water by activated carbon prepared from waste rubber tire. Water Res 45(13):4047–4055
33. Heinrich G., Vilgis T.A. Contribution of entanglements to the mechanical properties of carbon black-filled polymer networks. Macromolecules. 1993;26:1109–1119.
34. Hopewell J, Dvorak R, Kosior E (2009) Plastics recycling: challenges and opportunities. Philos Trans R Soc B: Biol Sci 364(1526):2115–2126
35. Jang J, Oh J, Stucky GD (2002) Fabrication of ultrafine conducting polymer and graphite nanoparticles. Angew Chem 41(21):4016–4019
36. Karger-Kocsis J., Mészáros L., Bárány T. Ground tyre rubber (GTR) in thermoplastics, thermosets, and rubbers. J. Mater. Sci. 2013;48:1–38. doi: 10.1007/s10853-012-6564-2.
37. Khan AAP, Khan A, Asiri AM (2019) Nanocarbon and its composites for water purification. In: Nanocarbon and its composites. Woodhead Publishing, pp 711–731
38. Kordoghli S, Paraschiv M, Tazerout M, Khiari B, Zagrouba F (2017) Novel catalytic systems for waste tires pyrolysis: optimization of gas fraction. J Energy Resour Technol 139(3)
39. Medina N.F., Garcia R., Hajirasouliha I., Pilakoutas K., Guadagnini M.,

Raffoul S. Composites with recycled rubber aggregates: Properties and opportunities in construction. Constr. Build. Mater. 2018;188:884–897.

40. Mente P., Motaung T., Hlangothi S. Natural rubber and reclaimed Rubber composites—A Systematic Review. Polym. Sci. 2016;2:1–19.
41. Mina M., Ania F., Balta Calleja F., Asano T. Microhardness studies of PMMA/natural rubber blends. J. Appl. Polym. Sci. 2004;91:205–210.
42. Mitra S., Jørgensen M., Pedersen W.B., Almdal K., Banerjee D. Structural determination of ethylene-propylene-diene rubber (EPDM) containing high degree of controlled long-chain branching. J. Appl. Polym. Sci. 2009;113:2962–2972.
43. Ramarad S., Khalid M., Ratnam C., Chuah A.L., Rashmi W. Waste tire rubber in polymer blends: A review on the evolution, properties and future. Prog. Mater. Sci. 2015;72:100–140.
44. Sienkiewicz M., Kucinska-Lipka J., Janik H., Balas A. Progress in used tyres management in the European Union: A review. Waste Manag. 2012;32:1742–1751.
45. Wypych G. Handbook of Polymers. 2nd ed. Elsevier; Amsterdam, The Netherlands: 2016.

Chapter **5**

Recycling of Thermosetting Plastics

Contents

INTRODUCTION

Large molecular weight polymers known as thermosets offer structural and aesthetic benefits, affordability, and workability that make them a desirable substitute for thermoplastics and other conventional materials like metals and wood.

To begin with, thermosets are typically polymers that undergo curing methods such as heat, ultraviolet rays, electron beam processing, or chemical reactions involving hardeners or catalysts. This curing process initiates irreversible chemical reactions, resulting in cross-linked polymer chains due to their inability to melt when subjected to high temperatures. This characteristic gives them superior mechanical strength. Additionally, they maintain their shape and do not deform even when exposed to cold temperatures. Consequently, they are well-suited for applications in environments marked by extreme temperature fluctuations.

The low cost of production of thermoset composites, which are made by mixing a thermoset polymer with aramid, carbon, or glass fibers, results in additional benefits and improved properties. Indeed, thermoset structural composites have found widespread application in the aviation, surface, and water transportation industries, guaranteeing excellent performance for the finished goods and permitting fuel savings due to their lightweight nature. Furthermore, their exceptional thermal, chemical, and mechanical stability renders them appropriate for use in protective and structural applications, including wind turbines and aerospace materials. Additionally, it was reported that a 25% reduction in CO_2 is possible when metals are replaced in airplane structures with fiber-reinforced plastic composite materials.

Thermosets, which primarily comprise polyesters, phenolics, silicones, epoxies, and polyurethanes, account for 44 million tons, or about 12 percent, of the world's plastic production volume today. It's also important to note that growth in the global thermosetting plastics market is anticipated over the next few years.

Epoxy resins in particular are widely used as electrical insulation, adhesives, coatings, and engineering composites; by 2030, at least 4 million tons of epoxy composites are expected to be produced. (Naheedand Jawaid, 2017)

Due to their chemical makeup, which includes urethane groups in the chains created by the reaction of an isocyanate with alcohol, polyurethanes are versatile polymers with stable physicochemical properties. They are primarily used in furniture, packaging, chemical-resistant coatings, urethane foams, and insulation foams. By 2030, 17 million tons of polyurethane foam should be produced.

As well as being highly flexible and resistant to heat, chemicals, sunlight, and aging, silicones are also used as lubricants and anti-foaming agents. Common uses for silicones include gaskets, heat insulation, and soft-touch surfaces. By 2030, the demand for silicon is expected to reach about 11 million tons.

The extensive and massive use causes thermoset waste to build up. The majority of thermoset and thermoset composite wastes are currently dumped in landfills because they are thought to be difficult to recycle. Currently, these wastes are ground up into fillers, burned, or digested using environmentally friendly technologies. Legislation in Europe has recently mandated recycling of this waste rather than landfilling; however, as of this writing, end-of-life products and production waste from thermoset composites have not yet found a truly satisfactory solution. Recycling is frequently a costly and low-rate process because of the cross-linked, three-dimensional chemical nature of the thermoset matrix, which makes it difficult to re-melt using heat or solvent as it can for thermoplastic matrices. Moreover, thermal and chemical recycling are more widely used and functional than incineration, which offers low energy efficiency and produces emissions; thermoset chemical recycling is a process of chemically catalyzed reactions that break down polymer chains into building blocks that can be reused in the same product as before the recycling process or in other products. Mechanical recycling, on the other hand, only permits the recovery of lower-performance reinforcements.

The problem of recycling thermosets has been extensively studied over the past 20 years, and while it remains unresolved, certain technologies can be recycled extensively. The thermosets that are most frequently utilized are epoxy composites and polyurethane foams. Thus, epoxies-based composites are recycled by catalyzed alcoholysis or, in certain cases, are transformed into recyclable thermoplastics using polyamine curing agents that can cleave at crosslinking sites. Polyurethane foams, on the other hand, are typically converted into polyol. These technologies create business opportunities by converting low-value product waste into high-value products, thus supporting the circular economy. Moreover, recycled virgin materials can be substituted with fiber-reinforced thermosetting composites, lowering the amount of raw thermoset matrix and raw fibers used and promoting a more sustainable society in the process. According to research, the process used to manufacture virgin carbon fiber has significant greenhouse gas emissions. Furthermore, virgin fiber should be recycled to improve resource efficiency because it requires about 14 times as much energy to produce than steel. (Pimenta & Pinho, 2011)

In order to become a cost-effective and environmentally beneficial waste management solution, thermoset recycling has become essential and needs more research.

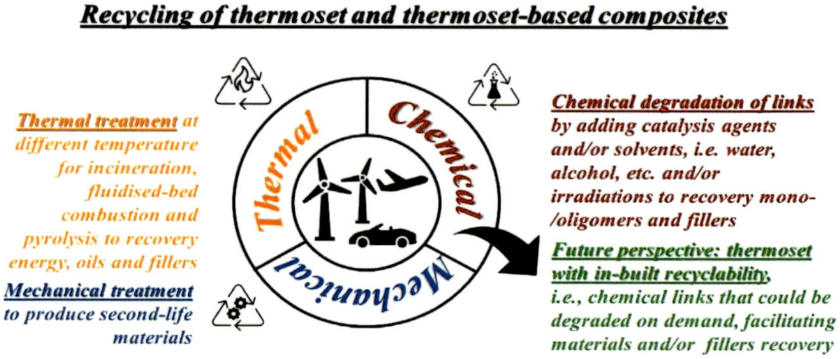

Figure 5.1. *Schematic diagram to summarize current recycling strategies for thermoset and thermoset-based composites, highlighting the future perspective in developing novel thermoset materials.*

Source: Morici, E.; Dintcheva, N.T. Recycling of Thermoset Materials and Thermoset-Based Composites: Challenge and Opportunity. Polymers 2022, 14, 4153. https://doi.org/ 10.3390/ polym14194153

Table 5.1. *Main advantages and disadvantages of different recycling processes*

Recycling Process	**(+) Advantages and (−) Disadvantages of Different Recycling Processes**
Thermal recycling	(+) Recovery of energy, oils and fillers (+) No use of solvent and catalysts agents (−) Reduced dimensions and properties of recovered materials (−) Production of emissions and CO_2
Mechanical recycling	(+) Recovery of fillers and matrix that can be used as additives to produce second-life materials (−) Long time pre-treatments to recover the materials (−) Low mechanical performance of recovered materials (−) Impossibility to re-manufacture the materials
Chemical recycling	(+) Recovery of mono-oligomers and fillers (+) Recovered fillers with unchanged shape, dimensions, composition and mechanical properties (−) Use of solvents and catalysts agents (−) Difficult to implement the process at a larger scale

Source: Morici, E.; Dintcheva, N.T. Recycling of Thermoset Materials and Thermoset-Based Composites: Challenge and Opportunity. Polymers 2022, 14, 4153. https://doi.org/ 10.3390/ polym14194153

5.1 THERMAL RECYCLING

Thermal recycling techniques, which entail processing at high temperatures, can be broadly categorized into two types: anaerobic combustion, or pyrolysis, which recovers simpler molecules used as feedstock for new chemical processing, and combustion or incineration, which recovers only energy and fluidized-bed combustion, which primarily recovers reinforcement fibers.

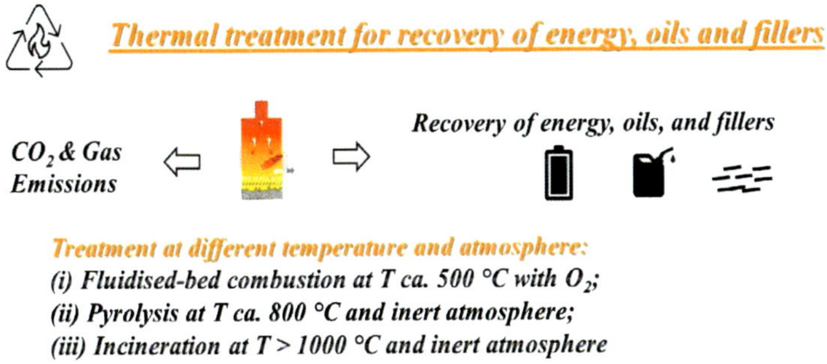

Figure 5.2. *Schematic diagram to illustrate the thermal treatment and recovery of energy, oils and fillers.*

Source: Morici, E.; Dintcheva, N.T. Recycling of Thermoset Materials and Thermoset-Based Composites: Challenge and Opportunity. Polymers 2022, 14, 4153. https://doi.org/ 10.3390/ polym14194153

5.1.1 Incineration

Energy recovery can be achieved by burning thermosetting polymers in an incinerator. The benefit of this method is that waste streams that are contaminated or mixed with other forms of municipal solid waste can be continuously processed. The disadvantages include the lack of material recovery, opposition from the public regarding the damaging effects of the produced gases on the environment, and legal restrictions that are already in place in some European countries. (Joshi & Ahmed,2016)

In some cases, co-processing thermoset composite waste in cement kilns can help solve financial and environmental problems. Because the thermoset matrix burns and releases heat energy, the waste materials become valuable new materials, and the incombustible residues and fibers are used as raw materials to make cement clinkers. Thus, co-processing entails recycling both materials and energy, enabling the recovery of natural resources and the preservation of fossil fuels. Nevertheless, certain mineral fillers or additives included in the thermoset matrix, like flame retardants and/or calcium carbonate, absorb energy or degrade throughout the process, reducing the amount of energy obtained through burning.

5.1.2 Fluidized-Bed Combustion

Fiber recovery is the primary goal of the procedure. The first to develop a fluidized bed process for recovering carbon fiber was the University of Nottingham. In a fluidized bed of silica sand, the organic matrix volatilizes and separates from the other ingredients at a temperature of about 500 °C. The air present guarantees that the polymer matrix oxidizes, resulting in clean fibers. However, to completely oxidize the organics, a secondary combustion chamber running at about 1000 °C is required. The fibers are extracted in a gas stream and subsequently collected in a revolving separator or cyclone. While the fillers can be used as raw materials, the organic components are used for the process's energy and heat recovery. The lower mechanical qualities of recycled fibers compared to virgin fiber, as well as their discontinuous and flocculate nature, high temperature, process pressure, and high energy consumption are the disadvantages of fluidized-bed combustion.

5.1.3 Pyrolysis

The foundation of the pyrolysis process is oxygen-free heating, which breaks down the polymeric matrix into low molecular weight liquids or gases while leaving the inorganic components intact. The fibers and other recovered materials can then be used as fillers or reinforcements for assembling the new products, while the organic portion can be recycled as fuel or as fresh resources for chemical processes.

A few industrial facilities are currently working on this throughout the world. It is important to note that some problems must be resolved. For instance, in sheet molding compounds, the optimal temperature for the process is thought to be around 600 °C; however, this could lead to unfavorable inorganic part degradation processes.

Furthermore, in order to ensure that the produced oils are safe for use as fuels or as feedstock for the petrochemical industry, it may be necessary to remove toxic compounds like bromine before pyrolysis.

For repurpose applications, post-treatments like fiber sizing might be necessary. Additionally, the recovered fiber's quality needs to be met for it to be put to good use. Automotive sheet molding and wind turbine blades are examples of industrial thermoset composites that were used in the investigation of a two-temperature step pyrolysis treatment for glass fiber recovery. The outcomes demonstrated that not only could performed fibers increase tensile strengths and failure strains, but that optimizing pyrolysis can enhance the quality of recovered fibers while also lowering the rate at which pre-existing surface damages grow during the process.

Additionally investigated was microwave pyrolysis, which is the breakdown of organic components of composite materials by microwave heating for rapid heating combined with reduced energy consumption. Even though the method has many benefits over traditional pyrolysis, including easy control and uniform internal heating, it is still primarily used in laboratories today. Technical obstacles must be addressed, such as regulating the feedstock's dielectric qualities, and further research must be done to determine the ideal operating conditions in order to avoid, for instance, char residue on the fiber surface impairing their mechanical qualities.

5.2 MECHANICAL RECYCLING

Because mechanical recycling is easy to do and affordable, it has been thoroughly studied. It is the process of reducing the size of waste so that it can be added to cement and asphalt or used as partially reinforced fibrous materials or powdered fillers to create new composites. (Dorigato, 2021)

In essence, they involve several stages: difficult and time-consuming collection and sorting are followed by a crushing procedure and additional pre-treatment that reduces the scrap components into smaller, more manageable pieces. A finer product is then produced by grinding the material in a high-speed mill. A blend of polymer, fibers, and fillers ranging in size from tiny fibrous materials up to 10 mm in length to fine powders smaller than 50 microns make up the recycled product. While the powder is matrix-rich and frequently used as filler, the fibrous product is fiber-rich and is used as reinforcement. Compared to other recycling methods, this one has advantages for the environment, society, and economy because it doesn't pollute the atmosphere.

Currently, recovered fillers can be economically competitive with conventional fillers like silicates and calcium carbonates, but mechanical recycling is not yet an industrially available process. Moreover, the recycled materials have low values because short and irregular fibers and fillers can be recovered. Consequently, the process's overall profitability could be maximized by using the energy saved to support large-scale development in the future. Shuaib and associates investigated the effects of granulator capacity and processing rates on lowering energy consumption in thermoset-based glass fiber recycling. Furthermore, they emphasize that the energy consumption and environmental impact of other recycling unit processes can be modeled using the utilized bottom-up approach. (Shuaib and Mativenga, 2016)

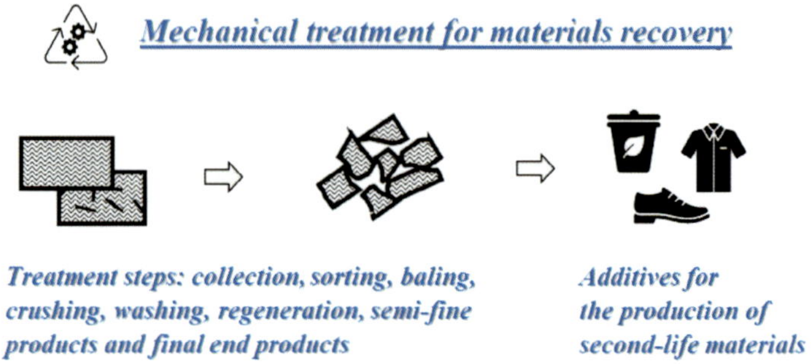

Figure 5.3. Schematic diagram to illustrate the mechanical recycling and recovery of materials.

Source: Morici, E.; Dintcheva, N.T. Recycling of Thermoset Materials and Thermoset-Based Composites: Challenge and Opportunity. Polymers 2022, 14, 4153. https://doi.org/ 10.3390/ polym14194153

Enhancing the interface adhesion between the fiber waste and resin matrix and finding an effective method to stop fiber agglomeration during reprocessing could lead to further improvements and more valuable recycled products. The field of short-fiber composite materials has seen recent successes with fused filament fabrication 3D printing; PLA reinforced with recycled glass fibers exhibits modulus and tensile strength of 18% and 19%, respectively, higher than samples reinforced with virgin glass fibers. This final outcome was explained by taking into account improved interfacial contact between recycled glass fibers that were partially coated in epoxy particles and the PLA matrix. Furthermore, the recycled glass fiber reinforced polymer (GFRP) was employed in a double-recycling process: when compared to fiber-reinforced cementitious materials, the incorporation of GFRP waste (20 weight percent) into a Fe-rich silicate slag, a byproduct of non-ferrous metallurgy production, allowed for a greater level of GFRP waste incorporation (20 weight percent) in the in-organic matrix and a 79 percent enhancement in flexural strength.

5.3 CHEMICAL RECYCLING

With the use of various techniques that break down the three-dimensional linkages between polymer chains, thermosets can be chemically recycled with success. This makes it easier to recover mono-oligomers and, in the case of composites, fillers. There are several ways to break down thermoset links: with solvents, with or without catalysis systems, and with the help of radiation. The development of new thermoset materials will likely focus on materials with inherent recyclability and degradability, or dynamic three-dimensional chemical linkages that can de- and re-crosslink "on-demand" in response to external stimuli like heat, pH, UV light, and mechanical stress. This is based on current understanding.

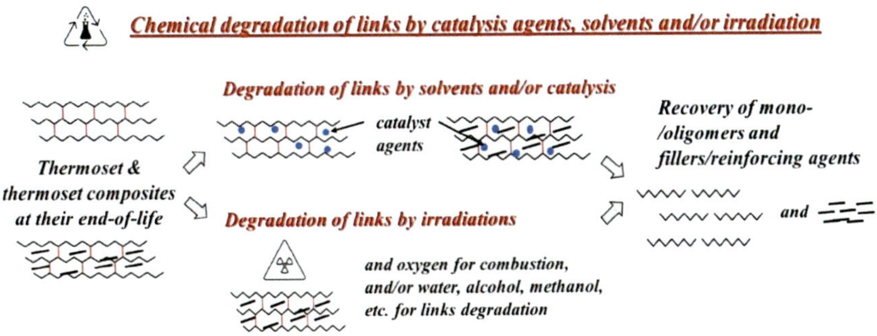

Figure 5.4. Schematic diagram to illustrate the chemical recycling and recovery of materials.

Source: Morici, E.; Dintcheva, N.T. Recycling of Thermoset Materials and Thermoset-Based Composites: Challenge and Opportunity. Polymers 2022, 14, 4153. https://doi.org/ 10.3390/ polym14194153

As a result, the recovery of thermoset mono-oligomers is a risky process since it often calls for the addition of catalysts and requires the use of potentially hazardous solvents. In some instances, the ecological and financial sustainability of the chemical recycling of these materials may not be a convenient issue, particularly for producers of second-life materials, even though the management of thermoset waste streams must be appropriately addressed. Governments must provide sufficient financial support and education to the public to reduce the impact of thermoset materials and ensure proper waste management. On the other hand, the recovery of valuable fillers from thermoset composite materials, like carbon and aramid fibers, may be seen as a sustainable and profitable matter, particularly if the fillers don't change in terms of composition, dimensionality, or structure. Naturally, the recovery of fillers with unchanged shapes, dimensions, and compositions could be advantageously and successfully accomplished by using the appropriate solvents (water, alcohol, methanol, ethanol, etc., also in their supercritical state) and/or catalytic agents (chemicals and/or irradiation absorbers) or irradiations (electron beam, UV irradiation, gamma irradiation, etc.).

However, as has been shown, solvolysis of cross-linked links is the traditional method for chemically recycling thermosets and their composites. The primary benefit of this approach is the potential to recover unaltered fillers; however, the large-scale scalability of this method is restricted by the solvents' use, and the supercritical solvents necessitate the use of expensive equipment. Okajima et al used supercritical methanol, which operates at 270 °C and 8 MPa to dissolve the matrix of the carbon fiber-reinforced sample and to sever the ester bonds holding the epoxy backbone together. The authors showed that while the C-C bond and the mechanical properties of the carbon fibers' shape were retained, the ester bonds could be selectively broken using supercritical fluid. This presents a chance to formulate new thermoset materials using recovered mono-oligomers and carbon fibers. (Okajima, I.; Hiramatsu, M.; et al.,, 2014)

Remarkably, Liu et al., suggested using ethanol as a solvent and $ZnCl_2$ as a catalyst to selectively break the C-N bonds in waste epoxy-based fiber-reinforced resins from the aerospace industry; the reaction was conducted at T = 190 °C. The authors demonstrate that the polymer fragments have a comparatively low molecular weight of 650 g/mol and terminal hydroxyl and amine groups that highlight the successful selective breaking of C-N bonds based on spectroscopy and spectrometry characterizations of productions. In amine-cured diglycidyl ether bisphenol A (DGEBA) filler reinforced resins, Wang et al suggested another method for selectively breaking C-N bonds. This method used acetic acid as a solvent, $AlCl_3$ as a catalyst, and relatively mild conditions at T = 180 °C. The C-C and C-O bonds were retained, but the N-H bonds had replaced the C-N bonds, according to the spectroscopy and spectrometry analysis.

Das et al., suggested adding an oxidizing agent to accelerate the cleavage of chemical bonds during the recovery of mono-oligomers from thermoset fiber-reinforced polymers. The authors combined acetic acid and H_2O_2 and found that under relatively mild conditions (T = 65 °C and 4 hours reaction time of approximately), peracetic acid was formed, exhibiting string oxidation action. The presence of aliphatic and aromatic compounds that suggested the simultaneous and un-selective breaking of C-N, C-C, and C-O bonds was highlighted by the final product characterization. All of these methods, albeit requiring the use of solvents, catalysts, and particular conditions, provide the possibility of recovering mono-oligomers and unmodified high-value fillers. While the innovative approach of creating dynamic networks will present more chances for thermoset and thermoset composites recycling to be successful, chemical recycling may still be a useful technique for recycling some thermoset filler-reinforced materials.

5.4 THERMOSET WITH BUILT-IN RECYCLABILITY

Recent years have seen the development of a novel strategy by researchers to get around the challenge of recycling thermosets: adding chemical linkers to the organic matrix to change it in a way that facilitates the materials' breakdown while maintaining their mechanical qualities.

The concept involves adding degradable crosslinkers or changing permanently crosslinked structures into dynamic ones so that cleavable bond exchange reactions can cause them to de- and re-crosslink. These final dynamic bonds are in fact stimuli that react to various stimuli, such as heat, radiation, acidity, and so forth. They might be applied to initiate structural modifications. Additionally, fiber-reinforced polymer-composites in which fibers could be readily recovered after the degradation of the resins were prepared using thermosets with cleavable bonds.

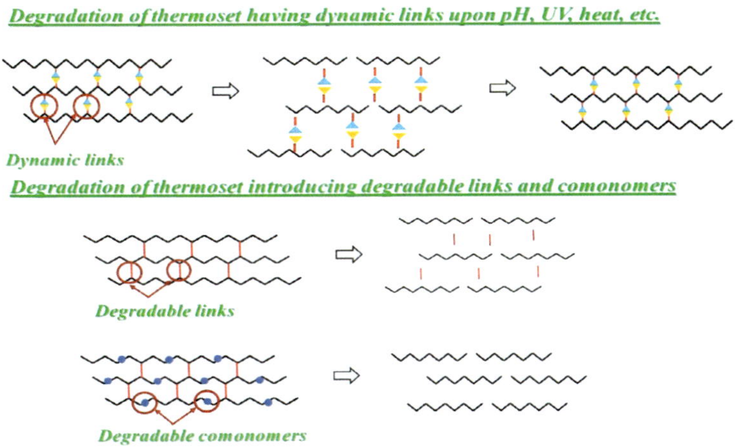

Figure 5.5. Schematic diagram to illustrate the perspective in developing the thermoset with built in recyclability.

Source: Morici, E.; Dintcheva, N.T. Recycling of Thermoset Materials and Thermoset-Based Composites: Challenge and Opportunity. Polymers 2022, 14, 4153. https://doi.org/ 10.3390/ polym14194153

Since the new dynamic covalent bond strategy allows for direct reshaping, recycling, and reprocessing, thermoset composites can also be made using traditional methods used for thermoplastic matrices, such as injection molding or hot pressing. In contrast, the presence of degradable crosslinkers forces recycled polymer to be re-synthesized or used in low-performing applications.

To create recyclable thermosets, a variety of cleavable bonds, including disulfide bonds, peroxide bonds, acetal linkages, ester bonds, B–O bonds, and structures containing phosphorus or nitrogen, have been thoroughly studied

It was found that the presence of a thermally labile tertiary ester linkage in an epoxy thermoset system decreased the matrix's breakdown temperature;

occasionally, the broken-down components create new anhydride crosslinks, which can be removed with the right solvents. The phosphotungstic acid aqueous solution was reported to be an effective catalyst system for the chemical degradation of anhydride-cured epoxy at a moderate reaction temperature of 190 °C. The ester bond in the crosslinked structure was selectively broken during the reaction to produce reactive group-containing oligomers, which were then utilized to create a fresh anhydride-cured epoxy system.

Niu and associates were the first to create self-repairing polymers by utilizing the B-O boronic esters bond, which underwent reversible depolymerization via hydrolytic cleavage. By crosslinking isocyanate-terminated prepolymers with boric acids, degradable polyurethane thermosets with high mechanical strength and toughness were created. The presence of a labile triple boron-urethane bond, which forms when hydroxyl groups in boric acids react with isocyanate groups in the prepolymers, greatly improves the mechanical properties of the degradable polyurethanes.

Well-proven cycloaliphatic diepoxides with cleavable acetal linkages that dissolve in solvent mixtures containing acid were discovered by IBM researchers. To create reworkable UV curing coatings, acetal groups were also added to methacrylates. By using a photoacid generator to produce acid from UV-cured methacrylates, thermal degradation was accomplished without the need for additional acids. (Shirai, 2007)

Dynamic disulfide bond-containing epoxy resin was synthesized and proved to be a great reprocessable resin. There were improvements in welding performance and reprocessability, and the mechanical properties, thermal stability, and service temperature were all on par with typical commercial epoxy resins. After three reprocessing cycles, 90% of the tensile strength was retained. Furthermore, fiber, resin, and solvent could all be recycled by performing chemical degradation in a solvent based on thiols in a closed loop. By simply replacing the traditional diamine hardener (DETDA) with 4-aminophenyl disulfide (AFD) as a dynamic hardener, fiber-reinforced composites based on epoxy resin with exchangeable disulfide crosslink were readily synthesized; the resulting composites displayed vitrimer behavior and could be recycled, reprocessed, and repaired using conventional methods and tools.

In many studies, active covalent bonds are introduced into the epoxy system in response to external stimuli like heat, light, or radiation. While a 100% recovery rate is ideal, the high energy required in the process causes degradation, which lowers the performance of the recovered material when compared to the original thermosets. Using dual dynamic networks—polymeric materials that combine two or more different crosslinkers in a single system—some researchers tested materials to better design a material and its properties for a necessary application. Recent research by Liang et al revealed a workable method for creating a dual dynamic network of waterborne polyurethane (WPU) using 1,3-dihydroxyacetone ketotriose as the chain extender. This condensation of carbonyl groups and ketohydrazine produced a covalent adaptive network. The newly developed WPU demonstrates strong adhesive properties to metal, even after four cycles of deconstruction and rebonding.

Lastly, adding a few cleavable co-monomers is a complementary strategy to the design of sustainable thermosets. Shieh and associates found that the silyl linkers in modified industrial polydicyclopentadiene facilitate chemical deconstruction and are more effective when degradable crosslinkers are introduced; this allows for the optimization of the cleavable bond location of a controlled thermoset degradation and improves the material properties of the original plastic thermoset.

5.5 AVAILABILITY OF WASTE PLASTIC AS RAW MATERIAL FOR PRODUCT DEVELOPMENT

More than 29 million metric tons of post-consumer plastic waste were collected in 2020; however, only about 34% of it was recycled and more than 23% was dumped in landfills or used in energy recovery projects. Municipal, commercial, and industrial waste are the sources of waste plastics. The main sources of industrial waste, also known as primary waste, are the packaging, manufacturing, and plastic processing sectors. Post-consumer waste from industries like packaging, building and construction, cars, electrical and electronic equipment, home, recreation, sports, and agriculture is commonly gathered from rural farms and gardens and is referred to as commercial waste.

Additionally, residential neighborhoods, streets, parks, collection depots, and waste dumps are sources of municipal waste plastics. The distribution of the percentages of plastic waste from various application areas in Europe is depicted in Figure 5.6, where it was found that building and construction, as well as packaging, account for the majority of plastic products. These products have caused significant environmental issues because of the high percentage (41%) of plastic products used for packaging and the careless disposal of products after use. Thus, there is a need to step up efforts to manage plastic waste originating from packaging.

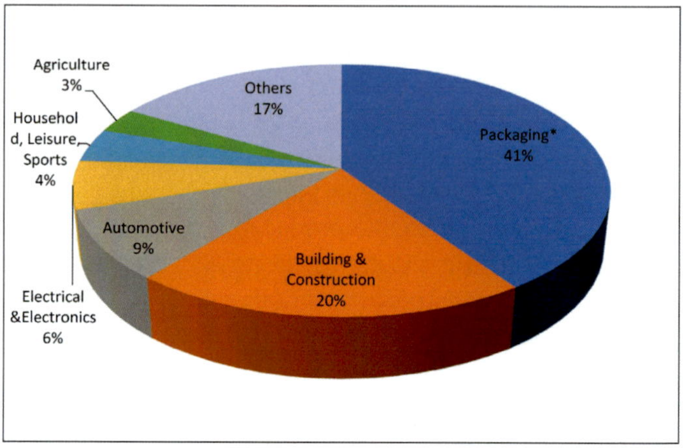

Figure 5.6. *Percentage distribution of waste plastics in Europe.*

*Including commercial and industrial packaging. Total 49.1 Mt.

Source: Oladele, I.O., Okoro, C.J., et al.,, Modern Trends in Recycling Waste Thermoplastics and Their Prospective Applications: A Review. J. Compos. Sci. 2023, 7, 198. https://doi.org/10.3390/jcs7050198

Plastic waste material is produced in large quantities as a result of the growing use of plastic products across many industries. Therefore, given their widespread availability, plastic waste materials present a potential source of raw materials for the development of new products in the future. Unlike the earlier belief regarding waste plastics, these materials require investigation and application to create products that are appropriate for a range of uses. Their easy availability has made this research necessary. Waste plastics, as opposed to metals, are useful materials for humanity because they are the most affordable, easily accessible, and long-lasting materials.

Benefits of Accelerating Waste Plastic Recycling

- *Environmental benefits*: By lowering the quantity of waste plastics that are dumped in landfills or released into the environment, recycling waste plastics helps to mitigate pollution and climate change. An estimated four to twelve million metric tons of waste plastic are reported to wash into rivers and into the ocean annually. The depletion of the ecosystem caused by this process poses a threat to marine life and the environment. Furthermore, since recycled plastics do not produce the same emissions as virgin plastics during production, recycling contributes significantly to the reduction of CO_2 emissions in the atmosphere.

- *Economic and social benefits*: By encouraging the expansion of recycling facilities locally, recycling waste plastic creates jobs and value. The establishment of recycling facilities encourages regional industrial endeavors aimed at extracting value from recycled materials.
- *Availability of raw materials*: Plastics that have been recycled could be used to make secondary goods. Waste plastic that has been processed can be appropriately adapted for use in a variety of fields.

5.6 PROPERTIES, APPLICATIONS, AND FUTURE OF THERMOSETTING PLASTICS

A class of polymers known for its special qualities and uses is thermosetting plastics. These materials, in contrast to thermoplastics, experience an irreversible chemical change during the curing process that turns them into a solid, rigid form. Heat or a catalyst causes this transition, which produces a three-dimensional network structure that is cross-linked. Electrical insulation, aerospace components, automotive parts, and other high-temperature applications frequently use thermosetting plastics because of their exceptional thermal stability and resistance to deformation. Due to their inherent strength, dimensional stability, and chemical resistance, they can be used in a variety of industries. These polymers differ from their thermoplastic counterparts in that, once cured, they cannot be remolded or reshaped. Modern manufacturing processes still rely heavily on thermosetting plastics because of their exceptional qualities and diverse range of applications.

5.6.1 Production of Thermosetting Plastics

Thermoset products are typically created using liquid molding techniques. After being fed into tanks or barrels, the polymers and additional agents are heated to a liquid state and combined. The liquid polymers are then injected into a mold cavity along with additional agents. The material goes through a curing process where the polymers cross-link together as it cools and hardens to the cavity's configuration. This procedure creates an unbreakable chemical bond that guards against the finished product's melting, softening, or warping in the event that it is exposed to extreme heat or corrosive conditions. Because of this, thermosets are ideal for outdoor use and high-heat applications.

There are two primary liquid molding processes used to produce thermoset plastics:

1) Reaction Injection Molding (RIM)

The A-component (polyol) and B-component (isocyanate) are two liquid agents that are heated to a high temperature in separate tanks before the RIM molding process starts. After that, a mixhead is fed under intense pressure. After being combined, the liquid mixture flows into a mold, where a chemical reaction occurs and a polymer is formed within the mold.

2) Resin Transfer Molding (RTM)

When RTM is a closed-molding procedure also referred to as liquid molding, the RIM molding process starts. To form and harden the part, resin formulations are combined with a hardener, also known as a catalyst, and then injected into a mold made of dry fibers, like fiberglass.

5.6.2 Properties of Thermosetting Plastics

Thermosetting plastics are widely used in many different industries because of a number of unique properties. Key characteristics of thermosetting plastics include the following:

High Thermal Stability

Thermosetting plastics are incredibly heat resistant and do not soften or distort at high temperatures. Because of this characteristic, they can be used in applications like electrical insulators and automotive engine components where exposure to high temperatures is required.

Chemical Resistance

Acids, bases, solvents, oils, and other chemicals are just a few of the many substances to which thermosetting plastics exhibit strong resistance. They are appropriate for chemical processing equipment, storage tanks, and chemical-resistant coatings because they are less likely to corrode or degrade in harsh chemical environments.

Dimensional Stability

Once cured, thermosetting plastics withstand changes in the environment without losing their shape or dimensional stability. Because of their low coefficients of thermal expansion, they are not greatly affected by changes in temperature. This characteristic guarantees accurate and stable component measurements across a broad temperature range.

Mechanical Strength

Thermosetting plastics give components structural integrity because of their exceptional mechanical strength and stiffness. Because of their high compressive, tensile, and flexural strengths, they can support heavy loads. The mechanical properties of materials can be further improved by adding reinforcing materials, like carbon fibers or fiberglass.

Electrical Insulation

Numerous thermosetting plastics are useful in electrical and electronics applications because of their superior electrical insulating qualities. They are resistant to electrical arcing, have a high dielectric strength, and low electrical conductivity. They are appropriate for circuit boards, insulating coatings, and electrical enclosures because of these qualities.

Flame Resistance

Since thermosetting plastics have cross-linking and an inherent chemical structure, they frequently have good flame resistance. They are appropriate for applications requiring fire-resistant materials, such as electrical connectors, switches, and fire safety equipment, because of their high ignition temperatures and low flammability.

Impact Resistance

Certain thermosetting plastics are capable of withstanding abrupt shocks and impacts without breaking because of their high impact resistance. This feature is crucial in applications like sports equipment and automobile bumpers where toughness and resistance to mechanical impact are essential.

It is significant to remember that the kind of resin and the particular formulation utilized can affect the unique characteristics of thermosetting plastics. The properties of various thermosetting plastics, such as unsaturated polyester resins, phenolic resins, epoxy resins, and melamine formaldehyde resins, can vary, offering a broad range of applications in various industries.

5.6.3 Applications of Thermosetting Plastics

Automotive Industry

The automotive industry uses thermosetting plastics for a variety of parts, including exterior body panels, interior trim, braking systems, engine parts, and electrical connectors. Because of their exceptional mechanical strength, dimensional stability, and heat resistance, they are perfect for withstanding the harsh conditions found in automotive environments.

Electrical and Electronics Industry

Because of their superior electrical insulation qualities, thermosetting plastics are widely used in electrical and electronics applications. Circuit boards, electrical enclosures, insulating coatings, connectors, and switches are among the products they are used in. Electrical equipment is guaranteed to operate safely and dependably thanks to their resistance to heat and chemicals.

Aerospace Industry

Thermosetting plastics are used in the aerospace industry to create strong, lightweight components. These polymers are utilized in electrical insulation systems, radomes, structural components, and interiors of airplanes. The important factors in fulfilling the strict safety and performance requirements are their high strength-to-weight ratio, flame resistance, and dimensional stability.

Construction Industry

There are several uses for thermosetting plastics in the building industry. They are utilized in composite panels, adhesives, coatings, decorative laminates, and insulation materials. For applications needing long-term performance in challenging environmental conditions, thermosetting plastics are a good choice because of their exceptional fire resistance, chemical resistance, and durability.

Consumer Goods

Thermosetting plastics are used in many consumer products, such as furniture, sporting goods, home fixtures, and kitchen appliances. In items like worktops, handles, casings, and laminates, they offer strength, resilience to impact, and visual appeal.

Industrial Equipment

Because thermosetting plastics have a high resistance to heat, chemicals, and wear, they are used in industrial machinery and equipment. Pumps, valves, gaskets, seals, bearings, and tooling components all use them. In industrial settings, thermosetting plastics' mechanical strength and dimensional stability guarantee dependability and a longer service life.

Medical and Healthcare

Medical and healthcare applications, such as parts of medical devices, surgical instruments, dental products, and laboratory equipment, involve thermosetting plastics. Because of their chemical resistance, sterilizability, and biocompatibility, they can be used in demanding healthcare settings.

It is crucial to remember that the precise uses for thermosetting plastics can differ based on the kind of resin utilized, the manufacturing technique employed, and the demands of each industry. Thermosetting plastics are a valuable material choice in many industries due to their advantageous properties and versatility.

REFERENCES

1. Hay, J.N.; O'Gara, P. Recent developments in thermoset curing methods. Proc. Inst. Mech. Eng. G J. Aerosp. Eng. 2006, 220, 187–195.
2. Boquillon, N.; Fringant, C. Polymer networks derived from curing of epoxidised linseed oil: Influence of different catalysts and anhydride hardeners. Polymer 2000, 41, 8603–8613.
3. Walczyk, D.; Kuppers, J. Thermal press curing of advanced thermoset composite laminate parts. Compos. Part A Appl. Sci. Manuf. 2012, 43, 635–646.
4. Fan-Long, J.; Park, S.J. Preparation and Characterization of Carbon Fiber-Reinforced Thermosetting Composites: A Review. Carbon Lett. 2015, 16, 67–77.
5. Caydamli, Y.; Heudorfer, K.; Take, J.; Podjaski, F.; Middendorf, P.; Buchmeiser, M.R. Transparent Fiber-Reinforced Composites Based on a Thermoset Resin Using Liquid Composite Molding (LCM) Techniques. Materials 2021, 14, 6087.
6. Gore, P.M.; Kandasubramanian, B. Functionalized Aramid Fibers and Composites for Protective Applications: A Review. Ind. Eng. Chem. Res. 2018, 57, 16537–16563.
7. Blanco, D.; Rubio, E.M.; Lorente-Pedreille, R.M.; Sáenz-Nuño, M.A. Lightweight Structural Materials in Open Access: Latest Trends. Materials. 2021, 14, 577.
8. Hagnell, M.; Kumaraswamy, S.; Nyman, T.; Åkermo, M. From aviation to automotive—A study on material selection and its implication on cost and weight efficient structural composite and sandwich designs. Heliyon 2020, 6, e03716.
9. Mishnaevsky, L.; Branner, K.; Petersen, H.N.; Beauson, J.; McGugan, M.; Sørensen, B.F. Materials for Wind Turbine Blades: An Overview. Materials 2017, 10, 1285.
10. Timmis, A.; Hodzic, A.; Koh, L.; Bonner, M.; Soutis, C.; Schäfer, A.; Dray, L. Environmental impact assessment of aviation emission reduction through the implementation of composite materials. Int. J. Life Cycle Assess. 2015, 20, 233–243.
11. Naheed, S.; Jawaid, M. Epoxy resin based hybrid polymer composites. In Hybrid Polymer Composite Materials; Thakur, V.K., Thakur, M.K., Pappu, A., Eds.; Woodhead Publishing: Kidlington, UK, 2017; pp. 57–82.
12. Gama, N.V.; Ferreira, A.; Barros-Timmons, A. Polyurethane Foams: Past, Present, and Future. Materials 2018, 11, 1841.

13. Zhang, Z.; Yuan, L.; Guan, Q.; Liang, G.; Gu, A. Synergistically building flame retarding thermosetting composites with high toughness and thermal stability through unique phosphorus and silicone hybridized graphene oxide. Compos. Part A Appl. Sci. Manuf. 2017, 98, 174–183.
14. Xue, X.; Liu, S.Y.; Zhang, Z.Y.; Wang, Q.Z.; Xiao, C.Z. A technology review of recycling methods for fiber-reinforced thermosets. J. Reinf. Plast. Compos. 2022, 41, 459–480.
15. Leszczyñski, S.; Brzychczyk, B. Waste management of half-finished products and thermosetting wastes. Polish J. Chem. Technol. 2007, 9, 122–126.
16. Morici, E.; Carroccio, S.C.; Bruno, E.; Scarfato, P.; Filippone, G.; Dintcheva, N.T. Recycled (Bio)Plastics and (Bio)Plastic Composites: A Trade Opportunity in a Green Future. Polymers 2022, 14, 2038.
17. Marson, A.; Masiero, M.; Modesti, M.; Scipioni, A.; Manzardo, A. Life Cycle Assessment of Polyurethane Foams from Polyols Obtained through Chemical Recycling. ACS Omega 2021, 6, 1718–1724.
18. La Rosa, A.D.; Blanco, I.; Banatao, D.R.; Pastine, S.J.; Björklund, A.; Cicala, G. Innovative Chemical Process for Recycling Thermosets Cured with Recyclamines® by Converting Bio-Epoxy Composites in Reusable Thermoplastic-An LCA Study. Materials 2018, 11, 353.
19. Pimenta, S.; Pinho, S.T. Recycling carbon fibre reinforced polymers for structural applications: Technology review and market outlook. Waste Manag. 2011, 31, 378–392.
20. Vieira, D.R.; Vieira, R.K.; Chain, M.C. Strategy and management for the recycling of carbon fiber-reinforced polymers (CFRPs) in the aircraft industry: A critical review. Int. J. Sustain. Dev. World Ecol. 2017, 24, 214–223.
21. Meng, F.; Olivetti, E.A.; Zhao, Y.; Chang, J.C.; Pickering, S.J.; McKechnie, J. Comparing Life Cycle Energy and Global Warming Potential of Carbon Fiber Composite Recycling Technologies and Waste Management Options. ACS Sustain. Chem. Eng. 2018, 6, 9854–9865.
22. Joshi, R.; Ahmed, S. Status and challenges of municipal solid waste management in India: A review. Cogent Environ Sci. 2016, 2, 1139434.
23. Gonçalves, R.M.; Martinho, A.; Oliveira, J.P. Recycling of Reinforced Glass Fibers Waste: Current Status. Materials 2022, 15, 1596.
24. Fu, B.; Liu, K.C.; Chen, J.F.; Teng, J.G. Concrete reinforced with macro fibres recycled from waste GFRP. Constr. Build. Mater. 2021, 310, 125063.
25. Utekar, S.; Suriya, V.K.; More, N.; Rao, A. Comprehensive study

of recycling of thermosetting polymer composites—Driving force, challenges and methods. Compos. Part B 2021, 207, 108596.

26. Pickering, S.J.; Kelly, R.M.; Kennerley, J.R.; Rudd, C.; Fenwick, N.J. Fluidized-bed process for the recovery of glass fibres from scrap thermoset composites. Compos. Sci. Technol. 2000, 60, 509–523.
27. Torres, A.; de Marco, I.; Caballero, B.M.; Laresgoiti, M.F.; Legarreta, J.A.; Cabrero, M.A.; González, A.; Chomón, M.J.; Gondra, K. Recycling by pyrolysis of thermoset composites: Characteristics of the liquid and gaseous fuels obtained. Fuel 2000, 79, 897–902.
28. Vasile, C.; Brebu, M.; Karayildirim, T.; Yanik, J.; Darie, H. Feedstock recycling from plastic and thermoset fractions of used computers (I): Pyrolysis. J. Mater. Cycles Waste Manag. 2006, 8, 99–108.
29. Ginder, R.S.; Ozcan, S. Recycling of Commercial E-glass Reinforced Thermoset Composites via Two Temperature Step Pyrolysis to Improve Recovered Fiber Tensile Strength and Failure Strain. Recycling. 2019, 4, 24.
30. Hao, S.; He, L.; Liu, J.; Liu, Y.; Rudd, C.; Liu, X. Recovery of Carbon Fibre from Waste Prepreg via Microwave Pyrolysis. Polymers 2021, 13, 1231.
31. Dorigato, A. Recycling of thermosetting composites for wind blade application. Adv. Ind. Eng. Polym. Res. 2021, 4, 116–132.
32. Conroy, A.; Halliwell, S.; Reynolds, T. Composite recycling in the construction industry. Compos. Part A Appl. Sci. Manuf. 2006, 37, 1216–1222.
33. Haider, M.M.; Nassiri, S.; Englund, K.; Li, H.; Chen, Z. Exploratory Study of Flexural Performance of Mechanically Recycled Glass Fiber Reinforced Polymer Shreds as Reinforcement in Cement mortar. Transp. Res. Rec. 2021, 2675, 1254–1267.
34. Shuaib, N.A.; Mativenga, P.T. Energy demand in mechanical recycling of glass fibre reinforced thermoset plastic composites. J. Clean. Prod. 2016, 120, 198–206.
35. Rahimizadeh, A.; Kalman, J.; Henri, R.; Fayazbakhsh, K.; Lessard, L. Recycled Glass Fiber Composites from Wind Turbine Waste for 3D Printing Feedstock: Effects of Fiber Content and Interface on Mechanical Performance. Materials 2019, 12, 3929.
36. Zhang, Y.; Pontikes, Y.; Lessard, L.; Willem van Vuure, A. Recycling and valorization of glass fibre thermoset composite waste by cold incorporation into a sustainable inorganic polymer matrix. Compos. B Eng. 2021, 223, 109120.

37. Okajima, I.; Hiramatsu, M.; Shimamura, Y.; Awaya, T.; Sako, T. Chemical recycling of carbon fiber reinforced plastic using supercritical methanol. J. Supercrit. Fluids 2014, 91, 68–76.
38. Yu, H.; Potter, K.D.; Wisnom, M.R. A novel manufacturing method for aligned discontinuous fibre composites (High Performance-Discontinuous Fibre method). Compos. Part A Appl. Sci. Manuf. 2014, 65, 175–185.
39. Ning, H.; Lu, N.; Hassen, A.A.; Chawla, K.; Selim, M.; Pillay, S. A review of Long fibre thermoplastic (LFT) composites. Int. Mater. Rev. 2020, 65, 164–188.
40. Kuroyanagi, M.; Yamaguchi, A.; Hashimoto, T.; Urushisaki, M.; Sakaguchi, T.; Kawabe, K. Novel degradable acetal-linkage-containing epoxy resins with high thermal stability: Synthesis and application in carbon fiber-reinforced plastics. Polym. J. 2022, 54, 313–322.
41. Oliveux, G.; Dandy, L.O.; Leeke, G.A. Current Status of Recycling of Fibre Reinforced Polymers: Review of Technologies, Reuse and Resulting Properties. Prog. Mater. Sci. 2015, 72, 61–99.
42. Ma, S.; Webster, D.C. Degradable thermosets based on labile bonds or linkages: A review. Prog. Polym. Sci. 2018, 76, 65–110.
43. Bapat, A.P.; Sumerlin, B.S.; Sutti, A. Bulk network polymers with dynamic B–O bonds: Healable and reprocessable materials. Mater. Horiz. 2020, 7, 694–714.
44. Zhang, G.; Tian, C.; Feng, H.; Tan, T.; Wang, R.; Zhang, L. Thermal Reprocessing and Closed-Loop Chemical Recycling of Styrene-Butadiene Rubber Enabled by Exchangeable and Cleavable Acetal Linkages. Macromol. Rapid Commun. 2022, 43, 2100887.
45. Post, W.; Susa, A.; Blaauw, R.; Molenveld, K.; Knoop, R.J.I. A Review on the Potential and Limitations of Recyclable Thermosets for Structural Applications. Polym Rev. 2020, 60, 359–388.
46. Overcash, M.; Twomey, J.; Asmatulu, E.; Vozzola, E.; Griffing, E. Thermoset Composite Recycling—Driving Forces, Development, and Evolution of New Opportunities. J. Compos. Mater. 2018, 52, 1033–1043.
47. Chen, J.S.; Ober, C.K.; Poliks, M.D.; Zhang, Y.M.; Wiesner, U.; Cohen, C. Controlled degradation of epoxy networks: Analysis of crosslink density and glass transition temperature changes in thermally reworkable thermosets. Polymer 2004, 45, 1939–1950.
48. Liu, T.; Guo, X.; Liu, W.; Hao, C.; Wang, L.; Hiscox, W.C.; Liu, C.; Jin, C.; Xina, J.; Zhang, J. Selective cleavage of ester linkages of anhydride-cured epoxy using a benign method and reuse of the decomposed polymer

in new epoxy preparation. Green Chem. 2017, 19, 4364–4372.

49. Niu, W.; O'Sullivan, C.; Rambo, B.M.; Smith, M.D.; Lavigne, J.J. Self-repairing polymers: Poly(dioxaborolane)s containing trigonal planar boron. Chem. Commun. 2005, 34, 4342–4344.
50. Bao, C.; Zhang, X.; Yu, P.; Li, Q.; Qina, Y.; XinJ, Z. Facile fabrication of degradable polyurethane thermosets with high mechanical strength and toughness via the cross-linking of triple boron–urethane bonds. Mater. Chem. A. 2021, 9, 22410–22417.
51. Buchwalter, S.L.; Kuczynski, J.P.; Stephanie, J.G. Cleavable Diepoxide for Removable Epoxy Compositions; International Business Machines Corporation: Armonk, NY, USA, 1999.
52. Shirai, M. Reworkable UV curing materials. Prog. Org. Coat. 2007, 58, 158–165.
53. Memon, H.; Wei, Y. Welding and reprocessing of disulfide-containing thermoset epoxy resin exhibiting behavior reminiscent of a thermoplastic. J. Appl. Polym. Sci. 2020, 137, 49541–49550.
54. de Luzuriaga, A.R.; Martin, R.; Markaide, N.; Rekondo, A.; Cabañero, G.; Rodríguez, J.; Odriozola, I. Epoxy Resin with Exchangeable Disulfide Crosslinks to Obtain Reprocessable, Repairable and Recyclable Fiber-Reinforced Thermoset Composites. Mater. Horiz. 2016, 3, 241–247.
55. Yu, S.; Zhang, R.; Wu, Q.; Chen, T.; Sun, P. Bio-Inspired High-Performance and Recyclable Cross-Linked Polymers. Adv. Mater. 2013, 25, 4912–4917.
56. Amamoto, Y.; Kamada, J.M.; Otsuka, H.; Takahara, A.; Matyjaszewski, K. Repeatable photoinduced self-healing of covalently cross-linked polymers through reshuffling of trithiocarbonate units. Angew. Chem. Int. Ed. 2011, 50, 1660–1663.
57. Bowman, C.N.; Kloxin, C.J. Covalent adaptable networks: Reversible bond structures incorporated in polymer networks. Angew. Chem. Int. Ed. 2012, 51, 4272–4274.
58. Rekondo, A.; Martin, R.; Ruiz de Luzuriaga, A.; Cabanero, G.; Grande, H.J.; Odriozola, I. Catalyst-free room-temperature self-healing elastomers based on aromatic disulfide metathesis. Mater. Horiz. 2014, 1, 237–240.
59. Winne, J.M.; Leibler, L.; Du Prez, F.E. Dynamic covalent chemistry in polymer networks: A mechanistic perspective. Polym. Chem. 2019, 10, 6091–6108.
60. Liu, Y.; Yu, Z.; Wang, B.; Li, P.; Zhua, J.; Ma, S. Closed-loop chemical recycling of thermosetting polymers and their applications: A review.

Green Chem. 2022, 24, 5691–5708.

61. Peng, W.L.; You, Y.; Xie, P.; Rong, M.Z.; Zhang, M.Q. Adaptable interlocking macromolecular networks with homogeneous architecture made from immiscible single networks. Macromolecules 2020, 53, 584–593.
62. Hammer, L.; Van Zee, N.J.; Nicolaÿ, R. Dually Crosslinked Polymer Networks Incorporating Dynamic Covalent Bonds. Polymers 2021, 13, 396.
63. Liang, R.; Zhang, H.; Wang, Y.; Ye, J.; Guo, L.; He, L.; Li, X.; Qiu, T.; Tuo, X. Dual dynamic network system constructed by waterborne polyurethane for improved and recoverable performances. Chem. Eng. J. 2022, 442, 136204.
64. Shieh, P.; Zhang, W.; Husted, K.E.L.; Kristufek, S.L.; Xiong, B.; Lundberg, D.J.; Lem, J.; Veysset, D.; Sun, Y.; Nelson, K.A.; et al., Cleavable comonomers enable degradable, recyclable thermoset plastics. Nature 2020, 583, 542 547.

Chapter 6

Recycling of Marine Plastic Debris

Contents

INTRODUCTION

Interestingly, 80% of marine debris originates from trash and debris found in urban runoff, i.e., sources located on land. Litter, trash and debris from construction sites, ports and marinas, commercial and industrial buildings, and trash blown out of garbage trucks and landfills are important elements of land-based sources. The remaining 20% comes from ocean-based sources like abandoned fishing gear and ship discharges overboard.

The majority of the municipal solid waste stream (80 million tons, or 31.7 %) is made up of food containers and packaging. Other than particles smaller than 5 mm, like pre-production plastic pellets, fragments, and polystyrene pieces, these items and plastic bags also make up the majority of marine debris. In addition to being commonplace in marine debris, packaging and single-use disposable goods also represent an unsustainable use of finite resources (oil, trees, energy sources, and water). The amount of marine debris in the world's oceans is rising. The Central Pacific Gyre saw a five-fold increase in plastic debris between 1997 and 2007, according to research from the Algalita Marine Research Foundation. In 1997, the baseline revealed that plastic pieces outnumbered plankton on the ocean's surface 6:1.

The amount of floating pelagic plastic particles off the coast of Japan increased tenfold in ten years during the 1970s and 1980s, and then tenfold every two to three years during the 1990s. During the early 1990s, there was a 100-fold increase in plastic debris in the Southern Ocean. The global production of plastic fibers quadrupled concurrently with these increases in plastic debris. Fish, seabirds, and marine mammals are harmed and sometimes killed by plastic debris in the ocean.

At least 267 species have been affected by marine plastic pollution globally, including 86% of all sea turtle species, 44% of all seabird species, and 43% of all marine mammal species. The effects include entanglement, drowning, starvation, suffocation, infection, and fatalities from ingestion.

A dead California grey whale washed up on Puget Sound beaches in 2010. After autopsies, it was discovered that the stomach had a golf ball, a pair of pants, over 20 plastic bags, tiny towels, duct tape, and surgical gloves. Plastic debris that floats is particularly likely to be consumed by seabirds that feed at the ocean's surface. The growth and survival of the chicks are negatively impacted by the things that adults feed them.

According to one study, plastic was found in about 98% of the chicks sampled, and the amount of plastic being consumed was growing over time. It has been discovered that floating plastics in the oceans collect pollutants and carry them through ocean currents because persistent organic pollutants in the marine environment adhere to the surface of plastic debris. It has also been discovered that plastic debris that floats and migrates spreads invasive marine species. More and more studies demonstrate that marine life can absorb pollutants into their bodies when it consumes plastics coated in them. Debris made of plastic is contaminating the human food chain.

Researchers from Algalita discovered that fish are consuming trash and pieces of plastic during a 2008 expedition around the Pacific Gyre. 35% of the 672 fish that were caught on that trip had eaten plastic debris. The American Chemical Council (ACC), which leads the plastics industry, spends millions of dollars annually trying to persuade legislators and Californians that anti-litter campaigns that place the blame for marine debris on individual behavior are the best way to combat plastic pollution. However, they have spent far more money pushing laws that encourage the use of plastics more than they have on public education.

While more public education is necessary to discourage littering, managing litter does not address the unsustainable resource consumption associated with the production of packaging and single-use disposable items. Furthermore, significant and ongoing funding is needed for street and waterway cleanups as well as public education campaigns aimed at reducing littering due to the growing prevalence of disposable products and packaging. Reducing the amount of money required for trash and litter management can be achieved by minimizing the production of disposable products. Preventive measures are more economical and environmentally friendly.

6.1 TYPES OF MARINE DEBRIS

Marine litter, another name for marine debris, is solid human waste that has been unintentionally or purposefully dumped into oceans or seas. Often referred to as beach litter or tidewrack, floating oceanic debris tends to accumulate at the center of gyres and on coastlines, frequently washing ashore. Ocean dumping is the deliberate disposal of waste in the ocean. There is also naturally occurring debris, like driftwood and drift seeds. Human influence has become a problem due to the growing use of plastic because many (petrochemical) plastics do not biodegrade as quickly as natural or organic materials. Discarded and lost fishing nets are the main source of plastic pollution in the oceans, accounting for approximately 10% of all plastic pollution. Plastic that gets into the water can seriously harm boats and coastlines, as well as fish, seabirds, marine mammals, and reptiles.

Figure 6.1. Marine debris washed up on a beach at Sharm el-Naga, Egypt.

Source: By Vberger - Own work, Public Domain, https://commons.wikimedia.org/w/index.php?curid=12262113

This issue is exacerbated by litter that gets into storm drains and waterways, spilled containers, landfill waste carried by the wind, and dumping. Serious negative effects of the increased water pollution include animals being caught in discarded fishing nets, plastic debris accumulating in large marine garbage patches, and rising contaminant concentrations in the food chain. Plastic bags, balloons, buoys, rope, glass and plastic bottles, medical waste, cigarette stubs, cigarette lighters, beverage cans, polystyrene, lost fishing line and nets, and various wastes from cruise ships and oil rigs are among the many man-made items that can end up as marine debris. Particularly, six-pack rings are thought to represent the issue ((Cecil Adams, 1999)).

From at least 1919 until 1970, the US military used ocean dumping to dispose of unused bombs and weapons, including chemical weapons, landmines, UXO, and conventional bombs. Millions of pounds of ordnance were dumped in the Gulf of Mexico and off the coasts of at least sixteen states, ranging from New Jersey to Hawaii (the US is not the only nation to have done this, of course; these do not wash up onshore). Plastic makes up 80% of marine debris (Alan Weisman, 2007). Because plastics usually do not biodegrade as many other substances do, they accumulate. When exposed to sunlight, they undergo photodegradation; however, this process is limited to dry conditions as water obstructs photolysis. In a 2014 study, researchers from the 5 Gyres group estimated that 5.25 trillion plastic particles, weighing 269,000 tons, were distributed equally throughout the oceans of the Northern and Southern Hemispheres based on computer models.

6.1.1 Persistent Industrial Marine Debris

Certain substances and processes used in industrial processes have a tendency to accumulate over time, remain in the environment, and are not easily broken down. Activities that involve harvesting or using resources from the marine environment and possibly losing or discarding equipment, supplies, machinery, or solid waste from industrial processes into the water or onto shorelines include fishing, boating, and aquaculture. Anything from a Styrofoam lobster float particle to a fishing boat can be considered part of this. In the coastal waters and along the shores of Charlotte County, New Brunswick, a study was carried out in 2003 to identify the types, amounts, sources, and effects of persistent industrial marine debris. It also looked at any potential relationships between the types and numbers of industrial operations in the area and the amount and types of persistent industrial marine debris. Materials like plastic or foam can decompose into smaller particles, which may be eaten by wildlife like fish, birds, and cetaceans because they resemble tiny sea creatures. When indigestible material builds up in the gut, it can cause blockages, a delusion of fullness, and ultimately death from inadequate nutrient intake.

6.1.2 Ghost Nets

Fishing nets that have been lost, abandoned, or otherwise disposed of in lakes, rivers, and the ocean are known as ghost nets (Macfadyen, 2009). These nets can be left tangled on a rocky reef or drifting in the open sea. They are frequently almost invisible in the low light. They have the ability to entangle a variety of animals, including the occasional human diver, including fish, dolphins, sea turtles, sharks, dugongs, crocodiles, and seabirds. When the nets function as intended, they impede movement, leading to malnourishment, cuts, and infections, as well as asphyxia for individuals who must surface to breathe. If we exclude fishing gear that was discarded or abandoned, the annual total estimated amount of lost fishing gear is approximately 48 million tons (48,000 kt), some of which may remain in the oceans for a long period before disintegrating.

6.1.3 Macroplastic

Macroplastics are plastic debris larger than 5 millimeters. This category includes items such as plastic bags, bottles, fishing nets, and large pieces of broken plastic.

6.1.4 Microplastics

The use of microplastics is one area of increasing concern regarding plastic pollution in the marine ecosystem. Microplastics, which are plastic particles smaller than five millimeters in diameter, are frequently present in face cleansers, hand soaps, and other exfoliating products. The microplastics from these products enter the ocean after passing through the water filtration system, but their small size makes it likely that the wastewater plants' preliminary treatment screens won't catch them (Fendall & Sewell, 2009)). Because they can readily ingest the plastic and become ill, these beads are detrimental to marine life, particularly filter feeders. Because of their small size, microplastics are very difficult to remove from the environment. People can try to steer clear of these dangerous plastics by using products that use eco-friendly exfoliants.

Microplastics are now common in the marine environment due to the widespread use of plastic on the planet. For instance, microplastics are present in deep sea sediment, the water column, and sandy beaches and surface waters. In addition to dead biological material (tissue and shells) and certain soil particles, microplastics can also be found in a wide variety of other marine particles that are carried into the ocean by rivers and the wind. Density of population and proximity to urban areas have been identified as the primary determinants of microplastic abundance in the environment.

6.1.5 Deep-sea Debris

The Arctic Ocean floor contains marine debris. There is little data on plastic debris in the mesopelagic and deeper layers of the water column, despite the fact that an increasing number of studies have concentrated on the accumulation of plastic debris on the coasts, in offshore surface waters, and that consumed by marine organisms that reside in the upper levels of the water column. Previous research projects have used remotely operated vehicles (ROVs) for video observation, submersibles, and bottom sampling. Furthermore, the majority of them are restricted to one-time initiatives that don't last long enough to demonstrate the deep-sea debris's long-term effects. Thus far, research has demonstrated that human activity does, in fact, affect the debris in the deep ocean. Plastic has been regularly found in the deep sea, particularly in areas off the coast of densely populated regions like the Mediterranean (Chiba et al.,2018)).

Litter is made up of a variety of materials that are denser than surface water, including metals, glass, and some plastics. It has been discovered to spread over the sea floor and open oceans, where it can entangle itself in corals and interfere with other sea life. It can also become buried under sediment, which makes cleanup very challenging, particularly because it spreads over a larger area than shipwrecks. When other organic particles gather and adhere to the plastic, which is normally negatively buoyant, the plastic can sink. Major particle and debris transfers are aided by other oceanic processes like offshore convection and coastal storms that impact circulation. Additionally, downwelling currents may be enhanced by underwater topographic features, which could result in the retention of microplastics in particular areas.

In 2017, the Global Oceanographic Data Center of the Japan Agency for Marine-Earth Science and Technology (JAMSTEC) released a Deep-sea Debris database that featured thirty years' worth of images and samples of marine debris taken since 1983. 3,425 pieces of man-made debris were identified from 5,010 dives in the database, which included both deep-sea submersibles and ROVs. Out of all the debris found, macro-plastic accounted for 33% (89% single-use) and metal for 26%. These two types of debris were the most significant. At a depth of 10,898 meters, plastic debris and plastic bags entangled in hydrothermal vent and cold seep communities were discovered at the Mariana Trench.

6.1.6 Garbage Patches (Gyres)

A garbage patch is a gyre of particles from marine debris that has been brought about by human population growth and the effects of ocean currents. These man-made accumulations of plastic and other waste are to blame for environmental and ecological issues that harm marine life, contaminate oceans with hazardous substances, and increase greenhouse gas emissions. Marine debris becomes mobile once it gets into the water. Flotsam can be carried by the wind or follow the direction of the ocean currents, frequently landing in the center of gyres where the currents are least strong.

6.2 SOURCES

From greatest to least, the following ten countries contribute more than 90% of the plastic waste that ends up in the world's oceans: China, Indonesia, Philippines, Vietnam, Sri Lanka, Thailand, Egypt, Malaysia, Nigeria, and Bangladesh. Most of this pollution is released through the Yangtze, Indus, Yellow, Hai, Nile, Ganges, Pearl, Amur, Niger, and Mekong rivers. "

According to estimates, container ships lose 10,000 containers at sea annually, most often during storms (Janice Podsada (2001)). In 1992, a spill happened in the Pacific Ocean when a storm caused thousands of rubber ducks and other toys, now referred to as "Friendly Floatees," to go overboard. Since then, the toys have been discovered all over the world, improving our knowledge of ocean currents. Similar events have occurred in the past, such as the time Hansa Carrier dropped 21 containers, one of which notably contained

Nike shoes that floated (Marsha Walton (2003)).

The majority of the hundreds of containers that the MSC Napoli dumped during its 2007 beaching in the English Channel ended up washing up on the World Heritage Site that is the Jurassic Coast. According to a 2021 study conducted in response to the 2014 loss of a printer cartridge container, some cartridges had spread out at a speed of between 6 and 13 centimeters per second on average. Five million Lego pieces were lost in a cargo container that was involved in a 1997 collision involving the Tokio Express off the coast of Britain. Collectors who combed the beaches for Lego dragons came to value some of the pieces. It also offered insightful information about the degradation of marine plastic. 52% of the items in Halifax Harbour, Nova Scotia, came from people using an urban park for recreation, 14% came from people disposing of waste, and only 7% came from shipping and fishing (Walker et al, 2006)). Roughly 45% of the debris found in the ocean comes from urban runoff and waste that is blown into the water from landfills. According to certain studies, certain areas may have a predominance of marine debris. For instance, a 2016 study on Aruba discovered that the majority of the debris on the island's windward side was marine debris originating from far-off places. After debris from six Korean beaches was gathered and examined in 2013, it was discovered that 44% of the debris was "land-based" and 56% was "ocean-based" (Yong, 2013)). Medical waste washed ashore in New Jersey during the 1987 Syringe Tide after being blown from the Fresh Kills Landfill. Many Antarctic fur seals are entangled in fishing-related debris, which is mostly composed of plastics, on the isolated sub-Antarctic island of South Georgia (Walker et al, 1997)).

6.3 ENVIRONMENTAL IMPACTS

Not every human-made object dumped in the ocean is dangerous. Because they sink to the bottom and become immobile, iron and concrete structures usually cause little environmental harm. At shallow depths, they can even serve as scaffolding for artificial reefs. For that reason, ships and subway cars have been purposefully sunk. Furthermore, when a hermit crab cannot find a real seashell that fits their needs, they have been known to substitute bits of beach trash for the shell.

6.3.1 Impacts of Plastic Pollution

Many animals that live on or in the sea consume flotsam by mistake, as it often looks similar to their natural prey. Large plastic debris has the potential to lodge itself permanently in these animals' digestive tracts, obstructing food flow and leading to infection or starvation death. Little floating plastic particles can become part of the ocean food chain by being consumed by filter feeders because they resemble zooplankton. The sustainability of fish and shellfish species may also be impacted by plastic pollution of the marine environment and food chain.

6.3.2 COVID-19 Pandemic Impacts

The COVID-19 pandemic has affected the quantity of marine debris discovered on Kenyan beaches; pandemic-related trash makes up about 55% of the debris. The trash related to the pandemic has appeared on Kenyan beaches, but it has not entered the water. The pandemic-related beach closures and reduced mobility may have contributed to the decrease in ocean litter, as fewer pieces of trash ended up in the water. Disposable masks have ended up along the beaches of Soko's islands in Hong Kong, indicating additional effects of the COVID-19 pandemic. This could be explained by the fact that during the pandemic, there was an increase in the manufacturing of medical supplies (such as masks and gloves), which increased the need for unusual disposal of these supplies.

6.4 REMOVAL

Removing marine debris is a critical task to protect ocean ecosystems, marine life, and coastal communities.

6.4.1 Coastal and River Clean Ups

Debris skimmer boats are one method for gathering and getting rid of riverine or marine debris. These kinds of devices can be used in situations where floating debris interferes with navigation. In San Francisco Bay, for instance, the US Army Corps of Engineers removes ninety tons of "drifting material" each month. Since a seaplane carrying Admiral Chester W. Davis arrived in 1942, the Corps has been performing this work. The pilot lost his life when Nimitz struck a piece of floating debris and sank. The Interceptor is a vessel designed to remove debris from rivers as a result of the Ocean cleanup. Debris is gathered manually and with the help of specialized beach-cleaning equipment once it qualifies as "beach litter." Initiatives exist to encourage fishermen to remove any litter that they may unintentionally haul up in the process of catching fish.

In other places, "trash traps" are put in place on tiny rivers to catch trash in the water before it gets to the ocean. For instance, the city of Adelaide in South Australia runs several of these "trash racks" or "gross pollutant traps" on the Torrens River, which empties into Gulf St. Vincent during the rainy season. Hand removal is another option in areas with lakes or close to the coast. For instance, Project AWARE encourages dive clubs to pick up trash as a kind of diving exercise. Every year, Ghost Fishing UK, with funding from the

Fat Face Foundation and World Animal Protection, conducts a diving marine debris removal operation in Scapa Flow, Orkney. Inadequate coordination among governmental levels and a disorganized regulatory apparatus—wherein accountability frequently varies for the ocean's surface, seabed, and shore—can impede the cleanup of marine debris. For instance, it is estimated that British Columbia's waters are home to 1600 abandoned or derelict boats. In 2019, the federal government of Canada enacted laws making it unlawful to abandon a vessel. However, enforcement of these laws is hindered by the fact that it is frequently challenging to identify the owner of an abandoned boat because provincial governments are responsible for licensing. The non-profit Dead Boats Disposal Society in Victoria points out that because there is little enforcement, abandoned boats are frequently allowed to sink, increasing the cost of cleanup and posing a greater risk to the environment (due to seepage of fuel, oil, plastics, and other pollutants).

6.4.2 Mid-ocean Clean Ups

The elimination of man-made debris (i.e. plastics) is only getting started. Nonetheless, a few initiatives have been launched that use ships equipped with nets to collect plastics, mostly for scientific purposes (Ocean Voyages Institute/ Kaisei 2009 and 2010 and New Horizon 2009). There is also the SeaVax from Bluebird Marine System, which had a cargo hold and an onboard shredder and was powered by wind and solar energy. The Manta ship built by Sea Cleaners has a similar concept. Boyan Slat of The Ocean Cleanup has suggested another way to collect fake litter. He recommended gathering the debris inside the current of gyres using platforms equipped with arms. The design of the SAS Ocean Phoenix ship is somewhat similar. Using GPS trackers and already-existing maritime equipment, Ocean Voyages Institute carried out a cleanup in June 2019 in the North Pacific Subtropical Convergence Zone, setting a record for the largest mid-ocean cleanup carried out in the North Pacific Gyre and removing over 84,000 pounds of consumer plastic trash and polymer nets from the ocean. During a cleanup expedition in May/June 2020, Ocean Voyages Institute broke the record for the largest mid-ocean cleanup in the North Pacific Gyre by removing over 170 tons (340,000 pounds) of ghostnets and consumer plastics from the ocean. The Ocean Voyages Institute can precisely track and send cleanup vessels to remove ghostnets by using specially designed GPS satellite trackers that are deployed by vessels of opportunity. The capacity and efficiency of cleanup will be significantly increased by the combination of GPS tracker technology and satellite imagery, which will enable the real-time location of plastic waste and ghostnets. An additional concern is that clearing marine debris from the ocean may have unintended consequences. In addition to being the primary lower-level food group for the marine food chain and the source of more than half of global photosynthesis, plankton could inadvertently be removed by cleaning up microplastics. Reusable shopping bags, reusable shopping containers, avoiding single-use plastics, and purchasing goods in reusable packaging are some of the best and most economical ways to contribute to the reduction of plastic entering our oceans.

6.5 TECHNOLOGIES FOR RECYCLING OF PLASTIC LITTER

Polymer fractions that have been pre-processed, sorted, and segregated are the main limits on recycling pathways. Taking care of marine plastic litter presents additional challenges because the plastic is not only incredibly diverse but also greatly impacted by a number of environmental elements, such as deterioration from UV radiation, oxidation, and the general biofouling process, in addition to mechanical degradation from wave movement and sandy shores. Furthermore, external factors affect the leaching of chemical additives like plasticizers from the polymeric chains, which further modifies the material's properties. The creative InNoPlastic Project, which aims to prevent, remove, and reuse marine plastic litter, has created a number of clean-up technologies. These include utilizing blockchain technology, creating a robot that can collect macro plastic litter, and mixing flocculants with drum screens to agglomerate nanoparticles.

A schematic diagram of technologies for recycling of marine plastic litter is illustrated in Figure 6.2.

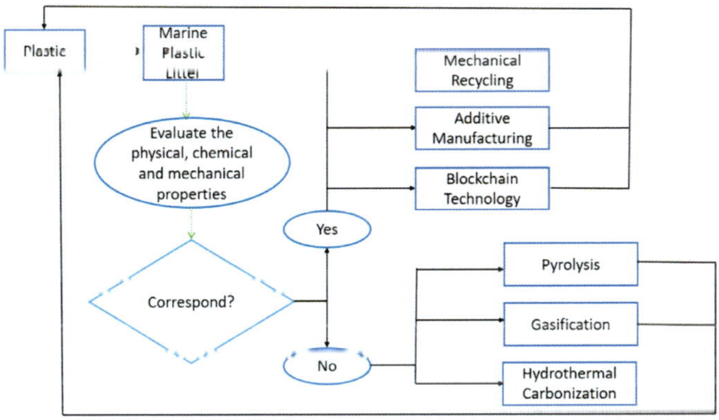

Figure 6.2. Scheme for recycling of marine plastic litter.

Source: Râpă, M.; Cârstea, E.M.; Şăulean, A.A.; Popa, C.L.; Matei, E.; Predescu, A.M.; Predescu, C.; Donţu, S.I.; Dincă, A.G. An Overview of the Current Trends in Marine Plastic Litter Management for a Sustainable Development. Recycling 2024, 9, 30. https://doi.org/10.3390/recycling9020030

6.5.1 Mechanical Recycling

Technology for mechanical recycling has already been put into place to handle plastic waste that comes from human activity. If marine plastic litter meets the requirements, it can be recycled provided that its mechanical, chemical, and physical properties do not significantly deviate from those of virgin polymers. On the other hand, it would be advised to downcycle the plastic waste if notable alterations were noticed. PET bottle waste, fishing gear waste consisting of PE, PA, and PP, buoys and pontoons made of EPS, and waste electrical and electronic equipment (WEEE) have all been reported to be mechanically recycled into marine plastic litter. The results have implications for improving marine plastic litter's capacity to be recycled.

The European Commission has set high goals in its ongoing initiatives to reduce the amount of lost fishing gear. These goals are to collect 15% of abandoned fishing nets and to recycle 15% of them; both are expected to be completed by 2025. Therefore, waste PE from fishing gear was recycled by adding it to gypsum-based materials in amounts ranging from 0.25% to 2.00% by weight as fiber reinforcement. The results of mechanical testing used to assess the characteristics of these gypsum-based materials showed improved post-crack performance. In a related study, high-performance cementitious composites were developed using ground waste fishing net (GN) and ground waste rope (GR) as fiber reinforcement, with varying fiber contents (1%, 2%, and 3%). The resultant composites' flexural strength was increased by up to 19%.

Four different plastic materials—PET from household waste on land, EPS from buoys, and PA and PP from abandoned fishing nets and ropes—were turned into fibers meant to strengthen concrete through the use of cryogenic grinding. When it came to fishing nets and ropes, the final concrete showed a 35% increase in elongation along with a strength value comparable to conventional concrete. A successful ~30% increase in strength was obtained in the PET case. Nevertheless, mechanical recycling of marine plastic litter has a significant negative environmental impact, even with potential benefits for resource recovery (see Table 6.1).

Table 6.1. *Advantages and disadvantages of mechanical recycling*

Main Findings	Limits	Solutions
• Resource recovery • Well-established technique for SUP • Economic benefit • Does not use toxic chemicals • Maturity at industrial scale	• Worse mechanical properties	• Virgin polymers are incorporated together with recycled plastics
	• Impact on the environment due to the various contaminants, such as salt water, marine organisms, biomass, and other pollutants, as well as plastic degradation process	• Washing of plastic waste • Sorting of plastic according to its composition • Innovative pontoon waste management strategy: on-site dissolution/precipitation using D-limonene (Xayachak et al, 2023)
	• Mechanical fragmentation of marine plastic litter (the estimated MPs counts in raw recycling wash water ranged between 5.97×10^6 and 1.12×10^8 MPs m^{-3}) (Brown, E., MacDonald, A.; Allen, S.; Allen, D., 2023)	• The specific filtration measures are necessary in the rotating drum and knife mill wash tanks, as well as in the compounding and wash tanks (Brown et al., 2023)

	• Adverse effects resulting from degradation	• Incorporating additives such as colorants (Ronkay et al, 2021)
	• The transportation, sorting and cleaning steps for the recycling are labor-intensive and costly	• Chemical recycling

Source: Râpă, M.; Cârstea, E.M.; ăulean, A.A.; Popa, C.L.; Matei, E.; Predescu, A.M.; Predescu, C.; Don u, S.I.; Dincă, A.G. An Overview of the Current Trends in Marine Plastic Litter Management for a Sustainable Development. Recycling 2024, 9, 30. https://doi.org/10.3390/recycling9020030

Because of a number of contaminants, including saltwater, marine life, biomass, other pollutants, and different types of plastics, as well as the process of their degradation, recycling marine plastic litter is challenging. For example, the recycling of waste electrical and electronic equipment (WEEE) collected in southwest England into consumer, industrial, and marine (fishing gear) plastics was thought to be a major way of releasing hazardous chemicals into the environment and a possible way of exposing wildlife to these materials. An analysis of WEEE (N = 264), particularly those colored black, using X-ray fluorescence (XRF) spectrometry revealed maximum concentrations of 43,400 mg kg^{-1} for Br, 2080 mg kg^{-1} for Cd, 662 mg kg^{-1} for Cr, and 23,800 mg kg^{-1} for Pb.

Reprocessing marine plastic litter is also linked to a decline in mechanical qualities. Typically, recycled plastics and virgin polymers are combined in melting machines. For example, in order to obtain new plastic products with adequate mechanical properties, 40 weight percent virgin HDPE was added during the injection-molding technology of PE and PP marine litter, collected from a sandy beach located in the northern coast of Latium Region (Central Italy). Additionally, marine litter made of PE and PP was converted into reduction agents or fuel by using pyrolysis technology, which was useful in the cement and metallurgical industries- see Figure 6.3 for an example.

In order to mitigate the effects of mechanical reprocessing on the environment, Xayachak et al., (2023) suggested EPS pontoons be precipitated or dissolved on-site as a creative waste management technique. One of the many benefits of using D-limonene—a solvent that is known to be environmentally friendly—instead of EPS is that it can be easily transported to the beach, allowing precipitation and dissolution to happen at room temperature. Another benefit is that the PS-limonene mixture can be easily transported to a recycling facility.

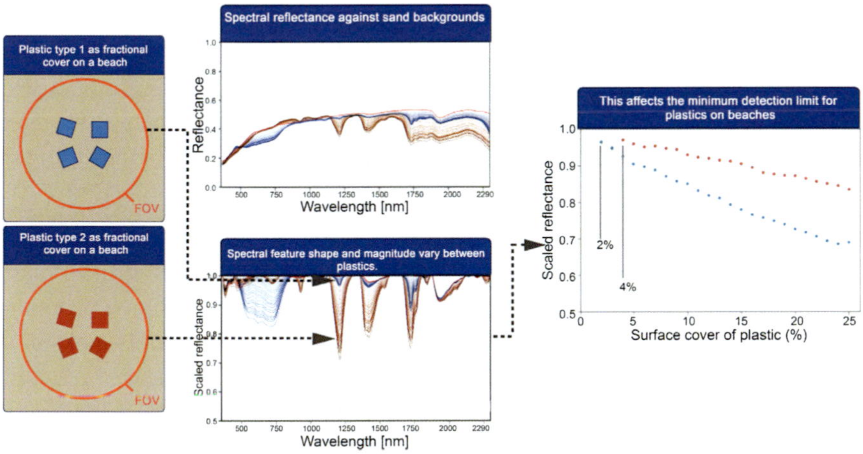

Figure 6.3. Studying marine plastic.

Source: Guffogg JA, Blades SM, Soto-Berelov M, Bellman CJ, Skidmore AK, Jones SD. Quantifying Marine Plastic Debris in a Beach Environment Using Spectral Analysis. Remote Sensing. 2021; 13(22):4548. https://doi.org/10.3390/rs13224548

6.5.2 Additive Manufacturing

One of the most widely used additive manufacturing techniques, fused deposition modeling (FDM) is a fundamental part of Industry 4.0 for quickly prototyping polymer and composite parts and is valued for its ease of use, speed, and affordability. The following procedures must be followed in order for PET made from bottles gathered from the seashore to be recyclable: gathering the bottles, cleaning them, cutting and grinding them, and then extruding the material to create a thin wire. This extruded material was then used to create 3D printed samples. It was possible to reduce the crystallinity of PET extrusion and facilitate its processing through 3D printing by optimizing the processing conditions.

The study conducted by Cañado et al., (2022) demonstrated that repurposing PA66 fishing nets collected from the Biscay Bay (Atlantic Ocean) through 3D printing reduced the environmental impact, as the Life Cycle Assessment (LCA) methodology indicated. The results showed that the global warming potential was 3.7 times lower than that of bio-based polyamide and 1.8 times lower than that of virgin petroleum-based polyamide.

6.5.3 Chemical Recycling

For marine plastic litter, which cannot be recycled mechanically, novel recycling technologies and procedures have been created. Fuel, pyrolysis oil, and a solid residue are produced through the chemical recycling of marine plastic litter. In addition to requiring less technical work during processing, this method offers the possibility of application in regions most impacted by marine pollution and lacking access to advanced engineering solutions. It also permits the storage or direct reuse of end products.

Given the significant yearly fuel consumption of maritime transportation worldwide (207 million metric tons in 2017 and more than 36 million metric tons in 2019 among EU member states), chemical recycling offers a promising avenue for the introduction of recyclable marine litter products within the context of a circular economy. Pyrolysis is a useful technique for recycling plastic marine litter, but it has high energy requirements. However, for nonrecycled plastics, hospital plastics that pose serious biohazards, and plastic marine litter, this method of chemical recycling is preferred. Pyrolysis oils must then be steam cracked in order to obtain base chemicals for the synthesis of second-generation virgin-quality polymeric materials. Consequently, pyrolysis of marine plastic litter "as is" and a distillation procedure enabled the production of a light, naphtha-like substance that could be steam cracked. A boiling point of approximately 240°C and higher concentrations of linear paraffins and α-olefins were found in the analysis of marine litter naphtha. The linear paraffins, α-olefins, a specific sequence of α,ω-diolefin, and n-paraffin examined by the two-dimensional gas chromatography (GC × GC) analysis indicated a higher concentration of PE in the collected marine litter due to the waste-derived naphtha fractions.

To ascertain the trace metal content, the CHNS/O analysis was performed using inductively coupled plasma-optical emission spectrometry (ICP-OES) in conjunction with inductively coupled plasma mass spectrometry (ICP-MS) analyses following microwave-assisted acid digestion. The following concentrations were highlighted by the results: silicon 1.900 ppm, potassium and sodium under LOD (50 ppb), calcium ~3 ppm, vanadium and tungsten metals 5.9 ppm, oxygen ~2500 ppm, and chlorine ~297 ppm (21). Kusenberg et al., reported the maximum allowable concentrations of a few significant contaminants found in pyrolysis oils. (2022) and consist of the following: 2 ppm for calcium, 20–200 ppm for potassium, 0 ppm for zinc, < 1 ppb for iron, 3 ppm for chlorine, 1 ppb for iron, 125 ppb for sodium, 50 ppb for vanadium, 0 ppm for nitrogen, nickel, and oxygen, and 100 ppm for each of these elements. It is clear that there is a chance for corrosion because the pyrolysis oil made from marine plastic litter has a composition that is higher than what is required for industrial steam crackers. Using marine plastic waste for "as received" pyrolysis recycling, meaning it is used after the plastic separation process. In an investigation conducted by Faussone et al., (2021) saw the collection of about 100 kg of marine debris from the Venice Lagoon, which was then processed "as is" into products such as marine gas oil (MGO), intermediate fuel oil (IFO), mixed floating litter, mussel nets, eco-leather, mooring lines, and trawl fishing nets. In comparison to the 95 and 85 values, respectively, required by EN228:2012, the characteristics for VN reported as research octane number (RON) and measured octane number (MON), estimated in accordance with the ISO 5164 and ISO 5163 standards, showed low values of 68.8 and 65.3, respectively. Lower values may be related to the relatively higher presence of low-octane hydrocarbons like hexane and heptane, offset by the presence of aromatics, given that the VN is in line with the gasoline's distillation range. Approximately 50% by volume of this output was found to be MGO. The increase in yield might be attributed to the removal and sorting of non-plastic components from marine plastic litter. Sorted PET waste doesn't need to be cleaned before pyrolysis because the final products' quality stays the same.

Through a comparison between gasoil obtained from pyrolysis of marine litter and gasoil obtained from commercial sources, it was discovered that the gasoil derived from marine litter fully complies with ISO 8217. On the other hand, BTX, ethylbenzene, styrene, and alpha olefins—all important recoverable platform chemicals for industrial upcycling—are noticeably abundant in VN. Gas chromatography-mass spectrometry (GC-MS) is used to identify the 2,4-dimethyl-1-heptene, which is a chemical component of all marine litter pyrolysis and allows for the distinction between commercial and pyrolysis marine gasoil.

In conjunction with the supply of multi-walled carbon nanotubes (MWCNTs), the environmental assessment of marine plastic litter recycling, which included PET bottle sorting, pyrolysis, and chemical vapor deposition (CVD) of seashore (SP) and underwater plastics (UP) waste, suggested that marine PET debris recycling could be a viable and sustainable strategy for scale-up.

Hee et al., (2021) reported an innovative approach to utilize the condensable fractions, generated by the pyrolysis process, of two heterogenous marine plastic litter batches collected from the North Sea and agricultural mulch foil, as reference, for biotechnological upcycling. The results of a thermogravimetric analysis (TGA) used to simulate the mass losses from pyrolysis of the examined marine plastic litter were 91 weight percent for mulch foil, 88 weight percent for Norderney, and 79 weight percent for Sylt marine plastic litter. The TGA and Derivative Thermogravimetry (DTG) signals consistently showed a progressive continuation of mass loss over 500 °C, indicating that all materials had migrated from the fixed carbon phase into the volatile phase. Due to the heterogeneity of the collected marine plastic litter or the calcination of contained carbonates at temperatures above 550 °C in TGA, different ash contents of 13.52 wt percent for marine plastic litter Sylt, 10.55 wt percent for marine plastic litter Norderney, and 0.83 wt percent in the case of mulch foil resulted after the incineration process. The results of the pyrolysis condensates analysis revealed that the carbon content of the marine plastic litter Sylt was 77.0%, that of the marine plastic litter Norderney was 72.7%, and that of the mulch sample was 56.6%. These findings suggest that the carbon source could be useful for biotechnological upcycling in the context of a circular economy concept. Marine plastic litter from Sylt had a caloric value of 41,500 kJ/kg, marine plastic litter from Norderney had a caloric value of 37,210 kJ/kg, and the mulch sample had a caloric value of 42,170 kJ/kg.

6.5.4 Hydrothermal Carbonization

A promising technology that supports the ideas of a circular economy and sustainable waste management is hydrothermal carbonization (HTC). The liquid fraction, solid hydrochar, and a tiny amount of gases are the end products of HTC. Fuel and adsorbent materials for the removal of dyes, heavy metals, and emerging pollutants could be made from hydrochar. HTC uses less energy and offers greater control over the released gases than pyrolysis and gasification techniques. It turns waste plastic into fuel by operating at a pressure and temperature that are reasonably moderate.

By Iñiguez et al., (2019), HTC developed an innovative approach, who for the first time employed marine water as a cheap and plentiful solvent in the process, enhancing the technology's overall sustainability and environmental impact. The increase in carbon and hydrogen contents was discovered by examining the hydrochar characteristics that came about as a result of processing four common types of polymers found in seawater at 200 °C and 250 °C, respectively. The authors proposed that 250 °C might be thought to be effective for a potential conversion of marine plastic litter into hydrochar if low CO emissions were observed with an increase in temperature, especially when the plastics cannot be recycled by other means.

6.5.5 Blockchain Technology

An additional cutting-edge strategy for managing marine plastic litter is the integration of blockchain technology (BCT) with an identity recognition system and digital token system. BCT shows to be crucial in overcoming obstacles associated with managing marine debris, raising public awareness of recycling, creating a global network for recycling, and stepping up societal and consumer oversight. Every unit of plastic waste that is gathered or recycled under this strategy is represented on the blockchain as a digital token. Tokens can be linked to the quantity of plastic collected and used as rewards for people or groups that take part in cleanup activities. By verifying the legitimacy and place of origin of the plastic waste, these tokens serve as a kind of digital product passport. This strategy's distributed ledgers and digital wallets have the ability to successfully replace cash transactions and paper documents in the traditional recycling chain, lessening the impact of global warming on regional economies and ultimately improving efficiency and safety. The blockchain ledger offers a thorough history for every plastic item from its point of collection to its processing or recycling stage, ensuring a transparent and traceable movement of tokens. A decentralized identity system allows participants in cleanup efforts to maintain control over their personal identity data, improving security and privacy.

A Chinese company called Waste2Wear, according to BCT, turned used plastic bottles that were collected from the ocean into textiles that were used for a variety of things, such as worker uniforms for hotels, hospitals, and schools, as well as home furnishings like pillows, bed linens, and curtains (Gong et al., 2022).

6.5.6 Cleaner Manufacturing Processes

The recycling of waste MPs from PET bottles, foam, and EPS with an average size less than 500 μm was the focus of only one study. Waste MPs were used as filler in a bio-based matrix to create a novel open-cell foam material. Due to its adaptability, this foam finds use in a variety of industries, including maritime, civil, and industrial, for thermal and acoustic insulation.

6.6 POLICY ISSUES

Many policies have implications for plastics recycling, which is frequently suggested as a solution to MPP and as part of a future circular economy. Therefore, how should the problem of MPP be addressed? Therefore, we will start by going over some of the technical difficulties that plastic recycling generally faces as well as the particular difficulties that plastic recycling presents for dealing with MPP. Next, we'll pay closer attention to extended producer responsibility (EPR), another policy, and how it might help control the pollution caused by plastics.

6.6.1 Challenges of Plastic Recycling

Recycling plastic materials is challenging due to their complexity. More precisely, a plastic product is made up of thousands of chemical compounds (of which hundreds are known to be toxic) in addition to the polymer itself (Groh et al., 2018). Because these chemical additives lower the quality and safety of materials, they can cause issues for recycling programs.

The Challenge of Maintaining Material Quality

Recycling initiatives must be tailored to each type of plastic in order to preserve material quality. This includes chemical additives (e.g. flame retardants, phthalates, etc.). Given the variety of plastics and the opaqueness surrounding the chemical makeup of materials, this presents a challenge. Consider Leslie et al., (2016) discovered that several toxic substances that are outlawed and found in older products are actually reintroduced into the consumer market. Unknown chemical mixtures will be present in marine plastics that are collected for recycling; products may have absorbed environmental toxicants (man-made toxic compounds) as well as leached additives into the nearby oceans.

When recycling mixed plastics, it can be especially difficult (i.e., composed of multiple distinct polymers), as this results in material properties that are unpredictable. As a result, there is little market for recycled mixed plastics. These difficulties mean that programs for recycling and beach cleanup should concentrate on a small number of polymers (e.g., PP, PE).

"Biodegradable" Plastics

In an attempt to address the issues and difficulties brought about by marine plastics, there have been initiatives to encourage the use of "biodegradable". But in most cases, these materials won't deteriorate in marine environments. Furthermore, it's unclear whether these materials are "safer" from an ecotoxicological standpoint, which could mislead customers and have detrimental effects on the environment (Haider et al., 2019). Because incorporating more than 5% biodegradable plastics can result in decreased material integrity and performance, biodegradable plastics are also problematic for recycling systems (Samper et al., 2018). Therefore, it seems improbable that these materials will be able to offer a sustainable solution to the issue of plastic pollution.

6.6.2 EPR

A producer's liability for a product is extended to the product's post-consumer stage by the Environmental Product Rule (EPR). EPR encourages producers to take the environmental effects of their product design into account by moving the burden of responsibility upstream, from municipalities (and taxpayers) to the producers. More precisely, EPR mandates that manufacturers provide funding for product collection, recycling, and/or safe disposal. Walls (2006) outlines a number of environmental goals for EPR, including a decrease in the use of virgin materials, more incentives for ecodesign, a decrease in production-stage pollution, a decrease in hazardous components, a decrease in waste volumes, and a decrease in waste disposal. From the extraction of virgin material to the creation of waste, product flows have historically been linear, as seen in the upper portion of figure 6.4 (which depicts a linear economy, from left to right). Theoretically, EPR induces a clockwise circular flow by altering this linear flow (see figure 6.4). EPR encourages producers to lower the costs associated with handling waste by transferring the burden of managing a product's post-consumer stage from the municipality to producers, who then pass along as much of the handling costs to consumers as possible. Ideally, this can accomplish each of the six EPR goals.

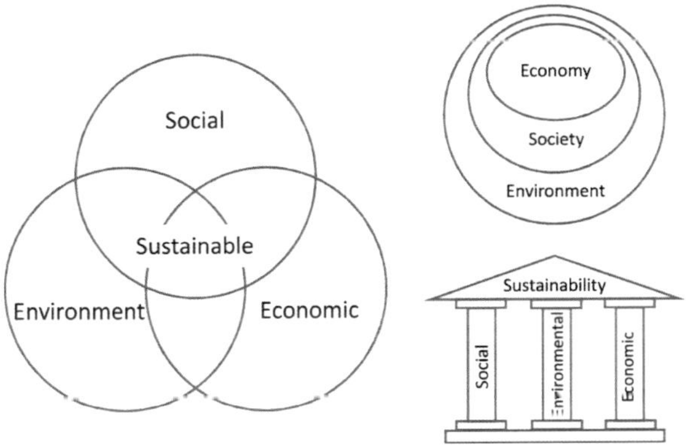

Figure 6.4. EPR as a policy toward creating a circular economy.

Source: https://en.wikipedia.org/wiki/File:Visualization_of_pillars_of_sustainability.webp

However, companies in a given industry frequently create producer responsibility organizations to split the cost of EPR requirements in order to cut costs. If businesses cooperate as well, this cooperation could lead to welfare losses (i.e. conspire) on other matters, like retail prices (Walls 2006). Cost-sharing among businesses also lessens the motivation for ecodesign, indicating a need for increased producer accountability on an individual basis.

In the European Union (EU), waste product recovery (EPR) is widely used for a wide range of products, such as electrical and electronic equipment, batteries, accumulators, cars, tires, waste oil, paper and card, and waste from construction and demolition. Indeed, the EU's goal of establishing a circular economy now depends heavily on EPR. EU regulations now include explicit targets for recycling rates for different waste types and mandatory EPR schemes for all packaging. By 2025, the overall goal for recycling packaging is set at 65% for all packaging and 50% for plastics. Starting in 2021, single-use plastic cutlery, plates, straws, plastic cotton bud sticks, and expanded polystyrene cups will be prohibited. If these goals are accomplished, perhaps this will inspire similar efforts around the globe, making MPP a more controllable issue.

REFERENCES

1. Alan Weisman (2007). *The World Without Us*. St. Martin's Thomas Dunne Books. pp. 112–128. ISBN 978-0312347291.
2. Brown, E.; MacDonald, A.; Allen, S.; Allen, D. The potential for a plastic recycling facility to release microplastic pollution and possible filtration remediation effectiveness. *J. Hazard. Mater. Adv.* 2023, *10*, 100309.
3. Cañado, N.; Lizundia, E.; Akizu-Gardoki, O.; Minguez, R.; Lekube, B.; Arrillaga, A.; Iturrondobeitia, M. 3D printing to enable the reuse of marine plastic waste with reduced environmental impacts. *J. Ind. Ecol.* 2022, *26*, 2092–2107.
4. Cecil Adams (16 July 1999). "Should you cut up six-pack rings so they don't choke sea birds?". *The Straight Dope*. Archived from the original on 6 October 2008. Retrieved 11 August 2008.
5. Chiba, S., Saito, H., Fletcher, R., Yogi, T., Kayo, M., Miyagi, S.,... & Fujikura, K. (2018). Human footprint in the abyss: 30 year records of deep-sea plastic debris. *Marine Policy*, *96*, 204–212.
6. Faussone, G.C.; Krzan, A.; Grilc, M. Conversion of Marine Litter from Venice Lagoon into Marine Fuels via Thermochemical Route: The Overview of Products, Their Yield, Quality and Environmental Impact. *Sustainability* **2021**, *13*, 9481.
7. Fendall, Lisa S.; Sewell, Mary A. (2009). "Contributing to marine pollution by washing your face: Microplastics in facial cleansers". *Marine Pollution Bulletin*. **58** (8): 1225–1228. Bibcode:2009MarPB..58.1225F. doi:10.1016/j.marpolbul.2009.04.025. PMID 19481226
8. Gong, Y.; Wang, Y.; Frei, R.; Wang, B.; Zhao, C.P. Blockchain application in circular marine plastic debris management. *Ind. Mark. Manag.* **2022**, *102*, 164–176.
9. Groh, K. J., T. Backhaus, B. Carney-Almroth, B. Geueke, P. A. Inostroza, A. Lennquist, M. Maffini, H. A. Leslie, D. Slunge, L. Trasande, M. Warhurst, and J. Muncke. 2018. Chemicals associated with plastic packaging: inventory and hazards. *PeerJ Preprints* 6: e27036v1.
10. Haider, T. P., C. Völker, J. Kramm, K. Landfester, and F. R. Wurm. 2019. Plastics of the future? The impact of biodegradable polymers on the environment and on society. *Angewandte Chemie International Edition* 58: 50–62.
11. Hee, J.; Schlögel, K.; Lechthaler, S.; Plaster, J.; Bitter, K.; Blank, L.M.; Quicker, P. Comparative Analysis of the Behaviour of Marine Litter in Thermochemical Waste Treatment Processes. *Processes* 2021, *9*, 13.

12. Iñiguez, M.E.; Conesa, J.A.; Fullana, A. Hydrothermal carbonization (HTC) of marine plastic debris. *Fuel* 2019, *257*, 116033.
13. InNoPlastic—Innovative Approaches towards Prevention, Removal and Reuse of Marine Plastic Litter. Available online: https://www.innoplastic.eu/ (accessed on 27 March 2024).
14. Janice Podsada (19 June 2001). "Lost Sea Cargo: Beach Bounty or Junk?". National Geographic News. Archived from the original on 6 April 2008. Retrieved 8 April 2008.
15. Kusenberg, M.; Eschenbacher, A.; Djokic, M.R.; Zayoud, A.; Ragaert, K.; De Meester, S.; Van Geem, K.M. Opportunities and challenges for the application of post-consumer plastic waste pyrolysis oils as steam cracker feedstocks: To decontaminate or not to decontaminate? *Waste Manag.* 2022, *138*, 83–115.
16. Leslie, H. A., P. E. G. Leonards, S. H. Brandsma, J. de Boer, and N. Jonkers. 2016. Propelling plastics into the circular economy — weeding out the toxics first. *Environment International* 94: 230–34.
17. Macfadyen, G. (2009). *Abandoned, lost or otherwise discarded fishing gear*. Rome: United Nations Environment Programme. ISBN 978-92-5-106196-1.
18. Marsha Walton (28 May 2003). "How sneakers, toys and hockey gear help ocean science". CNN. Archived from the original on 8 April 2008. Retrieved 8 April 2008.
19. Ronkay, F.; Molnar, B.; Gere, D.; Czigany, T. Plastic waste from marine environment: Demonstration of possible routes for recycling by different manufacturing technologies. *Waste Manag.* 2021, *119*, 101–110.
20. Samper, D. M., D. Bertomeu, P. M. Arrieta, M. J. Ferri, and J. López-Martínez. 2018. Interference of biodegradable plastics in the polypropylene recycling process. *Materials* 1110: 1886.
21. Veksha, A.; Ahamed, A.; Wu, X.Y.; Liang, L.; Chan, W.P.; Giannis, A.; Lisak, G. Technical and environmental assessment of laboratory scale approach for sustainable management of marine plastic litter. *J. Hazard. Mater.* 2022, *421*, 126717.
22. Walker, T. R.; Reid, K.; Arnould, J. P. Y.; Croxall, J. P. (1997), "Marine debris surveys at Bird Island, South Georgia 1990–1995", Marine Pollution Bulletin, 34 (1): 61–65, Bibcode:1997MarPB..34...61W, doi:10.1016/S0025-326X(96)00053-7.
23. Walker, T.R.; Grant, J.; Archambault, M-C. (2006). "Accumulation of marine debris on an intertidal beach in an urban park (Halifax Harbour,

Nova Scotia)" (PDF). *Water Quality Research Journal of Canada*. 41 (3): 256–262. doi:10.2166/wqrj.2006.029.

24. Walls, M. A. 2006. Extended producer responsibility and product design: economic theory and selected case studies. RFF Discussion Paper 06-08. Available at SSRN: https://ssrn.com/abstract=901661 or https://doi.org/http://dx.doi.org/10.2139/ssrn.901661.
25. Xayachak, T.; Haque, N.; Lau, D.; Emami, N.; Hood, L.; Tait, H.; Foley, A.; Pramanik, B.K. White spill: Life cycle assessment approach to managing marine EPS litter from flood-released pontoons. *Chemosphere* 2023, *337*, 139400.
26. Yong, C (2013). "Sources of plastic marine debris on beaches of Korea: More from the ocean than the land". *Ocean Science Journal*. 49 (2): 151–162. Bibcode:2014OSJ....49..151J. doi:10.1007/s12601-014-0015-8. S2CID 85429593.

Chapter 7

Recycling of Synthetic Fiber Plastic

Contents

INTRODUCTION

The majority of synthetic fibers, which are manufactured by humans, are made from petrochemicals, which are petroleum-based raw materials. Fibers are the basis for all fabrics, and fibers come from synthetic or man-made sources. They are composed of a monomer, which is a small unit or a polymer formed of several repeating units. They consist of polypropylene, nylon, acrylics, and polyurethane. These fibers are produced annually in millions of tons worldwide. Petrochemicals and other raw materials, such as petroleum, are used to make synthetic fibers. These substances are used to create a variety of fibers by polymerizing into a long, linear chemical with distinct chemical compounds. Almost half of all fibers are synthetic, and they are used in all areas of textile and fiber technology. Numerous fiber classes derived from synthetic polymers have been assessed as potentially valuable commercial products; the three most popular ones are nylon, polyester, and acrylic. While there are a few ways to make synthetic fibers, the melt-spinning method is the most widely used. The fiber must be heated until it starts to melt, and then it must be swiftly extracted from the melt using tweezers.

The molecules would then need to be arranged in a parallel configuration. As a result, the fibers can crystallize and orient and become closer together. A different technique is heat-setting. Compared to most natural fibers, synthetic fibers are more resilient and easily dyeable. Furthermore, a lot of synthetic fibers provide consumer-friendly features like stain resistance, stretching, and waterproofing. All fibers deteriorate and wear away due to sunlight, moisture, and oils from human skin. The sensitivity of natural fibers is generally higher than that of synthetic blends. The primary reason for this is the biodegradability of natural products. Synthetic fibers are typically made by extruding materials that form fibers through spinnerets into water and air to create a thread. Prior to the development of synthetic fibers, petrochemical-derived polymers were used to create synthetic fibers.

7.1 BASICS OF SYNTHETIC FIBER

A special kind of fiber that is not found in nature naturally is called a synthetic fiber. Humans go through a number of steps to produce synthetic fibers, starting in a laboratory. Synthetic fiber production processes generally entail the utilization of polymers derived from coal or petroleum, along with additional synthetic chemical materials. Many synthetic fiber varieties are made to resemble natural fibers like cotton, silk, or wool, and are widely used for their appealing aesthetics. The fact that synthetic fibers are frequently wrinkle-resistant, lightweight, cost-effective, and smooth to the touch makes them highly valued. Throughout the past 20 years, there has been an increase in interest in synthetic fibers such as glass, carbon, and boron. Glass-polypropylene and glass-epoxies, two synthetic composites, are making a healthy competition for natural fiber composites. Materials with high strength and stiffness that are commonly used as aerospace and automotive industry components are provided by synthetic fiber-reinforced polymer composites. In polymer composites for engineering materials, natural fibers are now used as a reinforced material instead of synthetic fibers due to environmental concerns. Often, the mechanical strength of polymer composites is increased by using hybrid composites made of both synthetic and natural fibers.

Throughout the past 20 years, there has been an increase in interest in synthetic fibers such as glass, carbon, and boron. Glass-polypropylene and glass-epoxies, two synthetic composites, are making a healthy competition for natural fiber composites. Materials with high strength and stiffness that are commonly used as aerospace and automotive industry components are provided by synthetic fiber-reinforced polymer composites. In polymer composites for engineering materials, natural fibers are now used as a reinforced material instead of synthetic fibers due to environmental concerns. Often, the mechanical strength of polymer composites is increased by using hybrid composites made of both synthetic and natural fibers.

7.1.1 Characteristics of Synthetic Fibers

Artificial materials known as synthetic fibers have unique properties that differentiate them from natural fibers. These fibers have a wide range of properties that make them useful in a variety of applications because they are chemically engineered. The remarkable strength and durability of synthetic fibers is one of their most notable qualities. Synthetic fibers frequently perform better than their natural counterparts because they are designed to withstand mechanical stress and wear, which makes them perfect for applications needing resilience and longevity. Synthetic fibers are not only strong but also remarkably resistant to environmental conditions. Synthetic fibers keep their structural integrity and integrity against degradation, unlike certain natural fibers that can break down when exposed to chemicals, sunlight, or moisture. Their popularity in outdoor and industrial settings, where harsh conditions are common, is partly due to their resistance. Furthermore, the ability to resist absorbing moisture frequently leads to rapid drying, which is a highly valued characteristic in textiles used for outdoor clothing and sportswear.

An additional important feature of synthetic fibers is their versatility. These fibers' properties can be customized by manufacturers to fulfill particular needs, opening up a wide range of applications. One of the most popular synthetic fibers, polyester, for instance, can be designed to be soft and opulent for bedding or to wick away moisture for sportswear. Because of their versatility, synthetic fibers can be used for a wide variety of products in the industrial, automotive, and textile industries. Notable are the thermal characteristics of synthetic fibers, as many exhibit superior heat resistance. Because of this property, they can be used in materials used in the aerospace industry or protective clothes where exposure to high temperatures is a concern. Furthermore, some synthetic fibers—like aramids like Kevlar—have the ability to withstand flames, which has made them useful in a variety of crucial applications, such as military and firefighter protective gear.

Many people commend synthetic fibers for requiring little upkeep. These fibers hold up better against mold, mildew, and pests than some natural alternatives, and they typically decompose more slowly. Synthetic fibers are widely used in consumer goods such as carpets, outdoor gear, upholstery, clothes, and upholstery because they are easy to maintain. Even though synthetic fibers have many benefits, environmental issues are becoming more and more important. Concerns regarding sustainability are exacerbated by the fact that petrochemicals are used in the production of some synthetic fibers. Still, in order to lessen the negative environmental effects of synthetic fibers, research and development efforts are concentrated on developing environmentally acceptable substitutes and putting recycling programs into place. (Rajak et al, 2022)

7.1.2 Historical Development and Evolution of Synthetic Fibers

Synthetic fiber development and evolution over time have been a fascinating journey filled with scientific advancements, technological innovations, and industry-changing effects. Synthetic fiber history dates back to the early 20th century, and a number of significant turning points have shaped its course.

- **Nylon Emergence (1930s):** When nylon was first introduced by the DuPont chemical company in the 1930s, synthetic fibers were born. DuPont chemist Wallace Carothers was instrumental in the creation of nylon, which was initially introduced as an alternative to silk. Upon their public debut in 1939, nylon stockings quickly gained popularity and demonstrated the potential of synthetic fibers in the textile sector.
- **Polyester Revolution (1940s-1950s):** Another important turning point came with the discovery of polyester, which followed the success of nylon. Terylene, the first polyester fiber, was created in

1941 by British chemists John Rex Whinfield and James Dickson in association with the Calico Printer's Association. But it wasn't until DuPont began selling polyester as Dacron in the 1950s that the material became widely used in textiles and, subsequently, a wide range of other applications, including industrial materials and packaging.

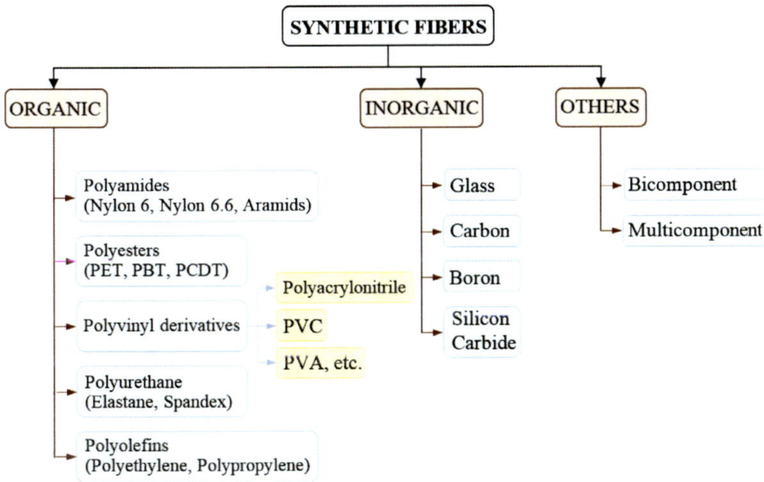

Figure 7.1. Details of fibers.

Source: *Rajak DK, Wagh PH, Linul E. A Review on Synthetic Fibers for Polymer Matrix Composites: Performance, Failure Modes and Applications. Materials. 2022; 15(14):4790. https://doi.org/10.3390/ma15144790*

- **Acrylic Fibers (1950s):** With DuPont's invention of Orlon in the 1950s, acrylic fibers became widely used. With qualities similar to wool, acrylics offered a substitute for a variety of textile applications. Their popularity in outdoor carpets and fabrics was aided by their resistance to mildew, moths, and sunlight.
- **Rayon and Viscose (Late 19th Century to Present):** Viscose and rayon are regarded as semi-synthetic fibers even though they are not totally synthetic. These fibers were first developed in the late 19th century, and the 20th century saw substantial advancements and commercialization. Rayon gained popularity as a textile and apparel material due to its silk-like properties. A rayon derivative called viscose was used to make a variety of goods, including clothes and packaging materials.
- **High-Performance Fibers (20th Century Onward):** High-performance synthetic fibers with specialized uses were developed in

the second half of the 20th century. In the 1960s, DuPont introduced amidam fibers, which are best represented by Kevlar and Nomex. Recognized for its remarkable durability and heat tolerance, Kevlar was used in bulletproof vests, aerospace parts, and other industrial applications.

- **Spandex (1959):** Chemist Joseph Shivers of DuPont created spandex, now marketed as Lycra, in the late 1950s. This elastic synthetic fiber, which offers unmatched stretch and recovery, transformed the textile industry and is now a mainstay in sportswear, undergarments, and other form-fitting clothing.
- **Environmental Considerations (Late 20th Century Onward):** Researchers started looking into eco-friendly substitutes as awareness of the negative effects synthetic fibers have on the environment has grown, particularly because petrochemicals are used in their manufacturing. The sustainability issues surrounding traditional synthetic fibers were intended to be addressed by advancements in recycling technologies and bio-based synthetic fibers.

The evolution of synthetic fibers throughout history is indicative of an ongoing search for materials with improved functionality, robustness, and adaptability. The development of synthetic fibers has had a significant impact on the development of contemporary textiles and materials, starting with nylon's introduction as a silk substitute and continuing with the introduction of high-performance fibers. The development of advanced and sustainable synthetic fibers is being propelled by ongoing research and innovation, which guarantees their continued relevance in the ever-changing demands of society and industry. (Rajak et al, 2022)

7.1.3 Environmental Impact of Synthetic Fibers

As synthetic fibers have become more and more widely used, the environmental impact of these materials has emerged as a significant and intricate issue. Since petrochemicals are the primary source of synthetic fibers like polyester, nylon, and acrylic, questions concerning their environmental sustainability have been raised. Fossil fuels must be extracted and refined during the production process, which results in energy consumption, carbon emissions, and environmental degradation. Because synthetic fibers are frequently not biodegradable, disposing of them at the end of their life cycle presents additional difficulties. Synthetic fiber production relies heavily on non-renewable resources, which is a major environmental concern. The main sources of energy consumption and greenhouse gas emissions for the production of synthetic fibers are the extraction of oil and natural gas. From the standpoint of resource management, this reliance on limited resources calls into question the long-term sustainability of synthetic fibers.

Furthermore, energy-intensive processes like spinning and polymerization are necessary for the manufacturing process itself. Concerns regarding the environmental impact of synthetic fibers are exacerbated by these processes, which add to their carbon footprint. The increasing demand for synthetic fibers makes it imperative to address the energy consumption during production in order to reduce the overall environmental impact of these fibers. An additional environmental challenge arises during the end-of-life phase. Synthetic fibers last longer in the environment than natural fibers, which biodegrade. Microplastics, which are tiny plastic particles that can infiltrate waterways and ecosystems and endanger aquatic life as well as possibly make their way up the food chain, are released when synthetic fibers break down. The environmental build-up of microplastics has sparked concerns about the wider ecological effects of disposing of synthetic fibers.

The development of bio-based substitutes and a greater focus on recycling are results of initiatives to mitigate the effects of synthetic fibers on the environment. Through recycling and conversion into new materials, synthetic textiles are kept out of landfills and contribute to less environmental impact. The difficulty of separating various synthetic fiber kinds and the potential for quality deterioration during the recycling process are still obstacles, though. Synthetic fibers made from biomaterials offer a promising solution to address environmental concerns. Because they are made from recycled materials or plant-based feedstocks, these fibers lessen the need for fossil fuels. Scholars and industry participants are investigating novel strategies for producing bio-based substitutes that provide equivalent performance attributes to conventional synthetic fibers while reducing ecological footprint.

7.1.4 Types of Synthetic Fibers

Fine filaments that can be produced artificially or naturally are called fibers. Man-made synthetic fibers are made from chemicals and are used to make clothing and other practical items. Polymerization is the process that creates these. There are two types of synthetic fibers: fully synthetic and semisynthetic. While natural polymers are used as a raw material to produce semisynthetic fibers like rayons, pure synthetic fibers like nylons, polyesters, and acrylics are made of chemicals.

Polyester Fibers

Unlike cotton, polyester is a synthetic fiber that is not derived from a plant. Polyester is instead made chemically from derived substances. There are five fundamental steps in the process of making polyester, though they differ based on the manufacturer and the various types of polyester.

- Create a monomer: Petroleum is separated into components that can be utilized for a variety of purposes at a refinery after it is extracted from the earth and transported there. The two monomers required to make polyester are ethylene and p-xylene, which are extracted from petroleum.
- Create a polymer: To create the polymer polyethylene terephthalate (PET), which is the plastic found in soda and water bottles, the monomers are combined with dimethyl terephthalate. PET has a consistency similar to honey in this form.
- Extrude the polymer: Once the PET is formed, it's extruded so it can dry and harden. Then, it's cut up into small pellets.
- Melt and spin the pellets: In order to create thread, the PET pellets are melted once more and the molten PET is forced through tiny holes called spinnerets. Upon cooling, these lengthy threads solidify into polyester fibers. Changes in spinneret shapes can result in varying fiber qualities depending on the intended use of the fibers.
- A yarn is born: In order to knit the fibers into fabric, they must first be twisted into polyester yarn. Now, in order to accomplish particular goals, the fibers can be further cut or altered with different chemicals.

Many variations on this production process are possible, depending on the desired result. The fibers, for instance, will be altered to form a thin filament in order to create fabrics that are smooth and soft. Additionally, the ethylene utilized in the procedure can be changed in a few ways. Instead of using petroleum to produce ethylene, sugarcane is typically used to produce plant-based polyester.

The Power of Polyester

Polyester has numerous attributes, many of which are unique to the fabric. These benefits are why polyester became so popular and is still widely used today. Some important polyester characteristics are:

- Durable: Compared to other natural fibers, polyester fibers are substantially stronger. Polyester is abrasion-resistant and won't rip or stretch. It is an excellent material for boat covers and high sun exposure applications because of its exceptional resistance to heat, light, and UV radiation degradation. Polyester is widely used to make home furnishings, safety belts, laundry bags, and outdoor gear like jackets, tents, and backpacks. It is also quick to dry, easily washable, and resistant to mildew and most chemicals. Polyester is so versatile that it's even used to replace artificial arteries.

- Moisture resistant: Unlike cotton, which absorbs liquid droplets, polyester wicks away moisture and repels it. Another feature that makes polyester perfect for outdoor apparel and equipment is this benefit. For instance, a polyester tent will withstand moisture and precipitation, keeping its contents dry. Because of this tendency to reduce the fabric's breathability, apparel manufacturers frequently combine polyester with breathable materials like cotton.
- Wrinkle resistant: Unlike cotton or linen, polyester keeps its shape longer between washings and ironings. Polyester's ability to resist wrinkles helps the material maintain its rigidity and shape, extending the amount of time that it can be worn or used between washings. When polyester was first offered as a substitute material for clothing, this was one of its most prominent features and advantages.
- Moldable: Polyester can be molded into shapes and is very malleable when heated. The amount of waste in our landfills can be reduced by repeatedly re-melting and using it again. Companies now specialize in separating blends of polyester and cotton to enable the original polyester to be reused repeatedly in textiles, thus enhancing the sustainability of polyester.

Similar to other textiles, polyester possesses attributes that could turn off potential buyers or render the material less suitable for specific uses. One of the problems with polyester is that it doesn't biodegrade like most other materials. However, scientists working on textile innovations are trying to create polyester that decomposes as quickly as materials like cotton. Furthermore, polyester isn't the most opulent material, so by itself it can feel rough. Polyester that is cheaply or poorly made will feel rough, especially against delicate skin. The process used to create or tether the fibers together can affect how polyester feels. Polyester can be smooth and silky in some cases, but it can also be coarse and scratchy in others. To create a more comfortable garment, apparel manufacturers can also combine polyester with other materials. (Rajak et al, 2022)

Nylon Fibers

A family of synthetic polymers known by the name of nylon is widely used to create a wide range of consumer goods and clothing. Nylon fibers are completely synthetic, meaning they have no organic basis, in contrast to other organic or semi-synthetic fibers. This kind of synthetic polymer was first used in clothes during World War II as a replacement for hemp and silk for parachutes.

Nearly all other textiles were made of wool at the start of the war, with cotton accounting for over 80% of textile applications in the US. However, by 1945, synthetic fibers, such as nylon, accounted for about 25% of the textile market share. Following the end of the war, producers looked for new strategies to promote this new category of synthetic fabrics. Some people made dresses out of recycled nylon parachutes when traditional dress materials like cotton and silk were hard to come by after the war. As a result, the notion of using nylon in women's clothing became more and more popular, and the industry of producing nylon lingerie and stockings quickly gained momentum. Simultaneously, nylon was becoming more and more popular in a range of other consumer and defense markets. The DuPont Corporation had been working on this material since the early 1920s, and the 1939 World's Fair was the official announcement of its creation. At the time, textiles were thought to be the primary use for this new polymer, and DuPont had no intention of using nylon for scientific or industrial purposes.

When nylon was first introduced, its current name was not yet known; however, DuPont had intended to sell it under the name no-run after realizing the potential of this polymer for use in textiles. Runs are types of abrasion to tights that result in the garments becoming unsightly. Nevertheless, the name was changed to nuron after it was quickly found that nylon was in fact susceptible to runs. But this name was also unsatisfactory, so it was changed to nilon, with a y in place of the i to make pronunciation easier.

When nylon was first introduced, it didn't go by its current name. DuPont had intended to sell this polymer under the name no-run because of its reputation for being able to withstand runs in fabric, which are types of deterioration to tights that make the clothing looks unattractive. The name was later changed to nuron after it was found that nylon was actually susceptible to runs. But this name wasn't good either, so it was changed to nilon and the i was changed to a y to make the pronunciation easier. In the early days of nylon fabric, users reported a number of problems with it. Although nylon was initially marketed as being stronger than steel, it was later discovered to be extremely prone to rips and tears. Additionally, the fabric's inability to wick away moisture raised questions. Nylon stockings would turn into coal and water in severe circumstances.

If manufacturers of nylon hadn't begun blending it with other textiles, this material would have been written off as a failed experiment. It was discovered that many of the negative characteristics of nylon fabric were eliminated while the positive qualities were preserved when combined with spandex, cotton, or polyester. The majority of nylon clothing made today is a combination of different materials. While nylon and other synthetic textiles have steadily lost popularity since the 1970s, this fabric was still in style during the 1940s and

1950s. Nylon's novelty began to fade with time, and people started to worry about the fabric's effects on the environment. Petroleum oil is the primary component of nylon fabric, which is not biodegradable. However, as of 2008, nylon fabric accounted for about 12% of global synthetic fiber production. Although nylon has become less common for use in consumer clothing, this polymer family is becoming more and more common in industry and research. For example, nylon can be processed into a very strong and flexible plastic. Nylon resins are widely used in toothbrushes, hair combs, machine screws, gun parts, and food packaging, among many other applications.

Acrylic Fibers

Acrylonitrile is a synthetic polymer used to make fabrics with acrylic fibers. Acrylic fabric is a type of fiber derived from fossil fuels because it is created through the reaction of specific chemicals derived from coal or petroleum with different monomers. Acrylic fabric is preferred for heat-retention applications because it is among the least breathable textiles available. For example, it's frequently utilized in sportswear, and tracksuits, hoodies, and athletic pants are frequently made of acrylic fabric. But, since acrylic fiber is thought to be carcinogenic, it might be wise to keep your skin away from it. Acrylic fiber needs to contain at least 85% acrylonitrile in order to be deemed authentic, even though it may also contain trace amounts of other synthetic compounds. The final fabric's characteristics will vary depending on the compounds that are combined with acrylonitrile.

Rayon Fibers

Made of purified and regenerated cellulose from plant sources, rayon is an artificial textile material. The first synthetic fiber was rayon, created in the late 1800s as an alternative to silk. Since the cellulose used to make rayon is extracted from soft woods or the short fibers (linters) that stick to cottonseeds, rayon is referred to as a regenerated fiber because it is transformed into a liquid compound, squeezed through tiny holes in a machine known as a spinnerette, and then transformed back into cellulose in the form of fiber. Attempts to work with the highly flammable compound nitrocellulose, which is created by treating cotton cellulose with nitric acid, represented the first practical steps toward producing such a fiber. British chemist Sir Joseph Wilson Swan displayed nitrocellulose fibers in London in 1884 and 1885. The nitrocellulose had undergone chemical treatment to return to nonflammable cellulose. The development of rayon as a useful fiber thus started in France with the work of industrial chemist Hilaire Bernigaud, comte de Chardonnet, who is often referred to as the father of the rayon industry, as Swan did not pursue the demonstrations of his invention. Fibers created by squeezing a nitrocellulose solution through spinnerettes, hardening the emerging jets in warm air, and then chemically reconverting them to cellulose were displayed by Chardonnet in 1889. The first commercially produced man-made fiber, "Chardonnet silk," was manufactured in a Besançon factory in 1891. It was an early form of rayon.

Chardonnet's method was straightforward and produced very little waste, but it was also costly, time-consuming, and possibly hazardous. A method for producing fibers from cuprammonium rayon was patented in 1890 by Louis-Henri Despeissis, another French chemist. This material was developed using the discovery made in 1857 by the Swiss chemist Matthias Eduard Schweizer that cellulose could be dissolved in an ammonia and copper salts solution and then regenerated in a coagulating bath following extrusion. German textile manufacturer J. –P. Bemberg started manufacturing rayon made of cuprammonium as Bemberg silk. The most widely produced form of cellulose today, a third kind, was created in 1891 from a syrupy, yellow, sulfurous-smelling liquid that three British chemists—Charles F. Edward J. Cross. By dissolving cellulose xanthate in diluted sodium hydroxide, Bevan and Clayton Beadle made their discovery. This fiber, which came to be known as viscose rayon (or just viscose), was being produced by the British silk company Samuel Courtauld and Company by 1905. US production of viscose began in 1911 with the American Viscose Corporation.

The fundamentals of viscose rayon production have not changed in modern times. The first step involves treating purified cellulose with sodium hydroxide, or caustic soda. Carbon disulfide is added to alkali cellulose after it has aged, resulting in cellulose xanthate, which is then dissolved in sodium hydroxide. Viscose, a viscous solution, is pushed through spinnerettes. After emerging from the holes, the jets coagulate to form a solid filament by reconverting to cellulose in a coagulating bath of acids and salts. During the manufacturing process, the filament can be adjusted and changed to achieve the desired levels of lustre, strength, elongation, filament size, and cross section. Rayon can be made to resemble silk and shares many properties with cotton. Water easily permeates the fiber, which weakens and swells when wet. It loses strength when exposed to strong alkalies, but it can be cleaned in mild alkaline solutions. Common solvents used in dry cleaning are safe. In applications where cotton is typically used, rayon can be used alone or in blends with other fibers to create clothing. Tire cord is made from high-strength rayon, which is obtained by drawing (stretching) the filaments during the manufacturing process to cause the cellulose polymers to crystallize. Tire cord is used in automobile tires. When making paper, rayon is also combined with wood pulp. Although production of rayon has decreased in industrialized nations due to environmental concerns related to the release of carbon disulfide into the air and salt byproducts into streams, rayon is still an important fiber. New rayon varieties like lyocell have been developed as a result of these worries. Wood cellulose is dissolved in a non-toxic amine oxide solvent to create lyocell, which is then recovered and washed from the regenerated fibers. (Rajak et al, 2022)

7.2 PROPERTIES OF SYNTHETIC FIBERS

The versatility of modern materials science has led to the development of synthetic fibers, which are widely used in a variety of applications due to their unique properties. Their remarkable strength-to-weight ratio is one of their key characteristics, which makes them highly valued for applications where durability and high tensile strength are critical. For example, polyester has exceptional strength, which guarantees durability and toughness in industrial materials, ropes, and textiles. Similar to this, aramid fibers, like Kevlar, exhibit remarkable heat resistance and tensile strength, making them indispensable in applications requiring strong, flame-retardant materials, like ballistic armor and protective clothing. The fact that synthetic fibers are resistant to chemicals, sunlight, and moisture is another noteworthy characteristic of these materials. They are perfect for outdoor applications because of their resistance, which also adds to their stability and durability under different circumstances. Because of their ability to wick away moisture and protect against UV rays, fabrics like polyester and nylon are frequently used in sportswear, outdoor gear, and marine applications. Furthermore, some synthetic fibers are better suited for industrial environments where exposure to corrosive materials is a concern because of their chemical inertness.

The ability of synthetic fibers to imitate and improve natural fibers is just one more example of their versatility. For instance, microfiber textiles offer improved moisture-wicking and quick-drying qualities while simulating the comfort and softness of natural textiles like cotton. Because of their versatility, textiles can be made to meet particular needs for comfort and functionality in apparel and home furnishings. Synthetic fibers are also characterized by their thermal qualities, which include differences in melting points and flame resistance. Modacrylic and some aramid fibers are examples of flame-retardant fibers that are used in protective apparel, upholstery, and automotive components where fire safety is crucial. On the other hand, synthetic fibers' capacity to insulate against heat makes them a valuable component of materials intended for use in cold climates, guaranteeing warmth and comfort in bedding and clothes.

Another benefit of synthetic fibers is their dimensional stability, which helps explain why they resist wrinkling, stretching, and shrinking. Because of this stability, fabrics are guaranteed to hold their shape and appearance over time, requiring less upkeep and extending the life of products. Customers looking for clothing that keeps a polished appearance with little effort can find convenience in wrinkle-resistant synthetic fiber clothing, such as blends of polyester. Moreover, a layer of versatility is added by the ease with which synthetic fibers can be engineered and manipulated during manufacturing processes. This makes it possible to create fibers with unique properties suited to certain uses. Applications of nanotechnology have resulted in the creation of smart textiles that have integrated features like temperature control, moisture sensing, and even antibacterial capabilities. (Rajak et al, 2022)

7.2.1 Strength, Elasticity, and Durability of Synthetic Fibers

The remarkable strength of synthetic fibers—which frequently outperforms that of natural fibers—is one of their main advantages. Items like aramids (e.g., Kevlar) and high-performance polyesters have exceptional tensile strength, which makes them perfect for uses that call for durable materials. These synthetic fibers are used to create lightweight, incredibly strong components like body armor, ropes, and composite materials for aircraft structures in industries like aerospace and defense where strength is a crucial factor. Another important characteristic that sets synthetic fibers apart is elasticity. For example, many synthetic fibers can be engineered to have desirable stretch and recovery properties, while some natural fibers might not have enough elasticity. The comfort and flexibility of materials in applications such as sportswear and activewear are enhanced by their inherent elasticity. For example, stretchable fabrics that adapt to body movements without losing shape can be made thanks to the remarkable elasticity of synthetic fibers like spandex. This characteristic is especially useful when making sportswear and undergarments.

One essential quality of synthetic fibers that makes them so desirable across a wide range of industries is their durability. Synthetic fibers are resistant to abrasion, wear, and tear, in contrast to natural fibers, which could be more vulnerable to environmental factors that cause degradation. Synthetic fiber fabrics, like polyester and nylon, are renowned for their toughness and can be used for a variety of purposes, such as industrial textiles, outdoor gear, and upholstery. These materials' durability adds to longer product lifespans and lower maintenance costs. Moreover, synthetic fibers are resistant to environmental changes, which adds to their durability. Numerous synthetic fibers exhibit resistance to chemicals, sunlight, and moisture, guaranteeing their stability and functionality in a variety of environments. This resistance is especially useful in outdoor settings where exposure to water and UV radiation is common, like marine environments. Thus, synthetic fibers are essential to the manufacturing of ropes, sails, and other nautical gear. Synthetic fibers are used for reinforcement in the fields of infrastructure and construction because of their strength, elasticity, and durability. For example, fiberglass is frequently used as a reinforcing material in the creation of reinforced concrete structures and composite materials. Synthetic fibers are added to materials to increase their tensile strength and general durability, which prolongs their useful life and preserves the structural integrity of infrastructure and buildings.

7.2.2 Resistance to Chemicals, Moisture, and Sunlight

Synthetic fibers' ability to withstand chemicals, moisture, and sunlight is one of their most important characteristics. It also greatly adds to their adaptability and suitability for a wide range of industrial applications. Synthetic materials, like nylon and polyester, show an amazing resistance to a variety of chemicals. They are extremely useful in industrial settings where exposure to corrosive substances is frequent due to their resistance. Synthetic fibers offer a strong barrier against the potentially harmful effects of acids, alkalis, and other harsh chemicals, ensuring both the wearer's safety and the materials' longevity in applications like protective clothing, industrial textiles, and chemical-resistant materials. Another important characteristic of synthetic fibers is their moisture resistance, which makes them ideal for applications where exposure to high humidity or water is a concern. Synthetic fibers retain their structural integrity in damp environments, in contrast to some natural fibers that may weaken or break down in certain situations. Polyester and other similar fabrics are frequently used in outdoor gear, such as tents and rain jackets, where the ability to withstand moisture is crucial to the wearer's dryness and the equipment's continued functionality. Furthermore, synthetic fibers' ability to withstand moisture is useful in settings like maritime ones, where they are used to make water-resistant ropes, sails, and other nautical gear.

A characteristic of synthetic fibers known as UV resistance is their notable resistance to sunlight. Because prolonged exposure to sunlight can cause material degradation, color fading, and strength loss, this feature is especially important for outdoor applications. For use in awnings, outdoor furniture, and automobile interiors, fabrics like polyester and acrylic are perfect because they are made to resist the harmful effects of ultraviolet (UV) radiation. Artificial fibers' ability to withstand ultraviolet radiation guarantees that goods maintain their structural integrity and visual appeal even after prolonged exposure to sunlight, which enhances the durability and functionality of materials in outdoor environments. Synthetic fibers are highly desirable in many fields, including agriculture, where they are used to produce weather-resistant and long-lasting crop covers due to their resistance to chemicals, moisture, and sunlight. Furthermore, synthetic fibers are utilized in the construction industry to create geotextiles, which are important for soil stabilization, drainage systems, and erosion control because these areas are subject to a variety of environmental conditions. (Rajak et al, 2022)

7.2.3 Thermal Properties and Flame Resistance

Synthetic fibers' thermal qualities and flame resistance are important features that are used in a variety of industries to ensure functionality, performance, and safety in a range of conditions. The various thermal behaviors exhibited by synthetic fibers are largely responsible for their extensive usage, as they are able to tolerate high temperatures and withstand flames. The melting point of synthetic fibers, which varies depending on the type, is one important feature of their thermal properties. High-performance fibers with a high melting point, such as aramids, which are best represented by Kevlar, are incredibly heat-resistant. This feature is especially useful for items like protective gear for firefighters, industrial workers, and military personnel where exposure to high temperatures is a concern. Certain synthetic fibers' capacity to keep their strength and structural integrity at high temperatures improves the materials' efficacy and safety in harsh settings.

The ability to withstand flames is a crucial component that guarantees the safety of goods and materials in a fire. Numerous synthetic fibers are designed to withstand flames, such as aramids, modacrylic, and some blends of fire-resistant polyester. These fibers minimize the chance of combustion and lessen the severity of burns by either self-extinguishing or having slow burn rates. In sectors where fire safety is crucial, like the manufacturing of protective clothes, drapes, upholstery, and car interiors, this flame resistance is essential. Additionally, flame-resistant synthetic fibers help to reduce the risk of fire-related mishaps by contributing to the development of materials used in public areas. Synthetic fibers are essential for thermal insulation in addition to being flame resistant. Thermal insulating materials are made in large quantities from fibers like nylon and polyester, which improves the energy efficiency of many products and buildings. Because of their ability to insulate, synthetic fibers are essential for making warm, cozy clothes, blankets, and home textiles—especially in colder climates. Additionally, the aerospace industry makes use of the thermal characteristics and flame resistance found in synthetic fibers. Aircraft interiors are made of fire-resistant synthetic materials, which guarantees passenger safety in the case of a fire. The general fire safety regulations in aviation are influenced by these materials' capacity to tolerate high temperatures and flames. (Rajak et al, 2022)

7.3 APPLICATIONS OF SYNTHETIC FIBERS

Because of its many benefits and attributes, synthetic fibers are now essential parts of many different industries' applications. Synthetic fibers such as polyester, nylon, and acrylic are widely used in the textile industry to produce apparel, accessories, and upholstery. Sporting qualities such as resilience to wrinkles, durability, and fast drying speed, polyester finds frequent use in sportswear, everyday clothing, and activewear. While acrylic fibers, which have a texture similar to wool and are resistant to mildew and insects, are popular in sweaters, blankets, and carpets, nylon, which is valued for its strength and elasticity, is used in hosiery, lingerie, and outerwear.

7.3.1 Textile Industry

Petroleum-based synthetic polymers are used to create synthetic fibers, which make up approximately half of all fiber usage. The many fiber varieties are made from a wide range of chemical substances. Every synthetic fiber has distinct qualities and attributes that make them suitable for a given set of uses. Across a wide range of sectors and industries, including welding, chemical processing, automotive and transportation, apparel, architecture and construction, filtration, marine, medical, and aerospace, synthetic fibers and fabrics are utilized.

In the textile industry, polyester, polyamide (nylon), acrylic, and polyolefin are the four most common synthetic fibers. The most common synthetic fiber, polyester, is used extensively in apparel of all kinds and to reinforce products like belts, hoses, and tires. High-performance outdoor clothing and pillows both use polyester fiberfill. Commonly used to make carpets, polyamide, or nylon, can also be combined with elastomeric fibers to create sportswear. While acrylic fibers don't absorb water, they are still similar to wool in certain aspects, making them a good choice for products like home furnishings and clothing for cold climates. Strong, colorfast, and resistant to abrasion, mildew, and stains are characteristics of polyolefin fibers, which are derived, for example, from polypropylene or polyethylene. They don't absorb water, just like acrylic fibers.

Numerous applications are made possible by these qualities, such as sailcloth, automotive textiles, industrial fabrics, indoor and outdoor carpeting, home furnishings, and safety gear like helmets, police and military vests, and armored vehicles. Contrarily, super-absorbent fibers have a high water retention capacity, which makes them perfect for a variety of uses, including self-healing concrete walls and diapers.

7.3.2 Functional Textiles are Continuously Developing

Textiles can now be given a multitude of characteristics and uses thanks to advancements in polymers and additives. Technical textiles are mainly functional fabrics that are produced for a variety of industries, including automotive, sports and protective apparel, agriculture, and home and industrial use. Either aesthetically pleasing or functional qualities are present in smart textiles. For example, they can change color or light up. These days, textiles can perform a wide range of tasks, such as monitoring and improving sports performance, controlling body temperature, monitoring the heart, and shielding the wearer from radiation and other environmental dangers. UV light resistance is not an intrinsic property of polymers. Degradation may eventually result from any component of the synthetic fiber system absorbing radiation.

Fiber additives can stop or limit radiation absorption and reverse deterioration. Due to this, they are necessary for synthetic fibers used in the textile industry, particularly those meant for use outdoors (like artificial turf and awnings) or indoors (like carpeting) that is exposed to light through windows. Synthetic fibers have completely transformed the textile sector and are now necessary for the creation of a vast range of goods and textiles. Polyester, a synthetic fiber that is widely used, has become essential to the textile industry. Its adaptability, which allows it to be designed to resemble natural fibers like cotton or wool and provide a wide range of applications, is what makes it so popular. Because of its reputation for being wrinkle-resistant, long-lasting, and easy to maintain, polyester fabrics are frequently used in sportswear, outdoor clothing, and daily wear.

Another well-known synthetic fiber that has had a big impact on the textile sector is nylon. Nylon is a fabric that is frequently used in sportswear, lingerie, and hosiery because of its strength, elasticity, and low weight. Its ability to mix invisibly with other fibers improves fabric performance, resulting in materials that are robust and cozy. Nylon is a great material for swimwear and intimate apparel because of its durability and quick-drying qualities. Acrylic fibers have made a name for themselves in the textile business thanks to their wool-like qualities. Warmth, softness, and resistance to mildew and moths are all provided by acrylic fabrics. Acrylic is a popular material for outerwear, blankets, and sweaters because of these characteristics. Acrylic textiles' durability and usefulness are further enhanced by their fast-drying qualities and color retention. The high-performance properties of synthetic fibers—particularly aramids like Kevlar—benefit the textile sector as well. Kevlar is used to make gloves, helmets, and bulletproof vests, among other protective gear, because of its extraordinary strength, heat resistance, and low weight. Because Kevlar is lightweight and durable, the aerospace industry uses it in aircraft components.

Stretchable fabric technology has been revolutionized by the synthetic fiber known as spandex, or lycra. Since spandex has unmatched elasticity and recovery, it is frequently used in sportswear, swimwear, and undergarments, guaranteeing a snug fit and flexibility. The design and comfort of form-fitting apparel have been transformed by its ability to stretch up to several times its original length. The development of environmentally friendly synthetic fibers has been prompted by the textile industry's increased focus on sustainability in recent years. Efforts to lessen the environmental impact of synthetic fibers are aided by innovations in recycled polyester, which is made from used plastic bottles, and bio-based substitutes that are constructed from renewable resources. The environmental impact of synthetic fibers remains a challenge for the textile industry, despite their extensive use. This is especially true when it comes to the disposal of these fibers after their useful life. The development of more sustainable synthetic fiber options, recycling programs, and circular economy principles are some of the main ways that ongoing research and technological advancements are addressing these issues.

7.3.3 Industrial Applications

With a variety of features that support their extensive use in a variety of industries, synthetic fibers are essential to many industrial applications. Synthetic fibers, including polyester, nylon, and acrylic, are widely used in the textile industry and are now essential parts of many fabrics. Because of their longevity, wrinkling resistance, and shape retention, these fibers are highly prized and are perfect for industrial textiles, upholstery, and clothing manufacturing. Awnings, tents, and outdoor apparel are among the outdoor applications that benefit from synthetic fibers' resistance to environmental elements like sunlight and moisture. Furthermore, synthetic fibers are widely employed in the automotive industry to produce carpets, seat covers, and interior parts. Fibers like nylon are well-suited to prolong the life and improve the appearance of car interiors because of their innate strength and resistance to deterioration. Furthermore, since fuel efficiency is a major factor in the design of contemporary vehicles, synthetic fibers' lightweight nature helps to improve it. Composite materials for aircraft components are manufactured using synthetic fibers in aerospace applications. The development of lightweight yet durable aircraft structures is aided by the high strength-to-weight ratios offered by carbon fiber-reinforced composites.

Synthetic fibers have many uses outside of textiles and transportation, such as in the building industry. Buildings and infrastructure's structural integrity are improved when reinforced concrete structures with materials like fiberglass and aramid fibers because they increase the concrete's tensile strength and durability. Geotextiles, which are used for soil stabilization, drainage, and erosion control, are another example of how synthetic fibers are used in construction. Synthetic fibers are vital to the medical technology industry because they are used in the creation of numerous implants and medical devices. Polypropylene and polyethylene fibers, for instance, are frequently utilized in the production of sutures because of their strength and biocompatibility. Synthetic fibers are also used to create artificial tissues and organs, which advances the field of regenerative medicine further. Moreover, the insulating qualities of synthetic fibers are advantageous to the electrical and electronics sectors. In order to produce cables and wiring that ensure the safety and dependability of electrical systems, insulation materials made of polyester, polyethylene, and other synthetic fibers are necessary. Certain synthetic fibers are especially well-suited for these applications due to their flame-retardant and thermal-resistant qualities. (Mamun et al, 2023)

7.4 RECYCLING PROCESS FOR SYNTHETIC FIBERS

Reusing synthetic fiber is becoming a more crucial step in the effort to lessen the impact on the environment and support environmentally friendly textile manufacturing. Owing to their affordability, adaptability, and durability, synthetic fibers like polyester, nylon, and acrylic are extensively utilized across multiple sectors, especially in the fashion and home textile industries. But there are serious environmental problems associated with their manufacture and disposal. Petrochemicals are the source of synthetic fibers, and producing them requires a lot of energy, which increases greenhouse gas emissions. Additionally, synthetic fibers contribute to the expanding problem of plastic pollution because they are non-biodegradable, meaning that, if improperly managed, they can remain in the environment for hundreds of years.

Synthetic textiles that have been used in manufacturing or as consumers are recycled into new, useful fibers. Synthetic fibers can be recycled in a number of ways, the most popular being mechanical and chemical recycling. In mechanical recycling, the waste textile is physically broken down into smaller pieces, which are subsequently processed again to create new fibers. Although this process is inexpensive and reasonably easy, it may cause the fibers' quality to deteriorate, which could limit their future use. Contrarily, chemical recycling entails dissolving synthetic fibers' polymers into their monomers so that they can be re-polymerized to create new fibers. Although more expensive and complicated, this process can yield high-quality fibers that are comparable to those made from virgin materials. Synthetic fiber recycling has a number of advantages for the environment. Because less virgin material is required, natural resources are preserved and the environmental impact of producing fiber is decreased. It also lessens environmental plastic pollution and helps divert textile waste from landfills. Furthermore, compared to producing fibers from raw petrochemicals, recycling synthetic fibers can result in significant energy savings because fibers made from recycled materials typically require less energy. Nevertheless, there are difficulties with recycling synthetic fibers. Textile contaminants, such as dyes, finishes, and mixed materials, can make recycling more difficult. Furthermore, the labor-intensive processes of gathering and classifying textile waste call for an effective infrastructure that is frequently deficient. (Mamun et al, 2023)

7.4.1 Mechanical Recycling

Involves cutting the material into small pieces, melting them down, and spinning new yarn. This process can also include shredding the fabric into fibers, carding the fabric to produce yarn, and blending the collected fibers with virgin fibers.

Mechanical recycling is the process of recovering plastic waste by mechanical processes such as sorting, washing, drying, grinding, re-granulating and compounding. Mechanical recycling does not change the chemical structure of the material, which permits multiple re-use/re-cycling of polymeric materials, creating a closed loop.

Mechanical recycling of synthetic fibers is a process that involves the physical reprocessing of waste materials, such as polyester, nylon, or acrylic, into new fibers or products without altering the chemical structure of the polymer. This method is commonly used for textiles, carpets, and other products made from synthetic fibers, and it plays a significant role in the circular economy by reducing waste and conserving resources.

The process begins with the collection and sorting of waste materials, which are then cleaned to remove any impurities such as dirt, dyes, or other contaminants. After cleaning, the materials are shredded or cut into smaller pieces, creating flakes or chips. These flakes are then melted down and extruded through a spinneret, a device that shapes the molten polymer into continuous filaments. These filaments are then cooled, solidified, and drawn to improve their strength and elasticity before being wound onto spools or further processed into yarns or fabrics.

One of the key advantages of mechanical recycling is its relatively low environmental impact compared to chemical recycling, as it requires less energy and does not involve the use of hazardous chemicals. However, the quality of the recycled fibers can be lower than that of virgin materials due to the degradation of polymer chains during the recycling process. This degradation can result in fibers that are weaker, less elastic, or less durable, which may limit their use in high-performance applications.

Moreover, the mechanical recycling of synthetic fibers faces challenges such as contamination from other materials, which can affect the quality of the recycled product. For example, if a polyester fabric is mixed with cotton, the resulting recycled fiber may have undesirable properties. To address this, advanced sorting technologies and pre-processing steps are essential to ensure that the feedstock is as pure as possible.

Despite these challenges, mechanical recycling remains a viable and increasingly popular method for managing synthetic fiber waste. It supports sustainability efforts by diverting waste from landfills and reducing the need for virgin resources. Additionally, advancements in technology and processes are continually improving the efficiency and quality of mechanically recycled fibers, making them more competitive with virgin materials. (Mamun et al, 2023)

7.4.2 Chemical Recycling

Chemical recycling of synthetic fibers is an advanced process that involves breaking down synthetic polymers, such as polyester, nylon, or acrylic, into their basic chemical building blocks, which can then be purified and re-polymerized to create new fibers. This method is distinct from mechanical recycling, as it addresses some of the limitations of mechanical processes by restoring the chemical integrity of the polymer, thus allowing the production of high-quality recycled materials that are often indistinguishable from virgin fibers.

The process typically begins with the collection and sorting of synthetic fiber waste, which can include textiles, carpets, and other products. The sorted materials are then subjected to chemical treatments that break the polymer chains down into their monomers or oligomers. For example, polyester can be depolymerized into its monomers, terephthalic acid (TPA) and ethylene glycol (EG), through processes such as glycolysis, hydrolysis, or methanolysis. These monomers are then purified to remove impurities and can be re-polymerized into new, high-quality polyester fibers.

Chemical recycling offers several significant advantages over mechanical recycling. One of the key benefits is the ability to produce recycled fibers with properties that are comparable to those of virgin materials, as the process effectively resets the polymer chains, eliminating the degradation that occurs in mechanical recycling. This makes chemical recycling particularly attractive for applications requiring high-performance materials, such as in the automotive, fashion, and technical textiles industries. Another advantage of chemical recycling is its ability to handle mixed-material waste streams that would be problematic for mechanical recycling. For instance, garments made from blended fibers, such as polyester-cotton blends, can be chemically separated into their constituent materials, each of which can then be individually recycled. This flexibility expands the range of materials that can be recycled and helps to address the growing issue of textile waste.

However, chemical recycling also faces challenges. The processes involved can be energy-intensive and costly, and they often require the use of hazardous chemicals, which must be carefully managed to minimize environmental impact. Additionally, the technology for chemical recycling is still in the early stages of commercial development, and scaling up these processes to handle large volumes of waste remains a significant hurdle.

Despite these challenges, chemical recycling is seen as a promising solution for the sustainable management of synthetic fiber waste. As the technology matures and becomes more economically viable, it has the potential to play a crucial role in creating a circular economy for textiles, where waste is continuously reprocessed into new products, reducing the reliance on virgin resources and minimizing environmental impact. (Mamun et al, 2023)

7.4.3 Biotechnological Solutions

Biotechnological solutions for recycling synthetic fibers involve the use of enzymes to selectively break down chemical bonds in polymers, such as polyester, nylon, and other synthetic materials, into their monomeric building blocks. This approach is a more environmentally friendly alternative to traditional chemical recycling methods, which often rely on harsh chemicals and energy-intensive processes. By harnessing the power of enzymes—biological catalysts that can operate under mild conditions—biotechnological recycling offers a promising path toward sustainable waste management in the textile industry.

Enzymes are highly specific in their action, meaning they can target and break down specific bonds within a polymer chain while leaving other components intact. For instance, certain types of enzymes, like cutinases or PETases, have been developed to target the ester bonds in polyethylene terephthalate (PET), a common synthetic polymer used in textiles and plastic bottles. These enzymes break down PET into its monomeric components, terephthalic acid (TPA) and ethylene glycol (EG), which can then be purified and re-polymerized into new, high quality PET.

The biotechnological approach to recycling offers several key advantages. Firstly, it operates under milder conditions compared to traditional chemical methods, typically requiring lower temperatures and pressures and avoiding the use of toxic chemicals. This reduces the overall environmental footprint of the recycling process, making it more sustainable and less harmful to the environment. Additionally, enzymes can be engineered to improve their efficiency, specificity, and stability, allowing them to be tailored to different types of polymers and recycling processes.

Another significant advantage is the potential for achieving higher purity in the recovered monomers. Because enzymes are highly selective, they can break down polymers more cleanly, resulting in fewer by-products and contaminants. This leads to monomeric building blocks that are purer and more suitable for re-polymerization into high-quality materials. Moreover, the use of enzymes can be particularly effective in recycling complex waste streams, such as textiles made from blended fibers, where traditional mechanical or chemical methods may struggle to separate and purify the different components.

Despite its potential, biotechnological recycling is still in the early stages of development and faces several challenges. The efficiency of enzymes, while improving, still needs to be optimized for industrial-scale applications. Additionally, the cost of enzyme production and the need for large quantities of enzymes can be a limiting factor. However, ongoing research in biotechnology and synthetic biology is focused on overcoming these challenges by engineering more robust and efficient enzymes and developing cost-effective production methods. (Mamun et al, 2023)

7.5 BIOTECHNOLOGICAL SOLUTIONS FOR RECYCLING SYNTHETIC FIBERS

The textile industry is facing mounting pressure to lessen its environmental impact as public awareness of sustainable practices rises. Due to their slow degradation, the widespread use of synthetic fibers has increased waste accumulation even though they help create a wide range of reasonably priced textile products. To lessen the quantity of textile waste that is burned or dumped in landfills, textile recycling techniques include mechanical recycling, chemical recycling (pyrolysis, enzymatic hydrolysis, hydrothermal technology), ammonolysis, gasification, glycolysis, and decolorization. Chemical treatment, which involves adding chemicals to break down the complex textile polymer into smaller polymer molecules, is an efficient way to recycle textiles. Degrading polymers, solution-purified polymers, oligomers, monomers, or raw materials in gaseous or liquid form—all byproducts of chemical textile recycling—is now a common practice through chemical recycling.

Even in textile waste, re-polymerization is a viable method for replenishing polymers like oligomers and monomers. Mechanical textile recycling, on the other hand, has been in use since the start of the industrial revolution. That's among the least expensive and easiest ways to recycle. A sustainable method of recycling synthetic fibers is not possible with all of these procedures, though. Because of this, biotechnological methods have surfaced as viable substitutes for recycling and breaking down synthetic fibers. Large-scale recycling can be achieved by using microorganisms that break down synthetic fibers in bioreactors. This brief analysis (refer to Figure 7.2) concentrates on biotechnological and mechanical approaches.

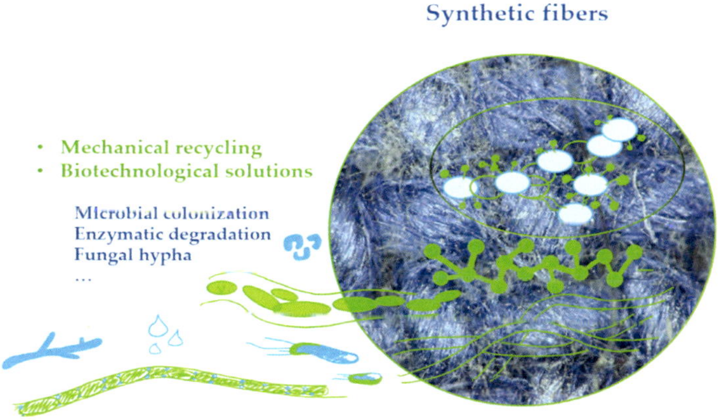

Figure 7.2. Synthetic fiber recycling options.

Source: Mamun A, Kuntz F, Golle C, Sabantina L. Biotechnological Solutions for Recycling Synthetic Fibers. Engineering Proceedings. 2023; 56(1):181. https://doi.org/10.3390/ASEC2023-16301

Consider Marqués-Calvo et al., outlined the outcomes of the enzymatic and microbiological biodegradation of PET and PET copolyester plastics using two commercial fungal lipases and two environmental isolate bacteria in a lab setting? Nevertheless, Mohanan et al., discussed the enzymatic and microbial deterioration of synthetic plastics used in the textile sector. There are regulatory aspects to integrating biotechnological solutions into the recycling strategies used today to support environmental preservation and facilitate the shift to a circular economy. This succinct overview investigates the possibilities of biotechnological approaches for recycling synthetic fibers. In order to create a more sustainable textile economy, it tackles the obstacles and emphasizes the applications that are essential. Enzymatic degradation is discussed, which is a basic aspect of biocatalysis. It is shown that enzymes can be used to specifically break down different types of synthetic fibers while producing the fewest byproducts possible. (Mamun et al, 2023)

7.5.1 Mechanical Recycling Processes and Limitations

For the purpose of managing man-made fibers sustainably, mechanical recycling techniques like mechanical sorting, mechanical shredding, and melt spinning offer practical and affordable solutions for different phases of the recycling process. These recycling techniques do have certain drawbacks, though. There are fewer options for use on the end market as a result of the mechanical recycling of textiles, which lowers quality because of the shorter fibers. Using recycled polyester (PES) fibers, which were likely mechanically recycled, a textile fabric composed of 80% cotton and 20% polyester is depicted in the microscopic image in Figure 7.3. With a digital microscope, pictures were captured. Although the surface of the yarn is uneven and has individual fibers protruding, the warp threads have relatively short fibers (see Figure 7.3b).

(a) (b)

Figure 7.3. Microscopic images of 20% polyester/80% cotton fabric (**a**) and warp yarn (yellow and black polyester fibers cover the white cotton fibers) (**b**).

Source: Mamun A, Kuntz F, Golle C, Sabantina L. Biotechnological Solutions for Recycling Synthetic Fibers. Engineering Proceedings. 2023; 56(1):181. https://doi.org/10.3390/ASEC2023-16301

Mechanical sorting is a crucial technique that effectively separates synthetic fibers from other materials according to physical attributes like size, density, or shape. It makes use of machinery like air sifters and conveyors. It works well with bulk materials but has limitations when working with similar properties, which could result in contamination. On the other hand, mechanical shredding is essential for reducing material size and getting it ready for additional processing, which makes it easier to handle and compatible with later recycling stages. Accurate sorting is crucial since mechanical sorting has trouble with similar materials and runs the risk of becoming contaminated. The use of mechanical shredding needs to be improved for sustainability reasons, as it poses issues with energy consumption and fiber damage. Melt spinning has a limited range of applications due to its energy-intensive nature and suitability only for certain synthetic fibers.

Applications of biotechnology provide a variety of creative ways to enhance the recycling of synthetic fibers. These methods make use of the inherent properties of microbes and enzymes to break down fibers, produce useful products, and support a more circular and sustainable textile sector. Sustained investigation and advancement in this field hold promise for transforming the handling of waste synthetic fibers and advancing ecological preservation.

7.5.2 Enzymatic Degradation

Enzymes that specifically target particular molecular bonds within the polymer structure cause the directed action of these enzymes to break down synthetic fibers. Because of their substrate selectivity, enzymes like lipases, cellulases, and proteases can break apart glycoside, ester, and peptide bonds, respectively. While other enzymes target different types of synthetic fibers, esterases, for example, have proven effective in breaking down synthetic fibers based on polyester. The fiber's molecular structure is broken down by enzymatic hydrolysis, which eventually forms shorter oligomers and monomers that can be taken up by microbes or utilized as building blocks for biochemical reactions. Synthetic fibers can be broken down enzymatically, which benefits waste management and creates opportunities for the circular economy. Reducing reliance on fossil fuels, the breakdown products that result, such as monomers and oligomers, can be used as feedstocks for biopolymers, biofuels, and other value-added products. Enzymatic degradation also contributes to lessening the environmental damage caused by the buildup of waste synthetic fibers. Furthermore, the polyester surface's hydrophilicity/hydrophobicity ratio plays a critical role in determining its enzymatic degradability. Scanning electron micrographs of polyester films (PES) deteriorated for 2, 4, and 10 hours, along with the alterations in surface morphology, are displayed in Figure 7.4.

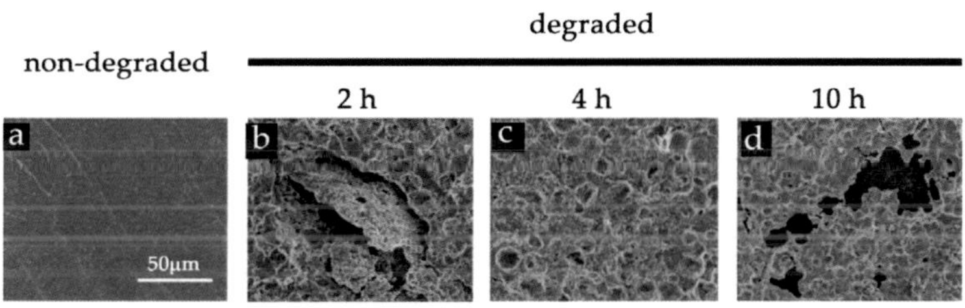

Figure 7.4. Scanning electron microscope images of poly(ethylene succinate) (PES) polyester films degraded for 2, 4, and 10 h.

Source: Mamun A, Kuntz F, Golle C, Sabantina L. Biotechnological Solutions for Recycling Synthetic Fibers. Engineering Proceedings. 2023; 56(1):181. https://doi.org/10.3390/ASEC2023-16301

The two-step, enzymatic application of keratinase in combination with a reducing agent, which allowed for the selective digestion of fiber composites of wool-polyester blends. The undigested polyester fibers were subsequently recovered. When comparing the enzymatic treatment to untreated samples, spectroscopic and mechanical analysis of the recovered synthetic fibers revealed no discernible change in the polyester's characteristics. Textile fiber separation using enzymatic methods for environmentally friendly waste management.

Polyethylene terephthalate (PET) is sequentially converted using a terephthalate-degrading pseudomonas umsongensis GO16 into two types of bioplastics: a medium chain length polyhydroxyalkanoate (PHA) and a novel bio-based poly(amideurethane) (Bio-PU). Due to the specificity of the enzymes, which renders them less useful for certain kinds of synthetic fibers, this enzymatic recycling method for synthetic fibers may have limitations. Its applicability may also be limited by the fact that the particular synthetic fiber and process variables may affect how well the enzymes degrade it. (Mamun et al, 2023)

7.5.3 Microbial Biodegradation

Microbial biodegradation makes use of microorganisms' innate capacity to disassemble complex synthetic polymers into simpler ones, such as bacteria and fungi. Enzymes found in a number of microorganisms have been shown to be capable of breaking down synthetic fibers. For instance, some bacteria are able to produce lipases and esterases, which are enzymes that attack the ester bonds found in polyester fibers. The potential of bioaugmentation and biostimulation techniques has also been investigated by researchers. Whereas biostimulation seeks to boost the activity of already-existing microorganisms by supplying nutrients or other growth-promoting elements, bioaugmentation entails the introduction of particular microorganisms into the environment. These methods seek to boost fiber degradation efficiency and quicken the process of degradation. A research was centered on the microbial biocatalysts that broke down synthetic polymers like PET, polystyrene, polyurethane, and polyethylene terephthalate.

Thanks to developments in genetic engineering, microorganisms can now be altered to increase their capacity to break down synthetic fibers. Scientists can modify microorganisms to effectively break down complex fiber structures and generate useful end products by introducing particular genes or pathways. The recycling of polyethylene terephthalate plastic through biocatalysis bacteria that have been isolated outside of a facility that recycles bottles to break down and digest plastic. Traditional synthetic fibers, which include acrylic, polyester, and nylon, are made from resources derived from petroleum and wind up in the environment or landfills, where they can linger for years and pollute the environment. Synthetic fibers can be broken down into their constituent molecules by biological processes in biochemical recycling, which can then be utilized to create new fibers or other materials.

Biotechnological processes can help convert depolymerized synthetic fiber monomers into useful chemicals and materials. The need for virgin resources can be decreased by using these monomers as feedstocks to create new polymers. This strategy can reduce waste production and aid in closing the loop in the production of synthetic fibers. The possible specificity of microorganisms and enzymes, which means they might not work as well on all kinds of synthetic fibers, is one of the drawbacks of enzymatic and microbial biodegradation techniques for recycling synthetic fiber. Furthermore, the particular fiber and process parameters affect how well the degradation process works. (Mamun et al, 2023)

7.5.4 Current Challenges and Limitations

Biotechnological solutions show promise, but there are still a number of obstacles to overcome. It's still difficult to find the right microbial strains or enzymes for a variety of synthetic fibers. Careful study is needed to optimize degradation rates as well as comprehend possible byproducts and their effects on the environment. Furthermore, it is important to give careful thought to how biotechnological processes can scale up to meet industrial demands. This field is still very young and has many obstacles to overcome. Synthetic fibers may not break down quickly enough by enzymes to produce an adequate supply of high-quality building blocks for the synthesis of new fibers. There are many formulations of synthetic fibers, and because the chemical structures of these fibers differ, it is difficult to develop enzymes that can efficiently break down every kind of synthetic fiber.

The quality of the recycled material may be compromised by contaminants or fragments that are produced during degradation and are challenging to separate from the degraded building blocks. Although experiments conducted on a lab scale show promise, industrializing enzymatic processes can be difficult and expensive. Because synthetic fibers are not biodegradable, the growing use of them in textiles has raised significant environmental concerns. There is an increasing amount of waste synthetic fiber, and traditional recycling techniques based on mechanical and chemical processes are not sufficient. Biotechnological methods have surfaced as a viable means of achieving efficient and sustainable recycling. (Mamun et al, 2023)

REFERENCES

1. Alagirusamy, R., & Das, A. (2018). Handbook of Properties of Textile and Technical Fibres (2nd ed.). Woodhead Publishing.
2. Rajak DK, Wagh PH, Linul E. A Review on Synthetic Fibers for Polymer Matrix Composites: Performance, Failure Modes and Applications. *Materials*. 2022; 15(14):4790. https://doi.org/10.3390/ma15144790
3. Mamun A, Kuntz F, Golle C, Sabantina L. Biotechnological Solutions for Recycling Synthetic Fibers. *Engineering Proceedings*. 2023; 56(1):181. https://doi.org/10.3390/ASEC2023-16301
4. Eichhorn, S. J., Hearle, J. W. S., Jaffe, M., & Kikutani, T. (Eds.). (2009). Handbook of Technical Textiles. Woodhead Publishing.
5. Hearle, J. W. S., & Mortimer, J. (2008). Atlas of Fibre Fracture and Damage to Textiles (2nd ed.). Woodhead Publishing.
6. Horrocks, A. R., & Anand, S. C. (2000). Handbook of Technical Textiles. Woodhead Publishing.
7. Kozłowski, R. M. (2017). Handbook of Natural Fibres (2nd ed.). Woodhead Publishing.
8. Liu, W. (2016). Nanofibers of Conjugated Polymers: From Fundamental Research to Applications. Springer.
9. Melliand Textile Reports. (Available through various issues and volumes, covering a wide range of topics related to textile fibers and materials.)
10. Morton, W. E., & Hearle, J. W. S. (2008). Physical Properties of Textile Fibres (4th ed.). Woodhead Publishing.

Chapter **8**

Thermal Analysis of Recycled Plastics

Contents

INTRODUCTION

An important field of research that explores how recycled polymers behave under different thermal conditions in comparison to their virgin counterparts is thermal analysis of recycled plastics. Thermogravimetric analysis (TGA) and differential scanning calorimetry (DSC) are two of the main methods used in this field to study important thermal properties, such as melting temperature, glass transition temperature, and thermal stability. Because of the effects of repeated heating and cooling, contamination from other materials, and degradation from prior use, recycled plastics frequently exhibit altered thermal characteristics. Repeated processing cycles, for example, may result in a drop in the melting temperature and an increase in the rate of thermal degradation, which may have a negative impact on the mechanical characteristics and general performance of the material.

Contaminants like paper or metal can affect the thermal degradation behavior of recycled plastics, making them a major challenge for thermal analysis. Contaminants frequently cause anomalies in the thermal profile, which makes recycling more difficult and lowers the quality of the finished product. In order to reduce contamination and enhance the uniformity of thermal properties, this problem must be resolved by optimizing the sorting and cleaning procedures.

The processing conditions of recycled plastics are a significant factor in their thermal behavior. The extrusion temperature, cooling rate, and reprocessing methods are a few examples of the variables that can have a big impact on recycled materials' thermal stability. To guarantee that recycled plastics fulfill the necessary performance requirements for the applications for which they are intended, effective management of these conditions is crucial. For instance, changing the processing parameters can improve the material's stability and lessen the harmful effects of thermal degradation. Recycled plastics are frequently treated with stabilizers and additives to enhance their thermal characteristics. These compounds have the potential to improve thermal stability and mitigate the effects of degradation, increasing the viability of recycled materials for a range of applications. More efficient recycling procedures and improved recycled goods can result from research into the kinds and concentrations of additives that work best.

Furthermore, the performance of recycled plastics in composite materials can be assessed using thermal analysis techniques. Developing advanced materials with desired properties requires an understanding of the thermal interactions between reinforcing agents and recycled plastic matrices. Expanding the uses of recycled plastics, enhancing their quality, and streamlining the recycling process all depend on thermal analysis. The sustainability and performance of recycled plastic materials can be improved by researchers and manufacturers by tackling issues with contamination, processing conditions, and material degradation. (ASTM, 2017, 2018; Graziano et al, 2019; Parameswaranpillai et al, 2019)

8.1 CHARACTERIZATION OF THERMAL PROPERTIES OF RECYCLED PLASTICS

Understanding recycled plastics' performance and suitability for different applications requires characterization of their thermal properties. In this process, important thermal characteristics like thermal stability, glass transition temperature, and melting temperature are analyzed. Thermogravimetric analysis (TGA) and differential scanning calorimetry (DSC) are frequently used methods in this analysis. By measuring the heat flow connected to phase transitions like melting or crystallization, DSC sheds light on how recycled plastics behave thermally in comparison to their virgin counterparts. TGA, on the other hand, monitors mass changes in response to temperature, providing important details about the material's stability and thermal degradation.

Because they are subjected to repeated cycles of heating and cooling, recycled plastics frequently display altered thermal properties that can result in chain scission and cross-linking. Reduced melting temperatures and faster rates of thermal degradation may follow from these modifications. Furthermore, the thermal profiles may become even more complex due to the presence of contaminants, such as leftovers from other polymers or foreign materials. Contamination, for example, can result in uneven melting behavior or early degradation, which impacts the usefulness and quality of recycled plastics.

It is crucial to take into consideration differences in the types of plastics and their prior processing histories in order to precisely characterize these thermal properties. Certain polymers, like high-density polyethylene (HDPE) and polyethylene terephthalate (PET), have unique thermal behaviors that affect how they are processed and used in final products. Researchers can determine the ideal circumstances for processing and using recycled plastics and gain a better understanding of how they respond to heat stress by methodically examining these properties. Comprehending the thermal properties of recycled plastics aids in evaluating their efficacy and directs advancements in material formulations and recycling technologies. Improved characterization methods can result in more effective quality assurance and recycling procedures, which will eventually help create high-performing and more environmentally friendly recycled plastic products.

8.1.1 Investigate how the Thermal Properties

Examining the behavior of recycled plastics under heat in comparison to virgin plastics is a necessary step in determining their thermal properties. Assessing the efficacy and appropriateness of recycled materials for diverse applications necessitates this inquiry. Methods like Thermogravimetric Analysis (TGA) and Differential Scanning Calorimetry (DSC) are frequently employed. While TGA tracks changes in mass as a function of temperature to reveal degradation patterns, DSC measures phase transitions such as melting and glass transition temperatures.

Due to multiple processing cycles that can break polymer chains and decrease thermal stability, recycled plastics frequently exhibit changed thermal properties. For example, when compared to virgin PET, recycled PET might have a lower melting point and a faster rate of degradation. Recycled plastics may contain contaminants that affect their thermal profiles and cause inconsistent performance.

Comprehending these thermal characteristics aids in streamlining the recycling procedure, enhancing the quality of the material, and guaranteeing that recycled plastics fulfill the necessary performance criteria. This information is essential for creating recycling methods that are more efficient and increasing the sustainability of plastic materials.

Melting Temperature of Recycled Plastics Compared to those of Virgin Plastics

Because of the effects of repeated processing and degradation, recycled plastics frequently have a different melting temperature than virgin plastics. Virgin plastics with well-defined melting points are essential for processing and use in materials like PET, HDPE, and PP. For instance, virgin PET usually melts between 250 and 260°C, HDPE between 130 and 140°C, and PP between 160 and 170°C.

On the other hand, recycled plastics usually have lower melting temperatures than virgin plastics. This decrease happens as a result of the polymer chains' tendency to break down with repeated heating cycles during recycling, producing shorter chains with lower melting points. Furthermore, the thermal profiles of recycled plastics can be further modified by impurities or additives. For example, recycled PET's melting point could be much lower than virgin PET's, which would make it less suitable for high-temperature uses.

Comprehending these fluctuations is essential for refining recycling procedures and guaranteeing that repurposed plastics uphold sufficient performance criteria. Precise thermal analysis facilitates enhanced quality assurance and modification of processing parameters to augment the effectiveness and practicality of repurposed materials across diverse uses.

Glass Transition Temperature of Recycled Plastics Compared to those of Virgin Plastics

Because of degradation and processing history, glass transition temperatures (Tg) of recycled plastic are frequently different from those of their virgin counterparts. The temperature at which a polymer changes from a stiff, glassy state to a more flexible, rubbery state is known as Tg, which is an important property. Tg is comparatively stable and well-defined for virgin plastics. For instance, the Tg of virgin PET is approximately 70–80°C, the Tg of HDPE is roughly -100°C, and the Tg of PP is approximately -10°C.

However, because of the effects of repeated processing cycles, recycled plastics often display variations in their glass transition temperatures, which can result in polymer chain scission and cross-linking. Tg frequently shifts as a result of these modifications; recycled plastics may exhibit a higher or lower Tg than their virgin counterparts. Due to polymer degradation, recycled PET, for example, may exhibit a lower Tg than virgin PET, which affects its performance in applications requiring thermal stability. On the other hand, because of the creation of cross-linked structures or the addition of specific additives during recycling, some recycled polymers may have a higher Tg.

Recycled plastics' additives and contaminants may also have an impact on Tg. Impurities may cause irregular Tg values by upsetting the polymer matrix. Even though additives are meant to improve or stabilize the material, they can also change Tg by changing how flexible or rigid the polymer is. Determining whether recycled plastics are suitable for a given application requires an understanding of these variations in Tg. Precise thermal analysis aids in adjusting processing parameters and enhancing the functionality of repurposed materials, guaranteeing they fulfill the requirements for the purposes for which they are intended. This knowledge also helps to improve recycling procedures so that recycled plastics are of higher quality and more useful.

Recycled plastics' additives and contaminants may also have an impact on Tg. Impurities may cause irregular Tg values by upsetting the polymer matrix. Even though additives are meant to improve or stabilize the material, they can also change Tg by changing how flexible or rigid the polymer is. Determining whether recycled plastics are suitable for a given application requires an understanding of these variations in Tg. Precise thermal analysis aids in adjusting processing parameters and enhancing the functionality of repurposed materials, guaranteeing they fulfill the requirements for the purposes for which they are intended. This knowledge also helps to improve recycling procedures so that recycled plastics are of higher quality and more useful. (ASTM, 2017, 2018; Graziano et al, 2019; Parameswaranpillai et al, 2019)

8.1.2 Different Types of Recycled Plastics

Based on the kind of polymer and the procedures they go through during recycling, recycled plastics are divided into different categories. To maximize their use and enhance recycling procedures, it is essential to comprehend the various types of recycled plastics.

Here's an overview of various types of recycled plastics:

1. Polyethylene Terephthalate (PET)

Properties:

- **Melting Point:** 250-260°C
- **Glass Transition Temperature:** 70-80°C

Applications:

- Recycled PET is commonly used in manufacturing new bottles, textiles, and packaging materials.

Recycling Process:

- **Mechanical Recycling:** PET is cleaned, shredded, and melted to form pellets or fibers.
- **Chemical Recycling:** PET can be depolymerized to its monomers and repolymerized into high-quality PET.

2. High-Density Polyethylene (HDPE)

Properties:

- **Melting Point:** 130-140°C
- **Glass Transition Temperature:** -100°C

Applications:

- Recycled HDPE is used in products like piping, plastic lumber, and containers.

Recycling Process:

- **Mechanical Recycling:** HDPE is cleaned, shredded, and reformed into new products.
- **Chemical Recycling:** Involves breaking down HDPE into its monomers or other useful chemicals.

3. Polypropylene (PP)

Properties:

- **Melting Point:** 160-170°C
- **Glass Transition Temperature:** -10°C

Applications:

- Recycled PP is used in automotive parts, containers, and industrial fibers.

Recycling Process:

- **Mechanical Recycling:** PP is cleaned, shredded, and melted for reuse.
- **Chemical Recycling:** Converts PP into its monomers or other chemical products.

4. Low-Density Polyethylene (LDPE)

Properties:

- **Melting Point:** 105-115°C
- **Glass Transition Temperature:** -125°C

Applications:

- Recycled LDPE is often used in products like floor tiles, rubbish bin liners, and landscaping materials.

Recycling Process:

- **Mechanical Recycling:** LDPE is cleaned and reprocessed into new materials.
- **Chemical Recycling:** LDPE can be depolymerized or converted into fuels and other chemicals.

5. Polycarbonate (PC)

Properties:

- **Melting Point:** 230-250°C
- **Glass Transition Temperature:** 150-160°C

Applications:

- Recycled polycarbonate is used in the production of electronics, automotive parts, and optical discs.

Recycling Process:

- **Mechanical Recycling:** PC is shredded, cleaned, and reprocessed.
- **Chemical Recycling:** PC can be broken down into its monomers for reuse.

6. Polystyrene (PS)

Properties:

- **Melting Point:** 240-270°C
- **Glass Transition Temperature:** 100°C

Applications:

- Recycled PS is used in insulation materials, packaging, and disposable cutlery.

Recycling Process:

- **Mechanical Recycling:** PS is cleaned and reprocessed into new products.
- **Chemical Recycling:** Involves breaking PS into its monomers or other useful compounds.

7. Acrylonitrile Butadiene Styrene (ABS)

Properties:

- **Melting Point:** 220-240°C
- **Glass Transition Temperature:** 105°C

Applications:

- Recycled ABS is used in automotive parts, toys, and electronic housings.

Recycling Process:

- **Mechanical Recycling:** ABS is ground, cleaned, and reformed into new products.
- **Chemical Recycling:** ABS can be depolymerized into its constituent monomers.

8. Polyamide (Nylon)

Properties:

- **Melting Point:** 220-265°C
- **Glass Transition Temperature:** 50-70°C

Applications:

- Recycled nylon is used in textiles, automotive parts, and industrial applications.

Recycling Process:

- **Mechanical Recycling:** Nylon is cleaned and reformed into new fibers or products.
- **Chemical Recycling:** Nylon can be depolymerized into its monomers or other useful chemicals.

Challenges and Considerations

- **Contamination:** Recycled plastics often face challenges with contamination from other materials, which can affect the quality and performance of the recycled product.
- **Processing Costs:** The cost of recycling can be high, and it's important to optimize the recycling processes to make them more cost-effective and efficient.
- **Quality Control:** Ensuring that recycled plastics meet quality standards for their intended applications is crucial for their successful use.

By understanding the different types of recycled plastics and their properties, industries can make informed decisions about material selection and recycling practices, ultimately contributing to more sustainable and effective use of plastic resources. (ASTM, 2017, 2018; Graziano et al, 2019; Parameswaranpillai et al, 2019)

8.1.3 Mechanical and Thermal Effects of Using Fine Recycled PET Aggregates in Common Screeds

The increasing amount of plastic that is consumed worldwide has led to a significant increase in plastic waste, which poses a serious risk to human health and the environment. Among the many plastic polymers, polyethylene terephthalate (PET) has become increasingly popular due to its remarkable qualities, which include strong durability, chemical resistance, and thermal stability. However, due to PET's poor biodegradability, it has accumulated in landfills and marine environments, underscoring the urgent need for quick action to guarantee effective management and long-term use of PET waste. One of the most intriguing ways to recycle PET waste in structural engineering and lessen the quantity of plastic that ends up in landfills is to mix it into building materials like screeds. A sustainable response to various environmental problems related to the use and production of cement is provided by the incorporation of PET into cementitious mixtures. In addition to lowering waste and CO_2 emissions related to the production of conventional materials, it helps with the recycling of plastic bottles. Because PET is lightweight, the mixture's density is lowered, resulting in a reduction in the amount of material used and the cost of transportation. Furthermore, the increased toughness of concrete that has PET added can prolong the useful life of buildings and lessen the frequency of repairs. Consequently, there are benefits to using PET in cement mixtures in terms of lessening the environmental effects of disposing of excess plastic in landfills. However, it also has additional advantages that are linked to the enhancement of reinforced composites' mechanical and physical performance.

All cementitious mixtures, including screeds, are composed of materials with low tensile strength but good compressive strength. To make up for this shortage of screeds, plastic materials made from recycled PET bottles could be a useful approach. Actually, compared to regular construction screeds, PET has a tensile strength that is about thirty times higher. Therefore, by creating a material that is more resistant to cracking, the use of PET fibers could raise the flexural strength of cementitious mixtures. Scientific literature has shown that these fibers' stitching effect can limit the onset of cracks by postponing the attainment of the ultimate deformation and enhancing ductility and toughness. When 4% of the weight of fine and coarse PET aggregates was added, the fracture toughness rose by up to 8.5 percent and 16.3 percent, respectively. Because of the resistance that the larger size of the aggregates provided, using coarse aggregates had a greater benefit than using fine ones. To further enhance the properties of cementitious mixtures, PET fibers can be mixed with other fibers or reinforcing materials. For instance, the mortar strength values affected by the combination of PET, PE, and PP are primarily those following the first crack. Put another way, the strength is primarily dependent on the properties of the mortar and is not affected by the fiber reinforcement until the cementitious mixture experiences its first crack. Cementitious mixtures used to construct pavement reinforced with 3 percent and 5 percent PET by mass were characterized by Arulrajah in. The reduction in resistant moduli was confirmed by using aggregate from recycled concrete in addition to plastic. On the other hand, the physical and strength properties of the 5 percent PET mixtures with RCA and CB, stabilized with cement, matched the requirements set by road authorities for the construction of pavement base and subbase.

The fiber-reinforced concrete's mechanical characteristics and resilience in a harsh maritime environment. The experimental findings demonstrated that the curing condition in seawater marginally altered the fiber-reinforced concrete's toughness and first-crack strength. Waste PET reinforcement has been demonstrated in numerous studies to reduce drying shrinkage and improve the cementitious mixture's compactness. Higher aspect ratio (AR) fibers are typically better at preventing cracking. For a PET fiber length of 50 mm, an increase in the volume fraction of the fibers in the range of 1 point 0 percent to 1 point 50 percent resulted in a significant reduction in the plastic shrinkage cracking of the mortar. Rather, at a fiber volume fraction of 1 percent and a fiber length of 50 mm, the plastic shrinkage cracks totally vanished. Greater resistance to cracking in screeds results in increased durability and material savings through thinner castings. Because PET aggregates are lightweight, using them also results in a reduction in screed thickness, which helps offset the material's weight loss. Because of this, screeds containing PET aggregates could be cast at slightly thinner thicknesses than conventional ones while maintaining the same flexural strength. By adding 20% of shredded PET by weight, polymer mortars containing binders, epoxy, and unsaturated polyester saw an average reduction in density of roughly 23%. There are various ways to incorporate PET into screeds; the reinforcement can take the form of fiber or aggregate. When regular, circular-shaped particles are applied, the compressive strength increases while the air content decreases in the first scenario. In the second scenario, PET fibers take on the same function as the steel bars that reinforce the concrete. Their geometry is what sets these PET reinforcements apart in terms of how well they work in cementitious mixtures. Better mechanical performance is actually ensured by having fibers with a high aspect ratio (AR), appropriate dosage, and small equivalent diameters.

Polyethylene Terephthalate (PET)

One thermoplastic polymer that is frequently used and valued for its strength, resilience, and adaptability is polyethylene terephthalate, or PET. It is a polymer that is frequently used in applications ranging from clothing fibers to beverage bottles. It is a member of the polyester family. To guarantee PET's performance and quality, especially when recycled, its characterization is crucial.

Thermal Properties of PET

- Glass Transition Temperature (Tg):
  - PET has a glass transition temperature around 70-80°C. At this temperature, PET transitions from a hard, glassy state to a softer, more flexible state. This property is crucial in determining the material's performance in various applications, especially when used in environments that experience temperature fluctuations.
- Melting Temperature (Tm):
  - The melting temperature of PET is typically around 250-260°C. This is the temperature at which PET transitions from a solid to a liquid. The high melting point makes PET suitable for applications requiring heat resistance, such as in food and beverage containers.

- Thermal Degradation:
  - PET begins to degrade at temperatures above its melting point, with significant weight loss typically starting around 300°C. Degradation can be accelerated by the presence of impurities or through repeated heating cycles, which can lead to changes in the polymer's properties and affect its overall performance.

Recycling and Thermal Analysis

- Effect of Reprocessing:
  - Recycled PET often undergoes multiple processing cycles, which can alter its thermal properties. Reprocessing can lead to a decrease in the melting temperature and an increase in the rate of thermal degradation. This is due to the breakdown of polymer chains and the formation of low molecular weight compounds.
- Contamination:
  - Contaminants present in recycled PET, such as residues from other materials or impurities, can affect its thermal behavior. For example, the presence of metal or paper can lead to irregular thermal profiles, impacting the quality of the recycled PET.
- Additives and Stabilizers:
  - Additives and stabilizers are often used to improve the thermal stability of recycled PET. These can help counteract the effects of degradation and enhance the material's performance by maintaining its thermal properties within desirable ranges.

Applications

- Packaging: PET is extensively used for beverage bottles, food containers, and other packaging materials due to its strength, clarity, and safety for contact with food and drink.
- Fibers: PET is also used in the textile industry for producing fibers for clothing, carpets, and upholstery, where its durability and resistance to wrinkling and shrinking are valuable.

To maximize PET's use and guarantee its quality across a range of applications, it is crucial to comprehend its thermal properties, especially when it comes to recycled PET. By evaluating the effects of contamination, reprocessing, and stabilizer addition, thermal analysis contributes to improved recycling procedures and improved PET product performance. (ASTM, 2017, 2018; Graziano et al, 2019; Parameswaranpillai et al, 2019)

8.1.4 Thermomechanical Properties of Virgin and Recycled Polypropylene—High-Density Polyethylene Blends

Just 9% of plastics in the world are recycled, which poses a serious threat to the environment. Because polyolefins—which include polypropylene (PP) and polyethylene (PE)—have good mechanical properties, are readily available, can be processed, and have low manufacturing costs, they are the primary constituents of municipal waste. Because of their close densities and structural similarities, it is not cost-effective to completely separate PP and PE during mechanical waste recovery; as a result, PP and PE typically remain mixed. Recycled PP (rPP) and recycled PE (rPE) have lower mechanical properties than virgin polymers as a result of irreversible thermomechanical degradation processes like chain scission that occur during the mechanical recycling process.

Blends of polypropylene and polyethylene (PP/PE) exhibit significant commercial interest due to their ability to mitigate the shortcomings of each material, namely low impact resistance at low temperatures and poor environmental stress cracking resistance, respectively. Thermodynamic immiscibility of blends of PP: Low-Density PE (LDPE) and PP: High-Density PE (HDPE) results in poor material performance because of the blends' phase-separated morphology and low interfacial adhesion between the phases. The final blend morphology and the crystallization behavior of the blend components determine the mechanical performance of immiscible blends. The development of morphology during polymer processing is influenced by a number of variables, including composition, the viscosity ratio of the constituents, interfacial characteristics, crystallinity, and processing conditions.

The mechanical characteristics, crystallization behavior, and morphology of PP:PE blends have all been documented in a number of studies. Comparative studies are important, even though determining and understanding the mechanical and thermal behavior of virgin polyethylene (vPE): virgin PP (vPP) blends is crucial to optimizing the properties of recycled blends. Since the literature primarily focuses on ternary systems (e.g., vPP:vHDPE and rPP:rHDPE blends), a comparison of the thermomechanical properties of these blends over a broad range of compositions is lacking. g. PP:HDPE as well as fillers, co-polymers, compatibilizers, and other polymers. Prior to adding more components, it's critical to comprehend how rPP:rHDPE blends perform differently from one another.

Degradation mechanisms during the recycling process have been found to affect the crystallinities of rPP, rHDPE, and rPP:rHDPE blends. Because of a decrease in molecular weight and an increase in chain mobility, rPP has a higher value of crystallinity than vPP. The ability of chains to fold into thicker lamella was enhanced by increased chain mobility, which consequently led to an increase in the rate of crystallization and crystallinity. Therefore, in order to maximize the mechanical performance of the recyclates, it is critical to comprehend the variation in crystallinity in rPP:rHDPE blends.

The impact of recycling cycles on the mechanical properties of polypropylene (PP), polypropylene (PE), and their blends has been studied and documented in the literature. As an illustration, after PP was put through ten consecutive injection molding cycles at 200 °C, it was discovered that the degree of crystallinity rose with each cycle. Yield stress and Young's modulus increased as a result. Who heated PP to between 175 and 190 °C for seven cycles in a row. After the third cycle, they saw a drop in yield stress and Young's modulus, which they attributed to a decrease in the number of tie molecules between the crystalline and amorphous phases. This not only emphasizes the significance of the reprocessing methodology but also shows how challenging it is to compare recycled materials' performance across the literature. Moreover, studies on the chemical analysis of recycled PE and PP have shown that these materials are variable, degrade over time, and contain contaminants and impurities.

PE can undergo a greater number of extrusion cycles before experiencing a decline in its mechanical qualities. The notable alteration in LDPE's crystallinity and, consequently, its mechanical characteristics up to the 40th extrusion cycle. But between the 40th and 50th cycles, there was a decrease in crystallinity, which could be attributed to crosslinking, short side branches in the side groups or backbone chain, or both. They discovered that crystallinity and Young's modulus decreased as a result of chain scission and branching, which happened up to the 60th cycle. Crosslinking prevented crystallinity and Young's modulus from changing after the 60th extrusion cycle, though. In comparison to the vHDPE, the rHDPE's Young's modulus had only decreased by 20% after the 100th cycle.

Research has been done to comprehend the mechanical characteristics of blends of polypropylene and ethylene that are recycled. The PP:LDPE blends' melt flow behavior was impacted by recycling, but the tensile properties were not significantly impacted (up to five recycling cycles). It's interesting to note that although PP experienced chain scission during each recycling cycle, the blend's general characteristics were preserved. They did not, however, look into the PP:LDPE up to 10 weight percent of LDPE. The range of blend compositions examined in the literature is usually narrow. But because waste streams vary so much, it's critical to comprehend the mechanical characteristics of every blend composition before introducing a third component. This would allow more recycled material to reenter the market and allow the recycling sector to respond to variations in the makeup of waste streams in various batches and locations. Consequently, the purpose of this paper is to comprehend the differences in thermomechanical characteristics between virgin and recycled PP:HDPE blends. (ASTM, 2017, 2018; Graziano et al, 2019; Parameswaranpillai et al, 2019)

8.1.5 Impact of Contamination on Thermal Degradation

Contamination has a major effect on the thermal degradation of recycled plastics and can negatively impact their quality and performance. Recycled plastics may contain impurities that affect the material's thermal behavior in a number of ways, including organic impurities, metal fragments, and leftovers from other polymers.

- **Altered Thermal Degradation Profile**: The homogeneity of the polymer matrix may be affected by contaminants, which could result in anomalies in the thermal degradation profile. For example, at lower temperatures, metal impurities can function as catalysts and speed up the plastic's degradation. Organic impurities may interact with the polymer to accelerate thermal degradation or result in unforeseen changes in the temperature at which degradation occurs.
- **Decreased Thermal Stability**: Contaminants have the potential to reduce recycled plastics' thermal stability. For instance, recycled PET that has foreign materials in it may degrade prematurely during processing or use if its thermal degradation temperature drops. The integrity of the material is impacted by this decrease in stability, which also restricts its use, especially in hot conditions.
- **Inconsistent Melting and Glass Transition Temperatures**: The melting and glass transition temperatures of recycled plastics can also be impacted by contamination. Impurities may interfere with the rigidity of the polymer, lowering its melting point, or they may change the glass transition temperature by upsetting the polymer's crystalline structure. These modifications may have an effect on the material's processing and final properties.
- **Decreased Mechanical Properties**: Because of uneven degradation or improper mixing, contaminants can cause physical defects in recycled plastic, such as weak spots or inconsistent textures. These flaws may lower the finished product's overall performance, mechanical strength, and flexibility.
- **Challenges in Quality Control**: While difficult to detect and

manage, contamination must be avoided if recycled plastics are to remain of high quality. To reduce the influence of contaminants on thermal properties and guarantee that the recycled material satisfies the necessary requirements, efficient sorting and cleaning procedures are essential.

Improving purification procedures, developing sorting technologies, and comprehending the interactions between contaminants and polymers are all necessary to address these problems. By using this method, contamination is lessened and recycled plastics are guaranteed to maintain their intended mechanical and thermal qualities.

Analyze how the presence of contaminants in recycled plastics affects their thermal degradation and stability.

The presence of contaminants such as paper, metal, or other polymers in recycled plastics can significantly affect their thermal degradation and stability, leading to various challenges in material performance and processing. Here's a detailed analysis of these impacts:

1. Altered Thermal Degradation Profile

- **Paper and Organic Contaminants:** Additional volatile compounds may be introduced into the recycled plastic matrix by paper and other organic contaminants. These organic contaminants may break down during thermal degradation at temperatures lower than those of the base polymer, which could cause degradation to start earlier. There might be more than one degradation step or less overall stability in the resulting thermal degradation profile.
- **Metal Contaminants:** Reactions involving thermal degradation may be catalyzed by metal contaminants. For example, oxidative degradation processes can be accelerated by metals like iron or copper. This catalytic activity may cause the plastic to break down at a lower temperature, decreasing the recycled plastic's thermal stability and possibly accelerating the breakup of the polymer chains.

2. Decreased Thermal Stability

- **Paper and Organic Contaminants:** The presence of paper or other organic materials can disrupt the uniform thermal degradation of the plastic. These contaminants often decompose at different temperatures, causing a non-uniform degradation of the polymer matrix. This results in decreased thermal stability and can lead to the production of by-products that further destabilize the plastic.

- **Other Polymers:** When recycled plastics contain residues of other polymers (e.g., a mix of PET and PVC), the different thermal properties of these polymers can create inconsistencies in the degradation process. For example, PVC releases hydrochloric acid during degradation, which can accelerate the degradation of adjacent PET, leading to a lower thermal stability of the mixed material.

3. Impact on Melting and Glass Transition Temperatures

- **Paper and Organic Contaminants:** Contaminants such as paper can introduce impurities that interfere with the polymer's crystalline structure. This can lower the melting temperature of the recycled plastic by disrupting the orderly arrangement of polymer chains. Similarly, the glass transition temperature (Tg) may be altered, affecting the plastic's flexibility and performance.
- **Other Polymers:** Residual polymers with different thermal properties can shift the melting and Tg values of the recycled plastic. For instance, the presence of a polymer with a lower melting point can reduce the overall melting point of the recycled material, affecting its suitability for applications requiring specific thermal performance.

4. Mechanical Property Degradation

- **All Contaminants:** Contaminants can cause physical inconsistencies within the recycled plastic, leading to defects such as weak spots or uneven texture. These issues arise from non-uniform degradation or improper mixing of different materials, resulting in reduced mechanical strength and durability.

5. Challenges in Quality Control

- **Identification and Removal:** Effective sorting and cleaning technologies are crucial for minimizing the impact of contaminants. Advanced techniques are required to detect and remove metal fragments, paper, and other polymers to ensure the recycled plastic maintains its thermal and mechanical properties.

Recycled plastics are highly susceptible to contaminants that alter their thermal stability and degradation profiles, changing melting and glass transition temperatures, and altering degradation profiles. To ensure high-quality recycled materials, addressing these issues entails developing technologies for sorting and cleaning materials, upgrading recycling procedures, and comprehending the interactions between contaminants and polymers. (ASTM, 2017, 2018; Graziano et al, 2019; Parameswaranpillai et al, 2019)

8.2 OPTIMIZATION OF RECYCLED PLASTIC PROCESSING

Improving the performance and increasing the applications of recycled plastics requires optimizing their processing. To effectively address the inherent challenges of recycled materials, such as degradation and contamination, processing conditions must be fine-tuned. To balance material quality and efficiency, critical factors include modifying temperature settings, cooling rates, and extrusion parameters. To preserve the mechanical properties of the recycled plastic, for example, adjusting the extrusion temperature can help reduce additional degradation of the polymer chains. Optimizing cooling rates also guarantees homogeneous crystallization and lowers the possibility of internal stresses that can compromise the functionality of the finished product.

It is also essential to use cutting-edge sorting and cleaning technologies to get rid of impurities like metal, paper, and leftovers from other polymers. The purity of recycled plastic can be increased, resulting in more consistent mechanical and thermal properties, by using high-efficiency separation systems and rigorous cleaning procedures. By enhancing their thermal stability and processing qualities, stabilizers and additives can further improve the performance of recycled plastics. Manufacturers can optimize the processing of recycled plastics, guaranteeing that they satisfy quality standards and are appropriate for a variety of applications, by regularly monitoring and modifying these parameters.

- Examine how different processing conditions (e.g., extrusion temperature, cooling rate) influence the thermal properties and performance of recycled plastics.
- Investigate the effects of reprocessing on the thermal stability and mechanical properties of recycled plastics.

(ASTM, 2017, 2018; Graziano et al, 2019; Parameswaranpillai et al, 2019)

8.2.1 Thermal Degradation Kinetics of Recycled Plastics

Studying the kinetics of thermal degradation in recycled plastics entails figuring out how quickly these materials degrade in the presence of heat. Understanding the stability, robustness, and processing behavior of recycled plastics in comparison to their virgin counterparts requires an understanding of this analysis.

1. Degradation Mechanisms

- **Chain Scission and Cross-Linking:** Recycled plastics often experience chain scission, where polymer chains break down into smaller fragments, and cross-linking, where chains form connections with each other. Both mechanisms affect the thermal degradation behavior. Chain scission typically lowers the molecular weight of the polymer, leading to a decrease in thermal stability. Cross-linking can sometimes enhance stability but may also result in brittleness.
- **Presence of Contaminants:** Contaminants such as metal particles or residues from other polymers can influence degradation kinetics. Metals can act as catalysts, accelerating oxidative degradation, while residues from other polymers may introduce additional degradation pathways.

2. Analytical Techniques

- **Thermogravimetric Analysis (TGA):** TGA is commonly used to study thermal degradation kinetics. It measures the weight loss of a sample as it is heated, providing data on decomposition temperatures and degradation rates. By analyzing the TGA curves, researchers can determine the onset temperature of degradation, the rate of weight loss, and the presence of multiple degradation stages.
- **Kinetic Models:** Various kinetic models, such as the Arrhenius equation, can be applied to TGA data to quantify the activation energy required for degradation. These models help predict how degradation rates change with temperature and can be used to compare the thermal stability of recycled plastics with that of virgin plastics.

3. Implications for Processing and Application

- **Processing Conditions:** Understanding degradation kinetics helps optimize processing conditions. For example, controlling the heating rate and maximum temperature during extrusion or molding can minimize further degradation and preserve the material's properties.
- **Material Performance:** Knowledge of thermal degradation kinetics is crucial for evaluating the suitability of recycled plastics for specific applications, especially those involving high temperatures. Accurate kinetic data ensures that recycled materials meet performance requirements and maintain their integrity throughout their service life.

Understanding the kinetics of recycled plastics' thermal degradation offers important insights into how stable and heat-resistant they are. Researchers and manufacturers can improve the performance of recycled plastics in a variety of applications by optimizing processing conditions and applying kinetic models in conjunction with techniques like TGA.

(ASTM, 2017, 2018; Graziano et al, 2019; Parameswaranpillai et al, 2019)

8.2.2 Role of Additives and Stabilizers in Thermal Performance

By reducing the effects of degradation, increasing thermal stability, and guaranteeing constant material quality, additives and stabilizers are essential for improving the thermal performance of recycled plastics.

Here's how they contribute to the thermal performance of recycled plastics:

1. Enhancing Thermal Stability

Antioxidants:

- **Function:** Antioxidants are added to prevent oxidative degradation during processing and service. They work by neutralizing free radicals that can cause polymer chain scission and cross-linking.
- **Effect:** By inhibiting oxidation, antioxidants help maintain the thermal stability of recycled plastics, reducing the rate of thermal degradation and extending the material's useful life.

Heat Stabilizers:

- **Function:** Heat stabilizers are used to protect the polymer from thermal degradation at high processing temperatures. They absorb or neutralize heat-induced chemical reactions that can lead to degradation.
- **Effect:** These stabilizers help maintain the integrity of the recycled plastic during processing, ensuring that it retains its desired properties even when exposed to elevated temperatures.

2. Preventing Degradation

UV Stabilizers:

- **Function:** UV stabilizers absorb or block ultraviolet light that can cause photo-degradation of the plastic. They prevent the breakdown of polymer chains caused by UV exposure.
- **Effect:** By protecting against UV-induced degradation, these

stabilizers enhance the material's durability, particularly for applications exposed to sunlight or other sources of UV radiation.

Flame Retardants:

- **Function:** Flame retardants are added to improve the fire resistance of recycled plastics. They work by either releasing chemicals that inhibit combustion or by forming a protective layer that shields the material from flames.
- **Effect:** Improved flame resistance contributes to better thermal performance and safety, especially in applications where fire resistance is critical.

3. Improving Processing Performance

Processing Aids:

- **Function:** Processing aids, such as lubricants and surfactants, are added to improve the flow and melt characteristics of recycled plastics during processing.
- **Effect:** These aids facilitate better processing conditions, reduce the risk of overheating or uneven melting, and help achieve a more uniform final product.

Compatibilizers:

- **Function:** Compatibilizers are used to improve the compatibility of different polymers or additives within the recycled plastic matrix.
- **Effect:** By enhancing compatibility, compatibilizers help maintain uniform thermal performance and reduce phase separation or inconsistencies in the material.

4. Balancing Performance and Cost

Cost-Effectiveness:

- **Function:** The selection of additives and stabilizers must balance performance improvements with cost considerations.
- **Effect:** While additives enhance thermal performance, their cost must be justified by the benefits they provide in terms of material stability and durability.

To maximize the thermal performance of recycled plastics, additives and stabilizers are essential. Through the improvement of processing, prevention of degradation, enhancement of thermal stability, and guarantee of compatibility, these materials aid in preserving the quality and performance of recycled plastics in a variety of applications. When properly chosen and implemented, they support sustainable manufacturing practices by facilitating the more dependable and efficient use of recycled materials.

(ASTM, 2017, 2018; Graziano et al, 2019; Parameswaranpillai et al, 2019)

8.3 THERMAL BEHAVIOR OF RECYCLED PLASTICS IN COMPOSITE MATERIALS

A crucial area of research that affects the functionality, robustness, and applications of composite materials is the thermal behavior of recycled plastics. Recycled plastics are frequently combined with additional materials, like fibers or fillers, to improve composite materials' mechanical qualities and broaden their range of applications.

Understanding the thermal behavior of these composites is essential for optimizing their processing and performance:

- **Thermal Properties and Performance:** Because of earlier degradation and contamination, recycled plastics in composites may display different thermal properties from their virgin counterparts. For example, recycled polypropylene (PP) or polyethylene terephthalate (PET) may have a different glass transition temperature or a lower melting point, which can affect the composite material's processing behavior and thermal stability. Recycled plastics may alter the composite's overall thermal profile, which may result in differences in the material's stability and rate of thermal deterioration.
- **Interaction with Reinforcements:** The type of reinforcement used also affects the thermal behavior of composites made of recycled plastic. Natural, carbon, and glass fibers are examples of common reinforcements. The thermal properties of the composite can be influenced by the interactions between these reinforcements and the recycled plastic matrix. For instance, natural fibers may have lower thermal stability and may affect the material's overall thermal behavior, whereas glass fibers typically have higher thermal stability and can enhance the composite's thermal performance. To achieve the desired thermal performance, the reinforcement and recycled plastic matrix must be compatible.
- **Impact of Processing Conditions:** The processing parameters have a big impact on how recycled plastic composites behave thermally. To prevent excessive degradation of the recycled plastics and to guarantee the best possible interaction between the plastic matrix and reinforcements, it is important to carefully control factors like

extrusion temperature, molding pressure, and cooling rates. For example, high processing temperatures can worsen the thermal degradation of recycled plastics, and incorrect cooling rates can cause the composite to have uneven thermal properties and internal stresses.

- **Additives and Stabilizers:** The thermal behavior of recycled plastic composites can be further influenced by the addition of stabilizers and additives. For instance, flame retardants can increase the composite's resistance to fire, while thermal stabilizers can improve the recycled plastic matrix's stability at high temperatures. Coupling agents can also strengthen the connection between reinforcements and recycled plastics, which will enhance thermal performance.
- **Applications and Performance:** The application of recycled plastic composites in a variety of industries, such as consumer goods, construction, and automotive, depends on an understanding of their thermal behavior. Composites that have been properly designed and processed can provide improved performance, such as increased resistance to thermal cycling or improved thermal insulation, which makes them appropriate for demanding applications.

The qualities of the recycled plastics, the kind of reinforcement utilized, the processing circumstances, and the addition of additives all affect the thermal behavior of recycled plastics in composite materials. Recycled plastic composites can have their thermal performance and functionality improved for a variety of uses by carefully controlling these variables.

8.3.1 Analyze the Interaction between Recycled Plastic Matrices

The way recycled plastic matrices interact with one another in composite materials is essential to determining the final product's overall performance and characteristics. Because of their previous processing and degradation, recycled plastics, which can originate from a variety of sources and types, frequently show variations in their molecular structure and thermal history. These differences may have an impact on how well various recycled plastic matrices combine and function.

Thermal properties, compatibility, and molecular weight are all important considerations when mixing recycled plastics. For instance, because recycled PET and recycled PP have different thermal behaviors and polarities, blending them can be difficult. These polymers' incompatibilities may cause phase separation or inadequate adhesion, which would impair their mechanical qualities and cause uneven performance.

Compatibilizers are among the additives that are frequently used to enhance the interplay between various recycled plastics. By strengthening the bonds at the interface between incompatible matrices, these additives can improve dispersion and lessen phase separation. Achieving a homogenous blend also heavily depends on processing parameters like temperature and mixing speed. The composite material's ability to achieve the intended mechanical, thermal, and aesthetic properties is dependent on how well the recycled plastic matrices interact with one another. This allows the material to be used in a variety of applications while maximizing the benefits of recycling.

Comparative Analysis of Thermal Analysis Techniques

A comparative study of thermal analysis techniques entails assessing various approaches to the investigation of a material's thermal properties, such as phase transitions, degradation patterns, and behavior when heated. Depending on the material and type of analysis needed, each technique has its own advantages and limitations as well as offering distinctive insights.

8.3.2 Thermal Cycling Effects on Recycled Plastics

Recycled plastics are significantly impacted by thermal cycling, which is the process of repeatedly heating and cooling materials. This has an effect on the materials' mechanical qualities, dimensional stability, and overall durability. Comprehending these impacts is imperative when assessing the appropriateness of repurposed plastics in settings where temperature fluctuations occur.

1. Mechanical Property Degradation

The mechanical characteristics of recycled plastics can alter significantly as a result of repeated heat cycling. Thermal expansion and contraction of the material due to alternating temperature extremes create stress inside the plastic matrix. This stress has the potential to cause weariness, microcracking, and failure over time. For instance, after multiple heat cycles, recycled polyethylene terephthalate (PET) may show decreased tensile strength and impact resistance as a result of internal voids or cracks forming in the polymer chains.

2. Dimensional Stability Issues

When thermal cycling is applied to recycled plastics, dimensional stability becomes a major concern. The accuracy and fit of components can be impacted by warping, warpage, or dimensional changes brought on by the repeated expansion and contraction. For example, dimensional changes may occur in recycled polypropylene (PP) used in automotive parts, which could affect assembly and performance. The inability of the material to completely regain its original shape following each heat cycle causes permanent deformation, which gives rise to these stability problems.

3. Impact of Previous Degradation

The way that recycled plastics react to heat cycling is largely dependent on their initial state. Plastics that have experienced significant degradation in the past—caused by contamination, improper recycling techniques, or previous use—are more vulnerable to further heat stress. Thermal cycling has the potential to accelerate the deterioration process of pre-existing damage, resulting in more noticeable effects on mechanical properties and dimensional stability.

4. Influence of Contaminants and Additives

Thermal cycling has an impact on recycled plastics due to the presence of contaminants and additives. Uneven thermal cycling effects can result from contaminants that act as stress concentrators or change the behavior of thermal expansion, such as metal particles or leftovers from other polymers. Stabilizers and impact modifiers, on the other hand, can increase a material's resistance to thermal stress and increase its overall durability, which can help lessen some of these effects.

5. Processing Conditions and Material Formulation

Conditions of processing and material formulation also have an impact on how thermal cycling affects recycled plastics. The ability of the material to tolerate heat cycling can be increased and internal stresses can be decreased under ideal processing conditions. The type and concentration of additives used in a formulation can improve mechanical performance and thermal stability, increasing the recycled plastic's resistance to temperature changes.

6. Practical Implications

Designing and utilizing recycled plastics in applications where temperature fluctuations are a possibility requires an understanding of the effects of thermal cycling. For instance, recycled plastic components used in automobiles need to be able to survive engine bay heat cycling without suffering appreciable performance deterioration. Similar to this, products used in consumer goods and construction must keep their structural integrity and shape when exposed to temperature changes in the environment.

Thermal cycling has a significant effect on recycled plastics, changing their overall durability, dimensional stability, and mechanical properties. The prior degradation of the material, the existence of contaminants, and the caliber of the additives used are some of the variables that determine how much of these effects are felt. The performance of recycled plastics under thermal cycling conditions can be improved, guaranteeing their suitability for a variety of applications, by carefully controlling these variables and optimizing processing conditions.

(ASTM, 2017, 2018; Graziano et al, 2019; Parameswaranpillai et al, 2019)

8.3.3 The Effects of Thermal Cycling (repeated heating and cooling) on the Physical and Chemical Properties of Recycled Plastics

The physical and chemical characteristics of recycled plastics are significantly impacted by thermal cycling, which is the process of repeatedly heating and cooling materials. These effects have the potential to greatly affect how well-performing, long-lasting, and useful recycled plastics are in a variety of applications.

Here's a detailed analysis of how thermal cycling impacts these properties:

1. Physical Properties

Mechanical Strength:

- **Degradation of Mechanical Properties:** Repeated thermal cycling subjects recycled plastics to alternating thermal expansion and contraction, inducing mechanical stress. This stress can lead to micro-cracking, fatigue, and weakening of the material. For example, recycled polyethylene (rPE) and polypropylene (rPP) often exhibit reduced tensile strength and impact resistance after multiple cycles. The cumulative effect of these stresses can result in a noticeable decline in the material's structural integrity.

Dimensional Stability:

- **Shape Retention Issues:** In recycled plastics, the constant expansion and contraction that occurs during heat cycling can lead to permanent deformation or warping. The material's incapacity to completely return to its initial dimensions following each cycle is the root cause of this problem. The fit and functionality of recycled plastic components, like construction materials or automobile parts, may be compromised by dimensional errors.

Surface Properties:

- **Surface Cracking and Deformation:** Repeated heat stress can cause recycled plastics to crack or distort on their surface. In addition to degrading the surface's aesthetic appeal, surface degradation can also increase a surface's vulnerability to outside influences like moisture intrusion.

2. Chemical Properties

Polymer Degradation:

- **Chemical Breakdown:** Through processes like oxidation and hydrolysis, hermal cycling speeds up the chemical deterioration of recycled plastics. Every heating stage has the potential to stimulate oxidative reactions that break down polymer chains and lower molecular weight. Degradation can produce volatile degradation products and results in the loss of mechanical properties.

Cross-Linking and Chain Scission:

- **Altered Chemical Structure:** Thermal cycling can cause changes in the chemical structure of recycled plastics. It is possible for cross-linking to take place, which could improve the material's thermal stability but also make it brittle. On the other hand, oxidative stress-induced chain scission lowers molecular weight and causes material deterioration.

3. Impact of Contaminants and Additives

Contaminants:

- **Exacerbation of Degradation:** Recycled plastics undergo alternating thermal expansion and contraction through repeated thermal cycling, which creates mechanical stress. This strain may cause the material to weaken, become fatigued, and microcrack. For instance, after several cycles, recycled polypropylene (rPP) and polyethylene (rPE) frequently show decreased impact resistance and tensile strength. The material's structural integrity may noticeably deteriorate as a result of these stresses acting cumulatively.

Additives.

- **Mitigation and Enhancement:** Processing additives have the potential to impact the thermal cycling performance of recycled plastics. Antioxidants and thermal stabilisers can counteract some of the adverse effects by preventing oxidation and degradation. Impact modifiers can increase a material's toughness and resistance to heat stress, which will increase the material's overall durability.

4. Practical Considerations

Performance in End-Use Applications:

- **Applicability Concerns:** Recycled plastics used in applications subject to thermal cycling, such as automotive components, electronic housings, or construction materials, must be carefully evaluated for their performance under these conditions. Ensuring that the recycled plastic can maintain its physical and chemical properties over time is crucial for reliability and safety.

Processing and Formulation Adjustments:

- **Optimization Needs:** To enhance the performance of recycled plastics under thermal cycling, optimizing processing conditions and material formulation is essential. This includes selecting appropriate additives, controlling processing temperatures, and improving material purification to minimize contamination.

The chemical and physical characteristics of recycled plastics are severely impacted by heat cycling. Repeated heating and cooling can cause changes in the chemical structure, dimensional stability problems, and mechanical strength degradation. To ensure that recycled plastics retain their functionality and durability under a range of thermal conditions, addressing these effects necessitates careful consideration of contaminants, additives, and processing conditions.

(ASTM, 2017, 2018; Graziano et al, 2019; Parameswaranpillai et al, 2019)

REFERENCES

1. Akahira T, Sunose T (1971) Method of determining activation deterioration constant of electrical insulating materials. Res Rep Chiba Inst Technol (sci Technol) 16(1971):22–31
2. Alani, A.H.; Bunnori, N.M.; Noaman, A.T.; Majid, T.A. Durability Performance of a Novel Ultra-High-Performance PET Green Concrete (UHPPGC). *Constr. Build. Mater.* **2019**, *209*, 395–405.
3. Al-Salem SM, Lettieri P, Baeyens J (2009) Recycling and recovery routes of plastic solid waste (PSW): a review. Waste Manag 29(10):2625–43.
4. Arulrajah, A.; Perera, S.; Wong, Y.C.; Horpibulsuk, S.; Maghool, F. Stiffness and Flexural Strength Evaluation of Cement Stabilized PET Blends with Demolition Wastes. *Constr. Build. Mater.* **2020**, *239*, 117819.
5. ASTM (2017) ASTM D6980-17, standard test method for determination of moisture in plastics by loss in weight. ASTM D6980-17.
6. ASTM E2160 04 (2018) Standard test method for heat of reaction of thermally reactive materials by differential scanning calorimetry, 04(Reapproved 2018), pp 1–5.
7. Barrall EM (1973) Precise determination of melting and boiling points by differential thermal analysis and differential scanning calorimetry. Thermochim Acta (Elsevier) 5(4):377–389.
8. Bazelatto Zanoni MA, Massard H, Ferreira Martins M (2012) Formulating and optimizing a combustion pathways for oil shale and its semi-coke. Combust Flame 159(10):3224–3234.
9. Coats AW, Redfern JP (1964a) Kinetic parameters from thermogravimetric data. II. J Polym Sci Part B: Polym Lett (Wiley Inc.,) 3(4914):917–920.
10. Coats AW, Redfern JP (1964b) Kinetic parameters from thermogravimetric data. Nature. Accessed 11 Aug 2020
11. Criado JM (1978) Kinetic analysis of DTG data from master curves. Thermochim Acta (Elsevier) 24(1):186–189.
12. Dollimore D et al (1992) The significance of the onset and final temperatures in the kinetic analysis of TG curves. Thermochim Acta (Elsevier) 196(2):255–265.
13. Duque JVF, Martins MF, Orlando MTDA (2020) Biodegradable masterbatch blends: The implications on thermal conversion and recycling stream of polyethylene. Open J Chem 6(1):008–010.
14. Elder JP (1996) Reconciliation of Arrhenius and iso-conversional analysis kinetics parameters of non-isothermal data. Thermochim Acta (Elsevier) 272(1–2):41–48.

15. Europe P (2019) Plásticos—Situación en 2019, Plastics—the facts 2018. An analysis of European plastics production, demand and waste data.
16. Faber J, Fawcett T (2002) The powder diffraction file: present and future. Acta Crystallograp Sect B: Struct Sci 58(3 PART 1):325–332.
17. Fan C et al (2014) The oxidation of heavy oil: thermogravimetric analysis and non-isothermal kinetics using the distributed activation energy model. Fuel Process Technol 119:146–150.
18. Flynn JH, Wall LA (1966b) General treatment of the thermogravimetry of polymers. J Res Natl Bureau Standards Sect A: Phys Chem.
19. Foti, D. Recycled Waste PET for Sustainable Fiber-Reinforced Concrete. In *Use of Recycled Plastics in Eco-Efficient Concrete*; Elsevier: Amsterdam, The Netherlands, 2018; pp. 387–410. ISBN 9780081026762.
20. Foti, D.; Lerna, M. New Mortar Mixes with Chemically Depolymerized Waste PET Aggregates. *Adv. Mater. Sci. Eng.* **2020**, *2020*, 8424936.
21. Friedman HL (2007) Kinetics of thermal degradation of char-forming plastics from thermogravimetry. application to a phenolic plastic. J Polym Sci Part C: Polym Symp.
22. Fujino J, Honda T (2009) Measurement of the specific heat of plastic waste/fly ash composite material using differential scanning calorimetry. Int J Thermophys 30:976–986.
23. Geyer, R.; Jambeck, J.R.; Law, K.L. Production, use, and fate of all plastics ever made. *Sci. Adv.* **2017**, *3*, e1700782.
24. Graziano, A.; Jaffer, S.; Sain, M. Review on modification strategies of polyethylene/polypropylene immiscible thermoplastic polymer blends for enhancing their mechanical behavior. *J. Elastomers Plast.* **2019**, *51*, 291–336.
25. Gražulis S et al (2012) Crystallography open database (COD): an open-access collection of crystal structures and platform for world-wide collaboration Nucl Acids Res (Oxford Academic) 40(D1):D420–D427.
26. Jomo KS et al (2019) Southeast asia's misunderstood miracle: industrial policy and economic development in Thailand, Malaysia and Indonesia, Southeast Asia's Misunderstood Miracle: Industrial Policy and Economic Development in Thailand, Malaysia and Indonesia.
27. Jones, H.; Saffar, F.; Koutsos, V.; Ray, D. Polyolefins and Polyethylene Terephthalate Package Wastes: Recycling and Use in Composites. *Energies* **2021**, *14*, 7306.
28. Jose, S.; Aprem, A.S.; Francis, B.; Chandy, M.C.; Werner, P.; Alstaedt, V.; Thomas, S. Phase morphology, crystallisation behaviour and mechanical

properties of isotactic polypropylene/high density polyethylene blends. *Eur. Polym. J.* **2004**, *40*, 2105–2115.

29. Joshi C, Seay J, Banadda N (2019) A perspective on a locally managed decentralized circular economy for waste plastic in developing countries. Environ Prog Sustain Energy.
30. Kumar, S.; Panda, A.K.; Singh, R.K. A review on tertiary recycling of high-density polyethylene to fuel. *Resour. Conserv. Recycl.* **2011**, *55*, 893–910.
31. Lerna, M.; Foti, D.; Petrella, A.; Sabbà, M.F.; Mansour, S. Effect of the Chemical and Mechanical Recycling of PET on the Thermal and Mechanical Response of Mortars and Premixed Screeds. *Materials* **2023**, *16*, 3155.
32. Liu G et al (no date) Glass transition temperature of polystyrene determined by DSC revisited. en.cnki.com.cn.
33. Maddah, H.A. Polypropylene as a Promising Plastic: A Review. *Am. J. Polym. Sci.* **2016**, *6*, 1–11.
34. Madi, N.K. Thermal and mechanical properties of injection molded recycled high density polyethylene blends with virgin isotactic polypropylene. *Mater. Des.* **2013**, *46*, 435–441.
35. Mendes, R.F.; Narciso, C.R.P.; Mendes, J.F.; Viana, Q.S.; Eugênio, T.M.C.; Vilela, A.P. Study of New Reinforcing Materials for Cementitious Panel Production. *Environ. Sci. Pollut. Res.* **2021**, *28*, 37217–37230.
36. Morîntale E et al (no date) Use of heat flows from DSC curve for calculation of specific heat of the solid materials. Elsevier. Accessed 12 Aug 2020.
37. Obeid F et al (2014) Thermo-catalytic pyrolysis of waste polyethylene bottles in a packed bed reactor with different bed materials and catalysts. Energy Convers Manag.
38. Ozawa T (1965) A new method of analyzing thermogravimetric data. Bull Chem Soc Jpn 38(11):1881–1886.
39. Pan R, Duque JVF, Debenest G (2021) Investigating waste plastic pyrolysis kinetic parameters by genetic algorithm coupled with thermogravimetric analysis. Waste Biomass Valoriz. Springer.
40. Parameswaranpillai, J.; Pulikkalparambil, H.; Sanjay, M.R.; Siengchin, S. Polypropylene/high-density polyethylene based blends and nanocomposites with improved toughness. *Mater. Res. Express* **2019**, *6*, 075334.
41. Qureshi MS et al (2020) Pyrolysis of plastic waste: Opportunities and challenges. J Anal Appl Pyroly (Elsevier B.V) 104804
42. Ragaert K, Delva L, Van Geem K (2017) Mechanical and chemical

recycling of solid plastic waste. Waste Manag.

43. Rodríguez-Carvajal J (1993) Recent advances in magnetic structure determination by neutron powder diffraction. Phys B: Phys Condens Matter (North-Holland) 192(1–2):55–69.
44. Ruland W, Smarsly B (2002) X-ray scattering of non-graphitic carbon: an improved method of evaluation. J Appl Crystallogr 35(5):624–633.
45. Ryberg MW et al (2019) Global environmental losses of plastics across their value chains. Resour Conserv Recycl.
46. Ryu HW et al (2020) Recent advances in catalytic co-pyrolysis of biomass and plastic waste for the production of petroleum-like hydrocarbons. Bioresour Technol (Elsevier Ltd.), 123473.
47. Salih, S.E.; Hamood, A.F.; Alsabih, A.H. Comparison of the Characteristics of LDPE:PP and HDPE:PP Polymer Blends. *Mod. Appl. Sci.* **2013**, *7*, 33–42.
48. Sarker M, Rashid MM et al (2012) Hermal conversion of waste plastic (HDPE, PP, PS) to produce mixture of hydrocarbons. Am J Environ Eng 2(5):128–136
49. Sarker M, Rashid MM et al (2012b) Low and high density polyethylene waste plastics conversion into liquid hydrocarbon fuel. chimica oggi/Chem Today 30(6)
50. Sheng C, Azevedo JLT (2005) Estimating the higher heating value of biomass fuels from basic analysis data. Biomass and Bioenergy Pergamon 28(5):499–507.
51. Singh RK et al (2020) A TG-FTIR investigation on the co-pyrolysis of the waste HDPE, PP, PS and PET under high heating conditions. J Energy Inst (Elsevier B.V.,) 93(3):1020–1035.
52. Soto JM et al (2018) A real case study of mechanical recycling as an alternative for managing of polyethylene plastic film presented in mixed municipal solid waste. J Clean Prod.
53. Starink MJ (2003) The determination of activation energy from linear heating rate experiments: a comparison of the accuracy of isoconversion methods. Thermochim Acta (Elsevier) 404(1–2):163–176.
54. Strapasson, R.; Amico, S.C.; Pereira, M.F.R.; Sydenstricker, T.H.D. Tensile and impact behavior of polypropylene/low density polyethylene blends. *Polym. Test.* **2005**, *24*, 468–473.
55. Tanzadeh, R.; Tanzadeh, J.; Hosseini, M.H.; Mirsepahi, M. Performance Evaluation of Reinforced Roller Compacted Concrete Pavement with Basalt Fibers and Recycled Polyethylene in Warm Regions of Iran.

In Proceedings of the International Conference on Transportation and Development 2016, Houston, TX, USA, 26–29 June 2016; pp. 1122–1131.

56. Tapia-Picazo, J.C.; Luna-Bárcenas, J.G.; García-Chávez, A.; Gonzalez-Nuñez, R.; Bonilla-Petriciolet, A.; Alvarez-Castillo, A. Polyester Fiber Production Using Virgin and Recycled PET. *Fibers Polym.* **2014**, *15*, 547–552.
57. Umasabor, R.I.; Daniel, S.C. The Effect of Using Polyethylene Terephthalate as an Additive on the Flexural and Compressive Strength of Concrete. *Heliyon* **2020**, *6*, e04700.
58. US EPA—United States Environmental Protection Agency (2019) Advancing sustainable materials management: 2017 fact sheet. United States Environmental Protection Agency, Office of Land and Emergency Management, Washington, DC 20460. EPA530F-18-004
59. Vyazovkin S et al (2011) ICTAC Kinetics Committee recommendations for performing kinetic computations on thermal analysis data. Thermochim Acta.
60. Vyazovkin S et al (2014) ICTAC kinetics committee recommendations for collecting experimental thermal analysis data for kinetic computations. Thermochim Acta (Elsevier) 590:1–23.
61. Wang C et al (2020) Structure of the global plastic waste trade network and the impact of China's import Ban. Resour Conserv Recycl.
62. Wojdyr M (2010) Fityk: a general-purpose peak fitting program. J Appl Crystallograp (International Union of Crystallography) 43(5 PART 1):1126–1128.

Chapter 9

Mechanical Recycling of Packaging Plastics

Contents

INTRODUCTION

Plastics are becoming more and more in demand worldwide. By 2030, 236 million tons of plastic are expected to be in circulation annually, up from 236 million tons currently. Reusing or recycling plastics that are already in use is crucial to reducing environmental pollution by preventing the unintentional or intentional release of more polymeric materials into the environment. Of the polymers in flow in 2016, only sixteen percent were collected for recycling, forty percent were disposed of in landfills, and twenty-five percent were burned (Figure 9.1). European nations have stepped up their efforts lately to raise recycling rates. 2018 saw the collection of 29.1 million tons of post-consumer plastic waste in Europe. Even though only about one-third of this was recycled, it was still a doubling of the amount recycled and a 39% decrease in plastic waste exports outside the European Union (EU) from 2006 levels. 39.9 percent of this plastic flow was used for packaging.

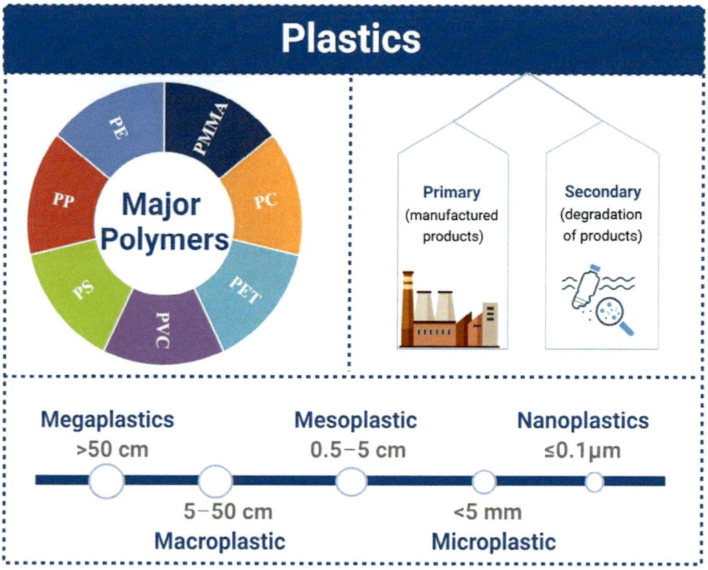

Figure 9.1. The main packaging polymers: PS (polystyrene), PE (polyethylene), PP (polypropylene), PVC (polyvinyl chloride), PET (polyethylene terephthalate), PC (polycarbonate), and PMMA (polymethyl methacrylate).

Source: *Chen Q, Zhao H, Liu Y, Jin L, Peng R. Factors Affecting the Adsorption of Heavy Metals by Microplastics and Their Toxic Effects on Fish. Toxics. 2023; 11(6):490. https://doi.org/10.3390/toxics11060490*

Since 42% of collected post-consumer waste in Europe is recycled, 40% is sent for energy recovery, and 19% is disposed of in landfills, packaging recycling is frequently more economically viable than other plastic market segments. In addition to promoting their use, plastics' stability also lessens their degradation potential. As a result, excess waste is dumped into the environment and landfill sites get saturated. Physical fragmentation has also resulted from the material's ubiquity and inconsistent disposal, introducing micro and nanoplastics into urban settings, water bodies, conservation areas, and our food chain. Landfill avoidance is largely achieved through dubious energy-from-waste strategies, but the EU Waste Directives' landfill taxes have helped slow this trend and raise recycling rates. The level of taxation varies by nation: of the 27 EU member states, 18 have landfill bans in place, while 24 have landfill taxes. Furthermore, according to the July 17–21, 2020, European Council conclusion, the use of non-recyclable packaging will be subject to taxes. In the UK, the cost of land-filling plastic waste is currently £94.15 ton^{-1}, an increase of 1345 percent from the landfill tax in 1996. In the UK, 47% of plastic packaging waste is collected for recycling, but only 43% of that waste is turned into useful feedstock, according to a 2018 report by The Waste and Resources Action Programme (WRAP).

Utilizing the United Kingdom as a model, forty percent of the waste gathered between 2015 and 2016 consisted of poly(ethylene terephthalate) (PET), twenty-two percent of polyethylene (PE), ten and a half percent of polypropylene (PP), and two percent each of polyvinyl chloride (PVC) and polystyrene (PS) (Figure 9.1). Table 9.1 lists the five polymers that are most frequently utilized in packaging. Food, cleaning supplies, and toiletries are packaged in bottles made of PET and high-density polyethylene (HDPE). PVC, low-density polyethylene (LDPE), and linear low-density polyethylene (LLDPE) are the main materials used to make packaging films. PET, HDPE, and PVC are the materials used to make plastic beverage bottles; however, PVC is being used less frequently due to legal pressure to outlaw it. LDPE and LLDPE are the common materials used to make single-use plastic bags. PS is mostly used in the food and consumer goods industries for packaging needs when it is in its expanded or solid state.

Table 9.1. *The five main packaging polymers by collection proportion and their main uses*

Polymer	Proportion of total waste collected from kerbside (%)	Applications in packaging
PET	40	Beverage bottles, trays, jam jars
HDPE and LDPE	22	Bottles, bags, bin liners, food wrapping material, squeeze bottles
PP	10.2	Bottles, straws, bottle caps

PVC	<2	Films, trays
PS	<2	Fast-food packaging, food packaging, disposable cutlery, consumer goods

Source: *Z. O. G. Schyns, M. P. Shaver, Mechanical Recycling of Packaging Plastics: A Review. Macromol. Rapid Commun. 2021, 42, 2000415. https://doi.org/10.1002/marc.202000415*

The plastics industry is mostly linear globally. More than half of the plastics that are produced, used, and disposed of are lost and never recovered. The need for increased production due to this disposal increases the planet's reliance on petroleum feedstock and the pollution that follows. A global push to convert the linear economy into a circular model is needed to meet consumption demands and protect the environment at the same time.

The creation of novel feedstocks for the plastics industry has received a lot of attention in the field of sustainable polymers; however, many of these novel polymers find it difficult to satisfy the demanding demands of being inexpensive, scalable, and possessing unique qualities. A circular economy model encourages the repurposing of plastics that are already in use while recommending the prudent use of our resources, including petroleum feedstocks. Reusing and reducing waste is essential, and biosourced feedstocks that don't harm our farming sector will keep expanding, but recycling plastics is essential to cutting down on plastic waste. Recycling needs to get better since the goal for complete circularity is zero land-filling of collected waste.

The necessity of single-use plastics has been brought to light by the COVID-19 pandemic. Concerns about virus-contaminated products and possible health hazards raise plastic consumption, instill reluctance in consumers to reuse products, and reduce recycling rates. Personal protective equipment (PPE) is starting to show up in municipal and institutional waste streams after previously being managed through specific medical waste streams. When mask use is encouraged or mandated, it is frequently impossible to prevent increases in PPE waste, which poses new difficulties for plastic production and recycling. Because PPE is more common and crude oil prices are almost at all-time lows, virgin plastic is preferred over more expensive recycled plastic. Predictions regarding the long-term effects of COVID-19 on our plastics economy are complicated by uncertainties regarding second spikes and long-term behavioral changes. However, managing plastic waste requires a systems approach due to these socio-material challenges. To keep the materials we rely on in supply, it is essential to preserve the polymers in their highest value state. Degradation, sorting, and plastic contamination continue to be the main obstacles to effective recycling. Primary, secondary, tertiary, and quaternary recycling are the four primary categories of recycling processes (Table 9.2). Extruding preconsumer polymer or pure polymer streams is the primary recycling process. Sorting polymer waste streams, reducing the size of polymer waste, and extrusion are the steps involved in secondary recycling. Many polymers can go through multiple cycles of primary and secondary mechanical recycling without experiencing performance loss if processing conditions are properly controlled. When polymers are no longer suitable for these simple mechanical recycling techniques, tertiary recycling is employed. When used in conjunction with conventional recycling techniques, chemical recycling can be very valuable if it is selective (returning the polymer to its monomeric feedstocks) as opposed to non-selective (pyrolytic or hydrocracking strategies). Plastics that are unsuitable for other forms of recycling are referred to as quaternary recycling and are used in pyrolysis energy recovery processes. Even though it has little value, quaternary recycling could produce greenhouse gases and have unforeseen societal repercussions.

Table 9.2. *Common definitions of plastic recycling*

ASTM D7209 definitions (withdrawn 2015)	ISO 15270:2008 standard definitions	Example
Primary recycling	Mechanical recycling	Bottle to bottle closed loop recycling
Secondary recycling	Mechanical recycling	Recycling into lower value plastic
Tertiary recycling	Chemical recycling	Depolymerization of polyesters
Quaternary recycling	Energy recovery	Pyrolysis

Source: *Z. O. G. Schyns, M. P. Shaver, Mechanical Recycling of Packaging Plastics: A Review. Macromol. Rapid Commun. 2021, 42, 2000415. https://doi.org/10.1002/marc.202000415*

Improved plastic circularity is obviously needed, and for several decades, chemists, materials scientists, and engineers have been addressing this issue. The development of biodegradable and bio-based plastics demonstrates the most concentrated effort to increase the sustainability of plastics. While bio-based plastics are frequently drop-in substitutes made from renewable carbon sources, biodegradable plastics are intended to break down naturally through enzymatic or hydrolytic degradation. The practice of selective chemical recycling of polymers has become increasingly popular in recent times due to the possibility of infinite recyclability through depolymerization into the original monomers. The field of biological recycling has expanded as well, generating downcycle feedstocks through enzymatic degradation and fermentation. Even though chemical and biological recycling are recognized as "green" recycling techniques, their sustainability cannot be determined without conducting thorough and impartial life-cycle assessments. According to our preliminary analysis, mechanical recycling will continue to be the most efficient way to recycle plastics in terms of cost, time, carbon footprint, and environmental impact.

Innovation in extrusion technologies has led to further improvements in polymer reprocessing. To enhance the quality of the polymer melt, extruders can be designed with sections for degassing, softening, drying, and filtering extrudate. Degassing sections allow several volatile compounds contained in polymer melts to escape by being vacuumed or by opening vents from the barrel. Eliminating these volatiles increases the value of recyclate by reducing acidolysis and hydrolysis and improving the odor of the polymer melt. In order to enhance blend homogeneity, mechanical, and optical qualities, and to eliminate larger, non-volatile impurities like dust or gel particles, polymer melts can also be filtered. Melt filters, which can include screen-changers like slide plates, woven screens, or filter cartridges, are selected based on a particular extrusion contamination. Extruder lengthening must be weighed against the potential increase in system dwell times that could worsen chain scission. Specific degradation mechanisms need to be taken into account when designing recycling systems. (Schyns and Shaver, 2021)

9.1 MECHANICAL RECYCLING OF POLY(ETHYLENE TEREPHTHALATE)

Virgin PET (vPET) possesses superior barrier, mechanical, and processability qualities. Consequently, a lot of PET is used for packaging, a lot of which is food grade. Virgin PET is ductile and has high elongation at break values of >80%, which, when mechanically recycled, have been found to rapidly reduce by a factor of 4 (Figure 9.2). The abrupt decline in the material's properties is caused by hydrolytic scission, thermo-oxidative, and thermo-mechanical chain degradation.

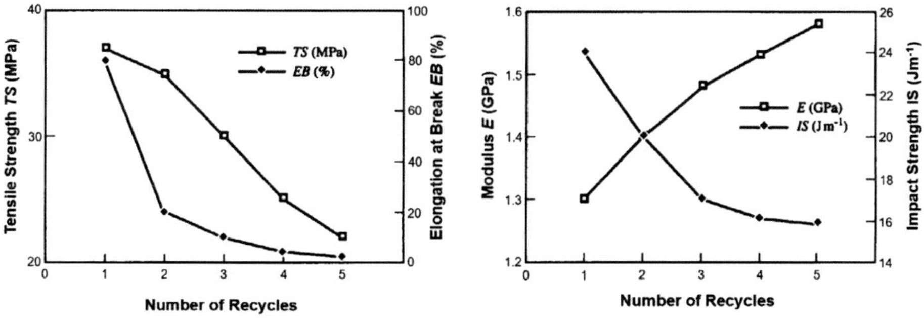

Figure 9.2. Left: Tensile strength and elongation at break versus number of recycles for PET. Right: Young's modulus and impact strength versus number of recycles for PET.

Source: *Z. O. G. Schyns, M. P. Shaver, Mechanical Recycling of Packaging Plastics: A Review. Macromol. Rapid Commun. 2021, 42, 2000415. https://doi.org/10.1002/marc.202000415*

9.1.1 Degradation Mechanisms of Poly(ethylene terephthalate)

In addition to producing possible byproducts like carbon dioxide, water, and carboxylic acid or aldehyde end groups, chain scission lowers the molecular weights of polymers. Shorter chain lengths embrittle the polymer, lessen viscosity, and decrease polymer elasticity. Processing equipment must therefore be calibrated in order to reduce melt viscosity. Furthermore, in post-consumer polymer melts, contaminants added during the PET's life cycle encourage chain scission reactions. Specifically, during extrusion, trace amounts of poly(vinyl alcohol) (PVA), poly(lactic acid) (PLA), and PVC can leach acids and encourage the hydrolysis or acidolysis of PET (Scheme 9.1). PET's chains may also crosslink, which has a negative effect on the quality of recycled materials. Large-scale recycling is hampered by the viscosity increases brought on by crosslinking reactions because higher torques can harm processing machinery. Due to the possibility of reactor damage during pyrolysis, PET is typically destined for landfill after undergoing irreversible damage from mechanical recycling.

PET chains break down as a result of carbon to hydrogen transfers or free radical attacks (Scheme 9.2). A macroradical is created by the extruder's stress and heat, which reacts with oxygen to form a peroxy-based radical. Radical hydrogen abstraction has the ability to reform macroradicals and generate hydroperoxide, which breaks down into two new radical species capable of starting new macroradical chains.

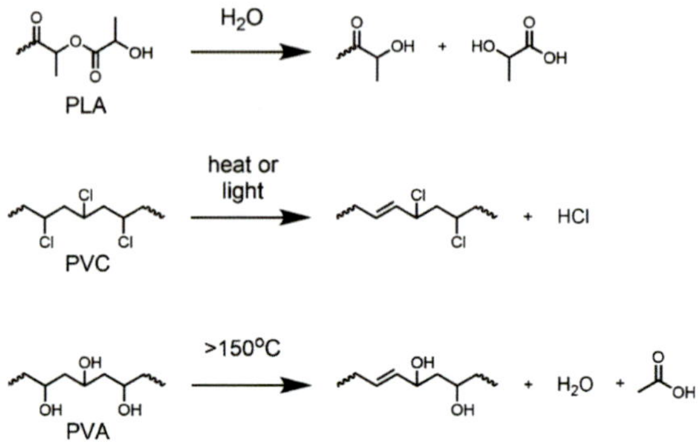

Scheme 9.1. Reaction pathways PLA, PVC and PVA can undergo during extrusion which can induce acidolysis or hydrolysis of PET.

Source: *Z. O. G. Schyns, M. P. Shaver, Mechanical Recycling of Packaging Plastics: A Review. Macromol. Rapid Commun. 2021, 42, 2000415. https://doi.org/10.1002/marc.202000415*

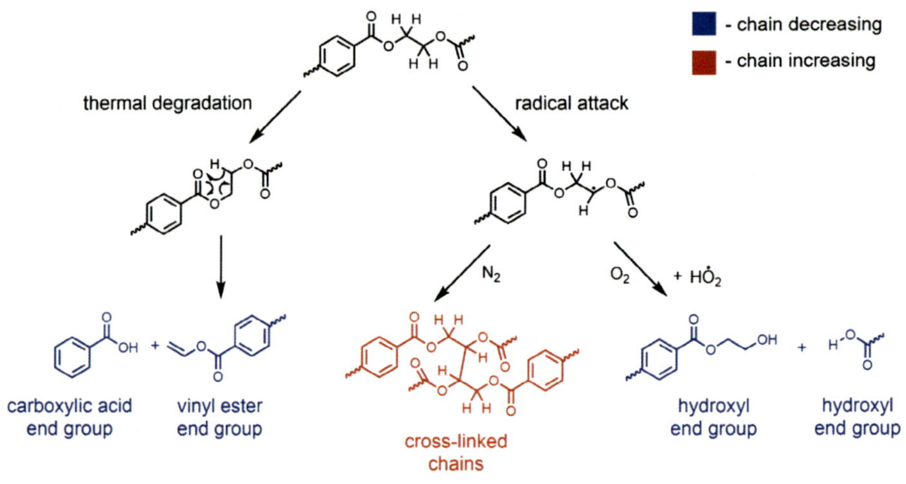

Scheme 9.2. Reaction pathways PET undergoes during mechanical recycling.

Source: *Z. O. G. Schyns, M. P. Shaver, Mechanical Recycling of Packaging Plastics: A Review. Macromol. Rapid Commun. 2021, 42, 2000415. https://doi.org/10.1002/marc.202000415*

Thermo-oxidation is facilitated by high carboxylic acid end group concentrations. Therefore, higher mechanical recycling cycles will result in faster chain scission rates because of thermo-oxidation. PET experiences hydrolysis, transesterification, esterification, and crosslinking reactions in addition to this free radical degradation (Scheme 9.2). Furthermore, radical crosslinking can happen when there is no oxygen present. The microstructure of the polymer is impacted by these degradation processes. Three fractions make up PET's microstructure: crystalline, rigid amorphous, and mobile amorphous fractions. Between the ordered crystalline and mobile amorphous fractions is the partially ordered rigid amorphous fraction. The most easily broken amorphous phase releases short chains that fold into intercrystalline domains and start the crystallization process. The crystalline domains are seen to thicken and reorganize, and as recycling increases, new, smaller-average-sized crystalline domains are created, increasing embrittlement and Young's modulus.

9.1.2 Improvements to Poly(ethylene terephthalate) Recycling

Extensive research has been conducted on enhancing PET mechanical recycling. A large body of current research stems from the pioneering work of La Mantia and Vinci. Their paper explores the number of extrusions PET can withstand before it becomes unsuitable for (re)manufacturing. Changes in the material properties become less noticeable after three recycling cycles (Figure 9.2) because degradation reactions take longer because of the size of the polymer. This finding supports the idea that the degradation kinetics are influenced by molecular weight, with the larger the polymer chain, the faster the degradation kinetics. Additives are frequently used in PET recycling in order to safeguard chains, encourage chain extension, enhance mechanical properties, and increase the value of the recycled material.

Extrusion in inert atmospheres and metal-based stabilizers like lead phthalate or tin mercaptide have been reported as ways to get around the effects of thermal oxidation. Reversing scission can also be achieved by reacting the hydroxyl and carboxyl end groups of broken and intact chains via solid state polymerization (SSP). Organic phosphates and other radical scavengers are used to stop the radical attack on chains. Gooneie et al.,stated that (1-oxido-2,6,7-trioxa-1-phosphabicyclo(2.2)) was effective.

An addition that also adds flame retardant qualities is 2)octan-4-yl) methoxy)dibenzo(c,e)(1,2)oxaphosphinine 6-oxide (DP). The authors do not address the possible effect of radical attack on β-scission; instead, they focus on how the breakdown products of DP function as radical quenchers and postpone the onset of PET crosslinking. Reactive chain extension holds the most promise for raising the quality of PET recycles out of all of these techniques. (Schyns and Shaver, 2021)

9.1.3 Chain Extender Use in Poly(ethylene terephthalate)

Chain extenders are a cheap and practical way to promote reactions inside processing equipment and undo some of the damage caused by PET chain degradation. These oligomers aim to fuse back together chains and increase molecular weight by targeting reactive chain ends.

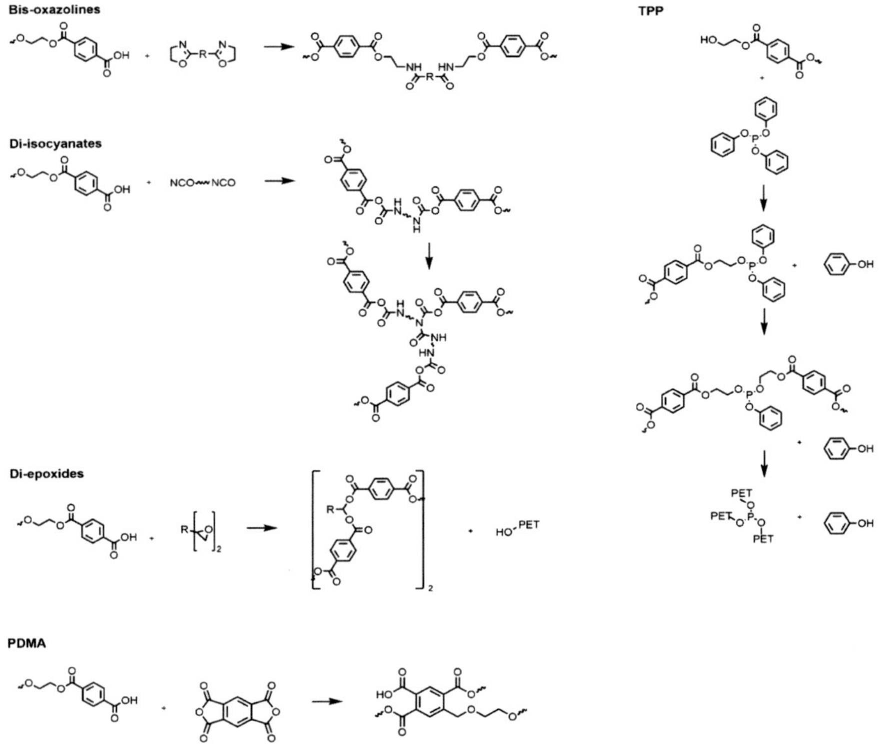

Scheme 9.3. Common reactions between chain extenders: bis-oxazolines, di-isocyanates, di-epoxides, pyromellitic dianhydride (PMDA), triphenyl phosphite (TPP) and PET.

Source: *Z. O. G. Schyns, M. P. Shaver, Mechanical Recycling of Packaging Plastics: A Review. Macromol. Rapid Commun. 2021, 42, 2000415. https://doi.org/10.1002/marc.202000415*

To maximize the number of reactions between chains, the extenders can have two or more reactive groups, resulting in both branched and linear topologies. Additionally, chain extenders can be used to make two different grades (i.e. PETs with notably distinct molecular weights or structures) by reacting the chains in tandem and undoing the chain scission brought on by thermo-oxidative processes. Oxazolines, isocyanates, epoxides, lactams, hydroxyls, carboxylic acids, and organic phosphates and phosphites are examples of common chain extenders. Scheme 9.3 illustrates some examples of widely used multifunctional extenders. Commercial extenders like Joncryl derivatives have these functionalities. Dynamically measuring in-situ length increases in chains can be achieved by tracking increases in torque values. Cavalcanti et al., describe effectively extending the chains through their carboxyl and hydroxyl terminal groups using triphenyl phosphite (TPP). Incarnato et al., explain how long chain branching of PET can be induced by using pyromellitic dianhydride (PMDA) as a chain extender in reactive extrusion (Figure 9.3).

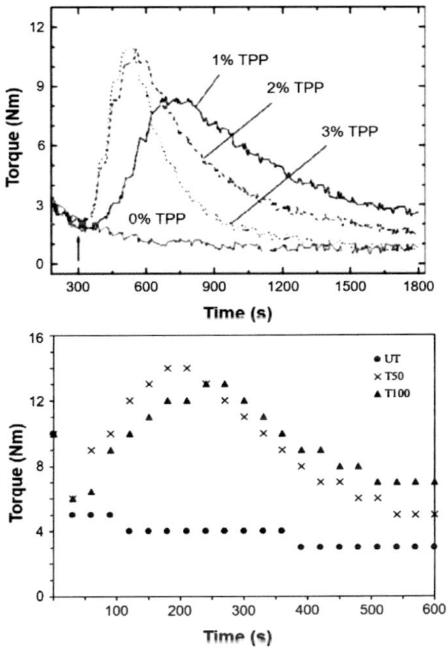

Figure 9.3. Top: The effect of TPP on the torque of PET processed at 260 °C and 60 rpm. Bottom: The effect of PDMA at 0.5% and 1% on torque for PET processed at 280 °C at 100 rpm.

Source: *Z. O. G. Schyns, M. P. Shaver, Mechanical Recycling of Packaging Plastics: A Review. Macromol. Rapid Commun. 2021, 42, 2000415. https://doi.org/10.1002/marc.202000415*

Increased intrinsic viscosity (*), a drop in melting temperature, and an increase in cold crystallization temperature are all facilitated by the existence of entangled networks resulting from PMDA branching. When PDMA is used, virgin PET's torque is increased more quickly than when TPP is used, by comparable factors (≈2-3) and at lower weight percentages (Figure 9.3). This discrepancy is a little unexpected because multifunctional extenders typically perform better than bifunctional extenders, indicating that extrusion conditions and extender reactivity are important factors. Moreover, PET, polyurethanes, PLA, and other speciality polymers have been chain extended using PMDA.

These days, chain extenders are frequently found in industrial settings. Tavares et al., show that adding the multifunctional epoxide oligomer Joncryl PR 002 to recycled PET works better than it does in virgin streams. They state that the reduction in molecular weight caused by extrusion-induced chain scission can be reversed with less than 1% of the Joncryl additive. Conversely, Duarte et al., claim that a combination of Joncryl 4368 and 4370 raises the molecular weight of vPET more successfully than post-consumer recycled PET; however, they also concur that the addition's 1% is needed to counteract process-induced degradation's effects. The increased number of reaction sites between end groups accounts for the enhanced efficacy of multifunctional chain extenders like PDMA, TPP, and Joncryl additives.

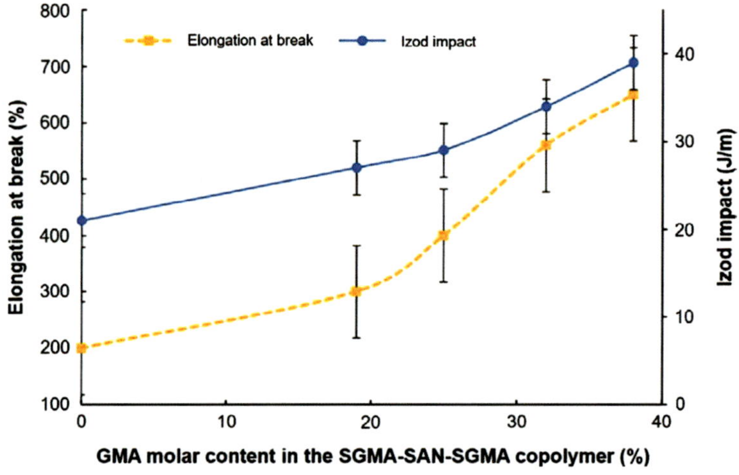

Figure 9.4. Elongation at break and Izod impact versus GMA content during chain extension of rPET. Reproduced by permission.

Source: *Z. O. G. Schyns, M. P. Shaver, Mechanical Recycling of Packaging Plastics: A Review. Macromol. Rapid Commun. 2021, 42, 2000415. https://doi.org/10.1002/marc.202000415*

Chain extension is also dependent on the type of reactive site: di-isocyanates have a higher degree of reactivity than di-epoxides with PET terminal groups. Benvenuta Tapia et al., describe the production of poly(styrene-co-glycidyl methacrylate-b-styrene-co-acrylonitrile triblock copolymers) via reversible addition-fragmentation chain-transfer (RAFT) polymerization. An epoxide-based chain extender raises the molecular weight and viscosity of rPET in reactive extrusion and matches similar thermal properties to vPET; mechanical properties and the concentration of glycidyl methacrylate (GMA) are directly correlated. (Figure 9.4).

Although encouraging, it's crucial to remember that these chain extenders might not be appropriate for the production of food-grade PET because of possible migration problems and regulatory concerns. Small oligomers that have not yet reacted, phosphites, or other molecules may seep through packaging materials and cause cytotoxicity in goods. Moreover, they might interfere with crystallinity or crystallization rates. The poor thermal stability of many chain extenders increases the risk of multiple extrusions. As most clearly seen in TPP-extended PET, chain extension occasionally results in an excess of acid groups being produced during the extension process, which in turn raises degradation rates.

The majority of packaging research endeavors center on enhancing the mechanical recycling procedure. While adding additives can enhance the recycled PET's qualities, it can also make recycling more difficult because it is unclear exactly what effects they will have, particularly near the end of the product's life. In fact, chain extenders are being used more commercially and are improving degraded recycled materials despite these risks. To comprehend the impact on a circular plastics economy, future research should concentrate on the applicability and mechanism of action of extenders over higher numbers of extrusion cycles. Reprocessing chain-extended PET continuously has not been thoroughly studied, but it may have an effect on recycling variability or degradation mechanisms. (Schyns and Shaver, 2021)

9.2 MECHANICAL RECYCLING OF POLYOLEFINS

When polyolefins are mechanically processed, they can degrade into a range of carbon-based molecules, including acids, cyclic ethers and esters, alcohols, aldehydes, ketones, and short-chain hydrocarbons. The three primary polyolefins used in packaging—HDPE, LDPE, and PP—are all made entirely of carbon and hydrogen bonds. They differ greatly from other packaging polymers and have distinct breakdown mechanisms from one another (Scheme 9.4). Along the polymer backbone, radical attack does produce macroradicals; however, after the initial attack, β-scission of the polymer chain produces other macroradicals. Shorter, branched, or crosslinked polymer chains can be formed by recombination or dissociation of these radicals, which can result in intramolecular radical transfer to produce a variety of other radical species. When polymer macroradicals react with any oxygen-based radical, carboxylic acid, carbonyl, and hydroxyl end groups are introduced, which may encourage additional degradation. While PP is more prone to chain scission, HDPE and LDPE exhibit higher rates of crosslinking. Polyolefin chains are mainly attacked during the amorphous phase of the polymer, when oxygen finds it difficult to permeate through strongly crystalline regions.

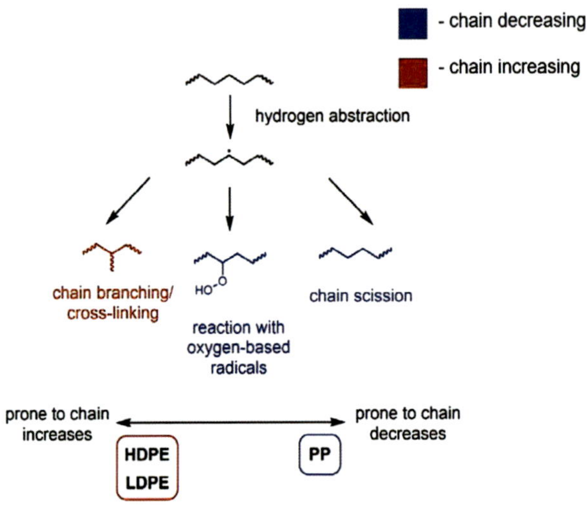

Scheme 9.4. General reaction scheme of polyolefins undergoing extrusion.

Source: *Z. O. G. Schyns, M. P. Shaver, Mechanical Recycling of Packaging Plastics: A Review. Macromol. Rapid Commun. 2021, 42, 2000415. https://doi.org/10.1002/marc.202000415*

9.2.1 Polyethylenes

PEs are structurally straightforward polymers whose extraordinarily high strengths have encouraged their widespread application in packaging. Compared to LDPE (0.924 g cm^{-3}), which is more loosely packed, HDPE (0.952 g cm^{-3}) is stronger but less elastic due to its high degree of crystallinity. Film production frequently uses LDPE because it is less crystalline than HDPE and is more transparent and elastic. As will be explained below, the densities of the two polymers can change how they degrade during extrusion.

High-Density Polyethylene

Depending on the extrusion conditions, HDPE exhibits competing mechanisms of chain branching and scission during extrusion. Pinheiro et al., report that the formation of stable carbonyl end groups through thermo-oxidative chain scission of the polymer backbones causes an increase in the carbonyl concentration in HDPE in oxygen-rich environments. In contrast, chain scission and the creation of double bonds take precedence over chain branching in low oxygen environments due to shear-induced degradation. The polymer chains gradually get shorter during continuous extrusion, which lessens their vulnerability to shear stresses.

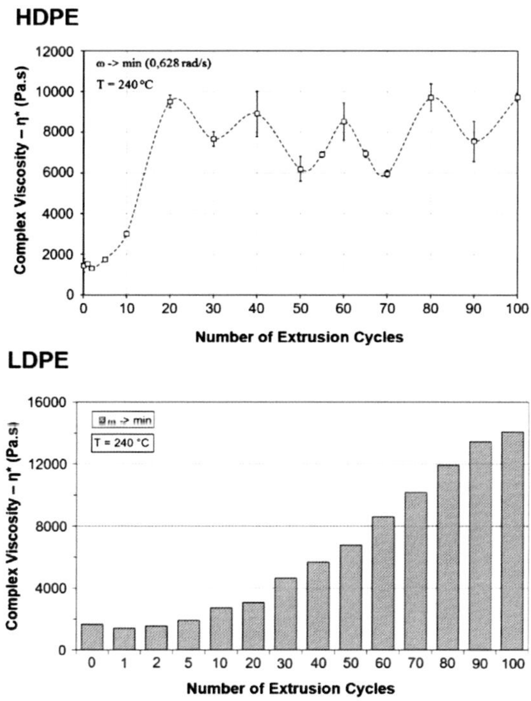

Figure 9.5. Complex viscosity versus number of extrusion cycles for top: HDPE and bottom: LDPE.

Source: *Z. O. G. Schyns, M. P. Shaver, Mechanical Recycling of Packaging Plastics: A Review. Macromol. Rapid Commun. 2021, 42, 2000415. https://doi.org/10.1002/marc.202000415*

The longer chain macroradicals can still attack the short chains at double bonds, though. Long chain branching is the result of a reaction between longer macroradicals and shorter polymer chains with double bonds. Using less direct experimental techniques, researchers have also confirmed simultaneous chain scission and branching. Oblak et al., executed a hundred consecutive HDPE extrusions (Figure 9.5). They observe that the material undergoes substantial structural alterations in the first thirty cycles, after which an oscillatory plateau in its properties is seen. They hypothesize that during the first thirty cycles of extrusion, chain branching is the predominant chain chemistry, followed by chain scission between 30 and 60 cycles, and above 60 cycles, chain crosslinking predominates, increasing viscosity by up to five times. Reducing screw speeds prevents this crosslinking, while increasing screw speed can improve chain scission reactions. As a result, HDPE recycling presents less of a challenge than PET recycling.

Although extruder control can reduce HDPE extrudate degradation, stabilizing additives are also frequently added. HDPE can be reactively extruded via grafting, crosslinking, or direct functionalization after polymerization. Through the use of radiation, moisture, and thermochemical techniques, crosslinking can be controlled. Peroxides have been utilized to minimize chain scission and preserve mechanical properties by causing crosslinking between chains. In order to reduce the chain scission and disproportionation side reactions caused by peroxides, Kim and Kim used trimethylol propane triacrylate (TMPTA) (Scheme 9.5).

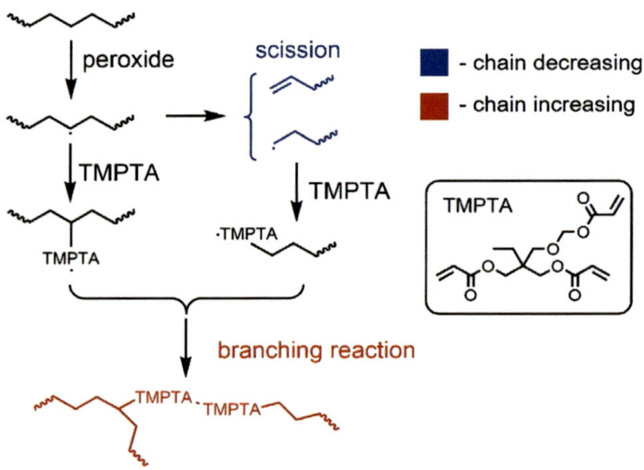

Scheme 9.5. Reaction scheme for controlled degradation of PE using peroxides and trimethylol propane triacrylate (TMPTA).

Source: *Z. O. G. Schyns, M. P. Shaver, Mechanical Recycling of Packaging Plastics: A Review. Macromol. Rapid Commun. 2021, 42, 2000415. https://doi.org/10.1002/marc.202000415*

While the melt flow index (MFI) decreased by more than a factor of 10 upon adding TMPTA, indicating decreased crosslinking, Su and Huang report increased branching with TMPTA use in PP. It is also reported that chain linking with epoxides, diamines, acrylates, phosphates, and silanes can be done in blended systems. Additional attempts to enhance HDPE involve grafting with polymerizable monomers or initiators, such as acrylics, vinyl silanes, maleic anhydrides, styrenics, oxazolines, and parabenzoquinone; however, in these cases, specialty PEs are frequently the goal rather than packaging polymers.

Low-Density Polyethylene

Because of its highly branched structure, LDPE reduces density and inhibits packing. Branched structures have more end groups and less thermal stability, making them more prone to crosslinking or chain branching reactions during extrusion. Rheology, thermo-rheology, GPC, and FT-IR studies show increases in LDPE molecular weights during recycling (Figure 9.5). There is still a competition between chain scission and chain branching during LDPE reprocessing, even though degradation mechanisms that increase molecular weight predominate. LDPE exhibits initial chain cleavage, recombination, and chain branching during the extrusion process. Radical chain reactions will increase scission rates when oxygen is present. Higher rates of chain branching can be induced by macroradicals cross-reacting in inert atmospheres and low oxygen environments. As it branches, LDPE's affinity for chain scission declines because of a reduction in the gyration radius. Since LDPE is mostly used in packaging with short circulation times, it is typically not stabilized against degradation during reprocessing, which causes issues for recycling and circular recovery systems.

Stabilizer Use in Polyethylenes

Stabilizers are frequently needed for PEs in melt reprocessing to prevent thermo-oxidative degradation and preserve recycle quality. These additions also strengthen the stability of UV radiation. Since ketones and peracids, which can form as degradation products under UV and thermo-oxidative stresses, form at different rates, combinations of UV inhibitors, thermal stabilizers, and anti-oxidants are essential. In order to reduce evaporation and volatization during processing, additives are selected based on their solubility, dispersion, and stabilization in the host matrix. Common antioxidants, phenols like Irganox 1010 serve as alkyl peroxy radical traps as well as hydrogen bond stabilization agents. Antioxidants based on phosphates and phenols work better together than when a phenolic compound is used alone. There may also be antagonistic interactions between stabilizing systems. Nitroxyl radicals generated from photooxidation react with the phenolic groups in systems stabilized with hindered amine light stabilizers (HALS) and hindered phenols, upsetting the redox system in charge of stabilization. Peña et al., cite the majority of antagonistic effects being mitigated by a three-additive system and recommend carbon black as a third additive because of its adsorption-desorption capabilities. Once more, choices made to lessen an immediate issue spread to other uses. For example, the use of carbon black in the previous example hinders sorting and darkens recycled materials, which lowers their quality and cost in later processing cycles.

Due to the migration of antioxidants and potential carcinogenicity, research efforts have focused on the development of safer, sustainable or natural antioxidant alternatives or the inhibition of migration through covalent bonding to the polymer chain. Many of these risks are unknown, with limited cytotoxicity data known for antioxidants or metal-based catalysts often used in their synthesis. Natural antioxidants include bio-derived phenols such as caffeic acid, ascorbic acid, α-tocopherol (vitamin E), curcumin, quercetin, and β-carotenes. Inclusion of 0.2 wt% vitamin E provides a higher level of protection than the same weight percentage loading of synthetic counterparts Irganox 1010 and 1076. Similarly, studies by Tatraaljai et al., and Zia et al., on the inclusion of curcumin during PE processing showed improved chain protection compared to Irganox with improved water barrier properties. In an attempt to promote the use of bio-derived antioxidants, Iyer et al., discuss the inclusion of very low cost antioxidants stemming from natural agro-wastes (Figure 9.6). The agro-waste stabilisers are not as effective as the commercial alternative Irganox 1010 but still protect against embrittlement of LDPE (Figure 9.6).

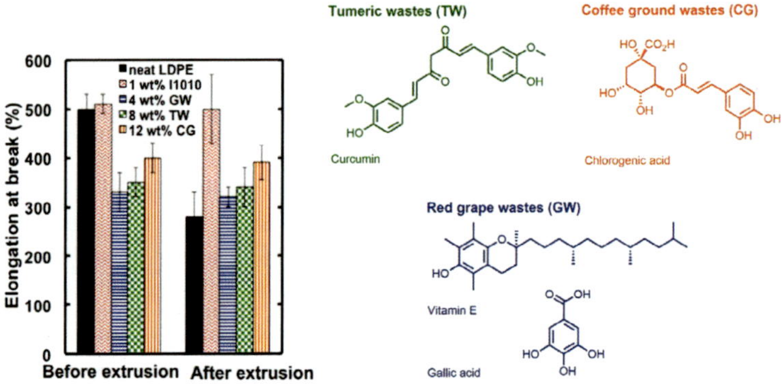

Figure 9.6. Elongation at break values for LDPE extruded with antioxidant agro-wastes.

Source: *Z. O. G. Schyns, M. P. Shaver, Mechanical Recycling of Packaging Plastics: A Review. Macromol. Rapid Commun. 2021, 42, 2000415. https://doi.org/10.1002/marc.202000415*

The use of stabilizers can increase the lifetime of plastic products and can provide protection against thermo-oxidation during mechanical recycling. As is now a common feature in this section, there is little understanding of the impact of these stabilizers over multiple processing cycles.

9.2.2 Polypropylene

Polypropylene (PP) is a necessary packaging polymer due to its high tensile strength, glossy appearance, and good optical qualities, but it also poses a problem for landfills due to its high volume to weight ratio. Tertiary carbons along the polypropylene backbone increase flammability, introduce a more reactive methine functional group, and make the material more susceptible to shear degradation. Due to thermo-mechanical and thermo-oxidative chain scission, repeated recycling of PP lowers molecular weights and increases crystallinity, which results in higher elastic moduli and less elongation at break.

The breakdown of PP radicals is remarkably similar to that of PE, as hydroperoxides remove hydrogen atoms from polymer chains. Separating catalyst residue and additives in the amorphous phase, the spherulitic structure concentrates degradation at these spherulite boundaries (Figure 9.7). The resulting macrorodicals combine or go through β-scission. The production of peroxyradicals is increased due to the higher reactivity of the methine proton, but the resulting radical is stabilized in comparison to PE. PP can withstand processing for up to five extrusion cycles at lower temperatures. In contrast to the chain branching and crosslinking seen in PEs, substantial chain scission limits performance beyond five cycles and/or at higher temperatures.

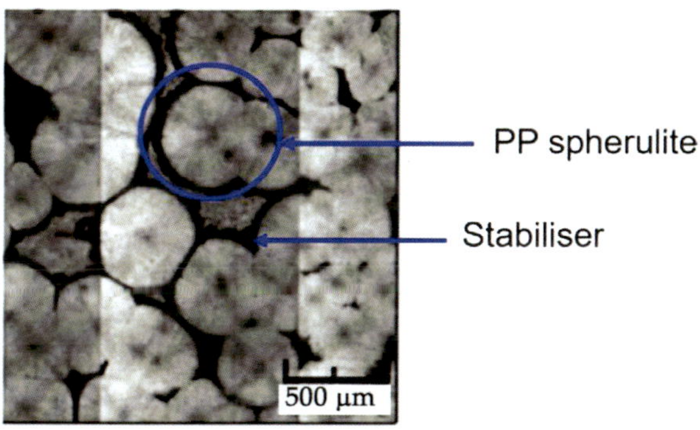

Figure 9.7. Optical image showing the spherulitic structure of PP.

Source: *Z. O. G. Schyns, M. P. Shaver, Mechanical Recycling of Packaging Plastics: A Review. Macromol. Rapid Commun. 2021, 42, 2000415. https://doi.org/10.1002/marc.202000415*

PP also needs stabilization in order to fend off the radical attack on the polymer chains during extrusion. Phenolic and hindered amine systems are both prevalent. The distribution of stabilizers and antioxidants is influenced by the spherulitic structure of PPs. The uniform distribution of stabilizers throughout the polymer matrix made possible by smaller spherulites enhances radical capture reactions and, consequently, efficacy. Taniike et al., reported that the distribution of the stabilizer within the PP polymer matrix is independent of the type of antioxidant or stabilizer. Virgin wine wastes, tannin wine seed extracts, and seed polyphenol extracts are examples of bio-derived agro-wastes that can stabilize the product and postpone the start of MFI oxidation degradation for up to 100 hours. PP and recycled PP (rPP) can both be successfully stabilized at concentrations of two to five weight percent by lignin, a naturally occurring phenolic polymer. In order to make the polymer matrix more rigid, lignin serves as a filler and an antioxidant. When lignin and Irganox 1010 are combined, there are reported synergistic effects because there is more commercial antioxidant that diffuses throughout the polymer matrix. The applicability of these stabilization techniques to protect over multiple recycling life cycles is unknown because they have not been tested in extrusion conditions.

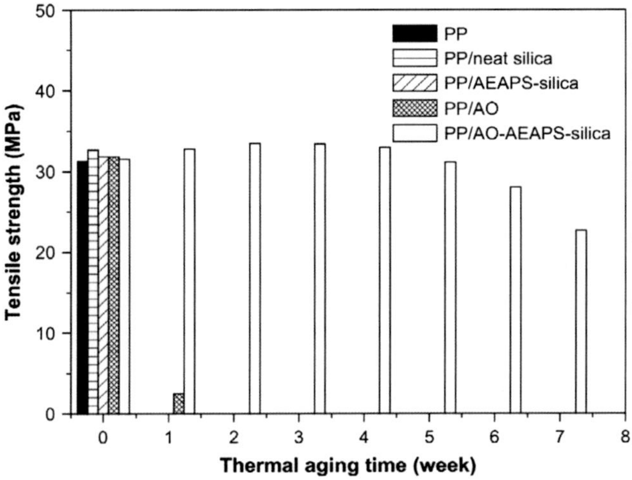

Figure 9.8. Effect of thermal aging on tensile strength of PP samples stabilized with AO and coupled to silica using AEAPS.

Source: *Z. O. G. Schyns, M. P. Shaver, Mechanical Recycling of Packaging Plastics: A Review. Macromol. Rapid Commun. 2021, 42, 2000415. https://doi.org/10.1002/marc.202000415*

Because stabilizers are localized in more mobile, amorphous regions, their migration in PP is also poorly studied, potentially posing health risks to humans and the environment. Gao et al., sought to stop the migration and volatization of antioxidants explain how antioxidants are grafted onto nanosilica particles. Following accelerated thermal aging, decreased antioxidant loss enhanced the stabilizing effects (Figure 9.8). Antioxidant migration from the polymeric material can also be intentionally utilized. Antioxidant defense of the polymer matrix and contained food products can be achieved through the controlled release of antioxidants from polypropylene matrices. Following their 8 weight percent incorporation into PP, thymol and carvacrol both released their active antioxidants into food-like simulants and stayed in the polymer for a period of 15 days.

9.2.3 Polyolefin Blends

The mixed polyolefin (MPO) waste fraction that is separated using flotation sorting of plastic wastes has made polyolefin blending an active research area for a number of years. Recycled MPOs are easier to make, require less sorting, and are far less expensive than pure PP and PE recycles. Nevertheless, due to inadequate adhesion between polymer phases, the generated recyclates have greatly reduced mechanical properties. Because of adhesion problems, these blends (iPP:PE blends in Figure 9.9) cannot be used for manufacturing because of their lower Young's moduli and decreased elasticity. The presence of noncrystalllizable oligomers at the interphases formed during synthesis can be used as an explanation for poor adhesion. Adhesion across MPO interphases can be promoted and the presence of these oligomers can be reduced by selecting the right catalyst during polyolefin synthesis. Research by Van Belle et al., demonstrates that changing the majority phase can change the mechanical behavior of polyolefin blends. Tensile behavior in blends of LDPE/PP, LDPE/HDPE, and LLDPE/HDPE is transitional; it will eventually transfer to the majority phase. PP/HDPE blends exhibit brittle behavior as opposed to the shear yielding observed in mono-polymers.

Recently, Eagan et al., report synthesizing a novel polyethylene-block-isotactic polypropylene (PE-*b*-iPP) copolymer to successfully "stitch" two polyolefin polymer phases together (Figure 9.9). Because of the previously mentioned interlocking entanglements between the two phases, the resulting copolymer stabilized blend exhibited significant improvements in its tensile properties when compared to conventional unstabilized PE-iPP blends. As blending agents for PP and PE, this research group has more recently introduced polyethylene grafted with isotactic polypropylene copolymers (PE-g-iPP), which has comparable compatibilizing effects to multi-block copolymers.

Rather than using living polymerization, the grafted compatibilisers are created by copolymerizing macromonomers with a hafnium pyridylamido catalyst. These novel copolymer compatibilizers offer compatible MPO wastes an economically advantageous solution. For the recycling industry, successful compatibilization of polyolefins is highly desirable. It is possible to improve the properties of one polyolefin by using those of another. Although HDPE is more affordable and has better mechanical qualities than LDPE, it produces films that are of lower quality and have more haziness and less toughness. It is still difficult to develop a polyolefin-wide blending technology that is composition-independent. To find the ideal ratios, research has to test a wide range of blend compositions, underscoring the need for methodical approaches to better polymer waste stream sorting procedures. (Schyns and Shaver, 2021)

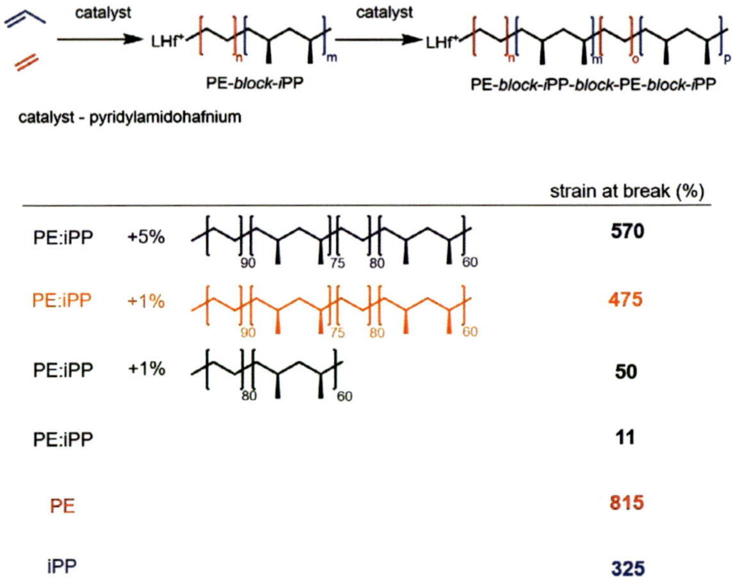

Figure 9.9. Synthesis of PE/iPP copolymers using an isoselective pyridylamido-hafnium catalyst. Tensile testing of PE/PP blends stabilized with different wt% of PE:iPP copolymer.

Source: *Z. O. G. Schyns, M. P. Shaver, Mechanical Recycling of Packaging Plastics: A Review. Macromol. Rapid Commun. 2021, 42, 2000415. https://doi.org/10.1002/marc.202000415*

9.3 MECHANICAL RECYCLING OF POLY(VINYL CHLORIDE)

PVC is widely used in the food and building industries due to its exceptional oxygen and water barrier qualities, strength, light weight, and flexibility. Nevertheless, PVC degradation nears the end of its useful life releases products that have been chlorinated into the environment. Degradation processes that take place during recycling include hydrodechlorination, crosslinking, functionalization of chains, and thermo-oxidation. Pyrolytic treatment of PVC is generally not recommended because it may release a highly toxic and corrosive gas called HCl, which could have an effect on the environment (i.e. acid rain) as well as plants (i. e. reactor corrosion). Heat- and photo-induced stress both cause PVC to become unstable. For PVC waste streams to be suitable for mechanical recycling, they must be pure and highly stabilized against undesired degradation.

9.3.1 Degradation Mechanisms of Poly(vinyl chloride)

The formation of internal double bonds during hydrodechlorination is facilitated by thermal and optical degradation, which considerably reduces mechanical strength. Polyene sequences are produced along the chain as a result of the first double bond formation, which stimulates subsequent reactions. Light is absorbed by these polyene segments, discoloring the polymer. Double bonds help to facilitate PVC chain crosslinking through reactions brought on by high degradation levels and an oxygen-deficient environment. Numerous additional chain structures and functionalities can be introduced by secondary reactions, particularly when oxygen is present. Radicals are produced when the C-Cl bond cleaves or when hydrogen atoms are abstracted. The oxygen in the surrounding environment reacts to produce oxygenated species, which ultimately form carboxylic acid end-groups. Additionally, the chain's polyene segments may react with oxygen. Nevertheless, PVC polyene formation takes place in both N_2 and O_2, making it difficult to control degradation and impractical to completely eradicate substantive degradation. Contaminants may act as a catalysts for the hydrodechlorination and mechanical performance losses, just like they do with other packaging polymers. (Schyns and Shaver, 2021)

9.3.2 Stabilization of Poly(vinyl chloride)

Stabilizers are typically added to PVC during extrusion to counteract the release of HCl and shield polymer chains from harm. In addition, certain stabilizers can restore chlorine atoms that have been lost from the polymer backbone. Stabilizers are used up in the polymer's processing during its first life, so recycling calls for additional stabilization. Metal-based stabilizers include lead, zinc-calcium, sulfur organotins, and metal soaps. Since a large portion of this legacy material is still in use, even though it is no longer in use (as of 2015), the presence of both lead-based and sulfur-based stabilizing systems react to form lead-sulfide, which appears as dark sports in the material and may be hazardous to humans and the environment. Adding cheap bases, like $CaCO_3$, can also stop HCl from escaping during processing. Incorporating $CaCO_3$ enhances the material's mechanical properties by acting as a filler (i.e. elongation at break, impact strength by up to 20 percent, and elastic modulus), and it is in fact maintained over several life cycles. Unfortunately, using too much of these fillers causes embrittlement, which makes it necessary to add more plasticizers, which further complicates the chemical makeup of PVC and calls for the use of fillers at low loadings. (Schyns and Shaver, 2021)

9.3.3 Plasticizer Use in Poly(vinyl chloride)

Even in the absence of stabilizers and fillers, PVC embrittlement occurs, making plasticizers necessary in almost all applications. By increasing free volume (Figure 9.10), plasticizers weaken the intermolecular interactions within materials, lowering the glass transition temperature (T_g), reducing stress at break, and increasing elongation at break. In 2006, 5.8 million versus 0.67 million tons of plasticizer were used in PVC, which is almost nine times more than stabilizer. Up until 2005, 80–90% of plasticizer use was accounted for by PVC, as PET and HDPE plasticizers are rarely used. The most popular phthalate plasticizers pose a health risk to people, which has led to a rise in calls to ban PVC from packaging polymers and many other uses. As phthalates migrate to external surfaces during thermal aging, phthalate plasticizer concentration rises with each recycling cycle, worsening these problems and eventually increasing environmental exposure.

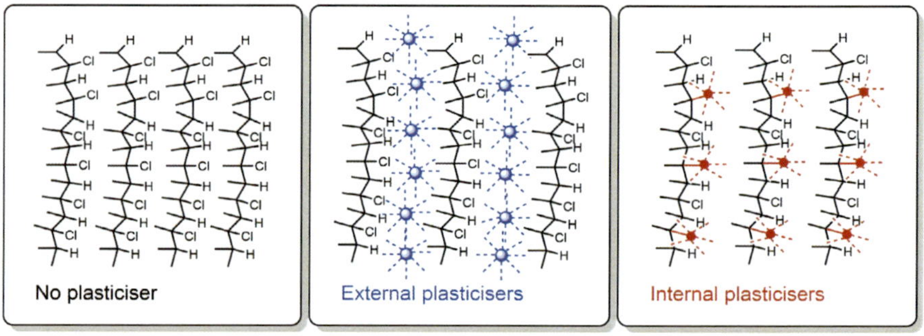

Figure 9.10. Schematic representation of internal and external plasticizers in PVC.

Source: *Z. O. G. Schyns, M. P. Shaver, Mechanical Recycling of Packaging Plastics: A Review. Macromol. Rapid Commun. 2021, 42, 2000415. https://doi.org/10.1002/marc.202000415*

Understandably, these worries have resulted in stringent limitations on the use of phthalates, which have presented significant difficulties for the mechanical recycling of PVC, including through the European PVC recycling program VinylPlus. Citrates, carboxylates, phosphates, sebacates, and epoxidized oils are some of the new, non-toxic plasticizers. Hexamoll DINCH, a 1,2-cyclohexane dicarboxylic acid diisononyl ester, is one of the most widely used and commercially available substitutes for phthalates for use in food-related packaging. It complies with international regulations in the toy, medical device, and food contact industries. Research is still being done on alternative phthalate-free plasticizers.

External plasticity can be substituted with internal flexibility by exchanging the chlorine atoms in PVC (Figure 9.10). Lu et al., describe how the polymer chain is reacted with the common plasticizer diisononyl phthalate to create a self-plasticizing polymer that has less phthalate migration. The system is also made more malleable by the nucleophilic substitution of thiophenol groups for chlorine atoms using K_2CO_3 or N,N-diisopropylethylamine, which results in lower T_g values. Jia et al., investigated PVC that self-plasticized as well. PVC has been made to self-plasticize through the use of phosphorous-containing castor oil derivatives (EAMR-DOPO), aminated tung oil methyl esters, cardanol-based groups, dehydroabietic acid derivatives, and mannich bases of waste cooking oil. Plasticizer migration and toxicity can be decreased by self-plasticized plastics, though the latter has not been proven beyond a reasonable doubt in practical systems. It is necessary to carry out additional research on the resilience of self-plasticized materials because it is anticipated that substituted PVCs will be less stable during mechanical recycling. However, self-plasticizing PVC presents a viable approach to improve recycling and reduce unintentional harm. (Schyns and Shaver, 2021)

9.4 MECHANICAL RECYCLING OF POLYSTYRENE

PS is used in packaging and comes in three different forms: solid, expanded foam, and polybutadiene reinforced polymer (HIPS). Because polystyrene has a low density, recycling and collecting it can be difficult. While expanded polystyrene (EPS) typically needs a solvent-based or mechanically-based method to de-foam and reduce its volume before reprocessing, solid PS can be mechanically recycled into other useful products. Although the chains are not harmed by the solvents used to de-foam EPS, caution must be exercised to reduce the solvent of choice's carbon and environmental impact. Noguchi et al., describe how to de-foam and shrink EPS using d-limonene, a naturally occurring vegetable oil derived from citrus fruits. This process yields a 20-fold reduction in the volume of EPS foam and shows promise for recycling due to the residual oil's antioxidant properties, which shield chains from scission caused by radicals. D-limonene is a costly process, but it dissolves EPS selectively while leaving expanded polyolefins and labels intact.

It is also derived from food and may be more environmentally friendly than other solvents like toluene, though its carbon footprint may be significantly higher. Other natural oils, such as star anise, eucalyptus, thyme, and chamomile oils, can be used as a substitute for d-limonene to effectively reduce the volume of EPS foam without lowering molecular weights. Trezek et al., aim to remove the impact and cost of solvents and explain how to use a machine to mechanically reduce the foamed polystyrene's size. Re-gassing recycled EPS (rEPS) can improve its foaming properties, but recycling rates are still lower due to the additional cost when compared to virgin polymer. (Schyns and Shaver, 2021)

9.4.1 Degradation Mechanisms of Polystyrene

PS is vulnerable to radical attack; a macroradical species is formed through thermal scission. Before the chain's β-scission through an unbuttoning reaction that forms tiny dimers and trimers, the radical reaction proceeds through intramolecular radical transfers.

PS chains can crosslink under specific reaction conditions, most involving low oxygen levels, just like any other polymer. Because polybutadiene chains are present, HIPS degrades differently. Degradation begins in the polybutadiene's unsaturated double bond and moves through the PS chains. PS is utilized in the packaging of electrical and electronic equipment even though it isn't used directly to make electric components. Efficient recycling of polymers used in electronics and their packaging is essential due to the growing amount of waste associated with electronics.

As PS is mechanically recycled, a decrease in molecular weight, viscosity, and elongation at break is seen. The main use of recycled polystyrene is in the production of disposable goods and cutlery. Remulliet al., reveal that the molecular weight of PS dropped by almost 50% by the tenth reprocessing cycle. Reprocessing is dominated by chain scission, as demonstrated by measurements of molecular weight and rheology. In HIPS reprocessing, chain scission and crosslinking are rival techniques.

The elongation at HIPS break decreases as chain scission takes over over crosslinking at higher temperatures or cycles. Vilaplana et al., report that the elongation at break of HIPS decreases by 38% after nine reprocessing cycles; this decrease is less than that of PET but is still significant for product manufacture. Since foam density variations and additives affect the recycling process and impede true circularity, mechanical recycling is not the main method used to recycle EPS. Because they can produce a high-quality rPS that is equivalent to the virgin material and reduce the volume of EPS, dissolution/precipitation techniques are frequently used with PS.

9.4.2 Filler Use in Polystyrene

Fillers are less researched than other packaging polymers, but they have the potential to strengthen the PS polymer matrix. Nagalakshmaiah et al., recently reported using twin screw extrusion to produce a cellulose nanocrystal (CNC)-PS composite material instead of the labor- and resource-intensive solvent casting methods. In order to adjust the CNC and reduce thermal and dispersion instabilities, mini-emulsion polymerization was employed. Styrene-co-2-ethylhexyl acrylate, a compatibilizing copolymer, was utilized to plasticize the polymer and reduce its T_g and mechanical performance.

Montmorillonite clay is added to HIPS to increase its viscosity and prevent chain scission reactions during degradation. However, because clay disperses poorly, adding clay again reduces the mechanical properties. On the other hand, clay agglomerates that undergo extrusion shearing shrink in size, distribute more evenly within the polymer matrix, and regain their mechanical characteristics. Furthermore, Remili et al., reveal that adding 5 weight percent of Cloisite 15A clay to PS keeps the chains from breaking down and preserves the mechanical qualities of the polymer after recycling (Figure 9.11).

Crosslinking from radical recombination reactions is responsible for the observed increase in PS molecular weight; the clay inhibits chain mobility and so restricts the movement of radical fragments created by chain scission. The Cloisite 15A clay in PS, as opposed to the nanoclay used in HIPS, improves the recycled composites' hardness and Young's modulus. This could be because the clay structures in the polymer matrix differ; montmorillonite clay tends to form random agglomerates, whereas layered Cloisite 15A is better able to trap fragments of radicals. Gutiérrez and Alvarez study the crosslinking reaction by employing zinc octanoate to catalyze the reactive extrusion of PS and native/oxidized corn starch. The compact structure of composite films resulted in higher elasticity; however, mechanical recycling has not yet investigated this.

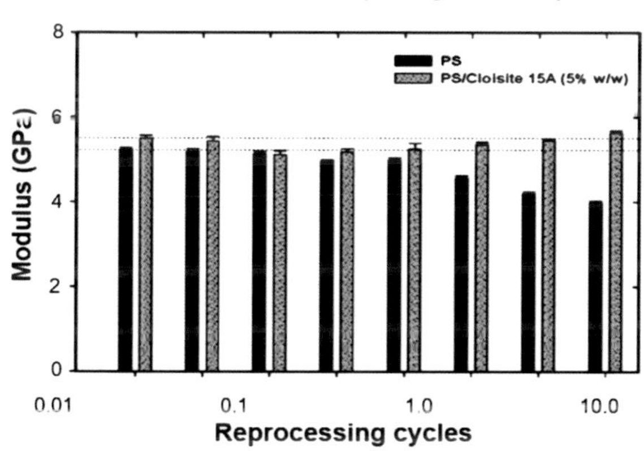

Figure 9.11. Young's modulus versus number of reprocessing cycles for PS compounded with and without 5%w/w Cloisite 15A.

Source: *Z. O. G. Schyns, M. P. Shaver, Mechanical Recycling of Packaging Plastics: A Review. Macromol. Rapid Commun. 2021, 42, 2000415. https://doi.org/10.1002/marc.202000415*

Similar to PVC, PS recycling is less common than that of other polymers because of operational difficulties and the possibility that future regulations will reduce the amount of material in use. (Schyns and Shaver, 2021)

9.5 APPLICATIONS OF COMPLEX POLYMER BLENDS IN MECHANICAL RECYCLING

Blending technologies have the potential to minimize contamination issues in polymer waste streams and recover value in systems where sorting polymer waste is impractical. It can be hard to blend polymers with different chemical polarities, and it gets even harder when there are several polymers in the system. Elevated entanglements and larger polymer blocks have a favorable impact on the material's mechanical characteristics.

Co-monomers such as those described by Eagan et al., can be used to compatibilize other blends. For example, poly(ethylene-*co*-vinyl acetate) and poly(styrene-*b*-ethylene/butadiene-*b*-styrene) are used to compatibilize blends of HDPE, PVC and PS, with gamma radiation then used to induce crosslinking reactions. Though the radiation improved the uniformity of blends, the elasticity and tensile strength of the blends suffers, making it difficult to justify the use of energetically costly radiation as a recycling solution, although important research in this field is continuing. PET-polyolefin blends are frequently made compatible by grafting polyolefins with functional groups—such as maleic anhydride, acrylic acid, ethylene-vinyl acetate copolymers, maleimides, or glycidyl methacrylate—that can interact with PET end groups. Mi et al., report blending polypropylene (PP) and polyester (PET) using multi-flow vibrate injection molding in conjunction with in situ microfibrillation and polyolefin grafted maleic anhydride (POE-g-MA) to regulate the blend morphology. For every polymer, two distinct morphologies were employed: PET spherules and fibers and PP spherulites and orientated structures known as "shish-kebabs" (Figure 9.12). Improved mechanical properties were obtained by blending the PP spherulites with the orientated microfibrillar PET, while improved impact strength was obtained by blending the orientated PP with the PET spherules. PP "shish-kebabs" and PET fibers cannot coexist because the PET fibers obstruct the flow of PP. By adding GMA grafted onto PP with styrene as a co-monomer, bottle grade PET can be made better. This will increase the impact strength of the material because of the improved chain interactions. PET is compatible with poly(styrene-co-(ethylene-butylene)-styrene) grafted with maleic anhydride (SEBS-g-MA) and exists as a minor phase in PP. The hydrophobic PP encapsulates the PET chains, preventing them from degrading. Although this has not yet been investigated for real-world systems, the PP-PET blend is reported to be stable up to three extrusion cycles, which is comparable to reports of mechanical recycling of pure PET streams. Another obstacle to maintaining material value is the lack of food-grade PP.

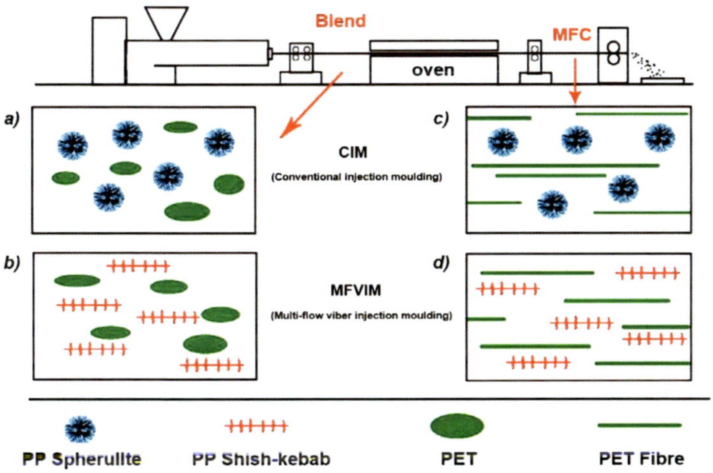

Figure 9.12. Experimental set-up used to produce a) blends b) microfibrillar composites c) blends, and d) microfibrillar composites via a,c) CIM and b,d) MFVIM. (a–d).

Source: *Z. O. G. Schyns, M. P. Shaver, Mechanical Recycling of Packaging Plastics: A Review. Macromol. Rapid Commun. 2021, 42, 2000415. https://doi.org/10.1002/marc.202000415*

It's not always possible to create high quality products using compatible blends. Ragaert et al., describe a secondary application that makes use of SEBS-g-MA as a compatibilizer for the waste sorting product that is sink fraction destined for incineration. They use infrared (IR) sorting to eliminate most of the PVC from this discarded fraction, leaving a mixture primarily composed of PET (73 percent) and PP (17.6%). The authors create a new roofing tile called "greentile," recovering value from a waste fraction that would otherwise be disposed of in a landfill or burned. They also report the properties of the blended material with and without SEBS-g-MA compatibilizer.

PLA and PET are problematic packaging partners, as they are difficult to separate due to their similar density (≈1.25 g cm^{-3} PLA and ~1.35 g cm^{-3} PET), physical appearance and IR signature from similar polyester backbones. As a result, improper waste sorting frequently results in PLA contamination of PET waste streams. At PET processing temperatures, PLA breaks down into the parent lactic acid, and the acids catalyze the breakdown and esterification of PET. Since PLA is one of the most widely used "biodegradable" plastics, it is not frequently gathered for automated recycling. According to Gere and Czigany, a compatibilized PET-PLA blend was created using an ethylene-butyl acrylate-glycidyl methacrylate (E-BA-GMA) terpolymer. The blends exhibit lower elastic moduli but higher toughness. The chain extender Joncryl ADR 4368 was also included by the authors, and it might be the real cause of the mechanical advancements noted. Although reprocessing was not investigated, the compatibilized blends exhibit greater thermal stability than their non-compatibilized counterparts.

The degrading mechanisms of various polymers can also be taken advantage of by blending technologies. Li et al., report utilizing tetraethyl thiuram disulphide (TETD) to create dithiocarbamate radicals that will react with the macroradicals produced by polymer degradation in order to suppress degradation reactions within the extruder and thus make PP and PS compatible. To efficiently regulate the polymer's degree of degradation, the authors combine the TETD with di-tert-butyl peroxide (DTBP), which removes hydrogen atoms from the PP backbone and speeds up the polymer's β-scission. A PP-g-PS copolymer is created in situ by the reaction, and this copolymer serves as a compatibilizer for the PS and PP fractions. By causing crosslinking of the polymer chains, dicumyl peroxide, another additive, was used in conjunction with an unnamed low molecular weight polyester (molecular weight of 3000 with 30 mol percent double bonds) to improve the stiffness of co-mingled plastic waste (65% PE, 11% PS, 10% PP, 7% PET, and 9% other plastics).

Reprocessing is rarely taken into consideration, despite the fact that polymer blends can provide improved mechanical properties and retain value in poorly sorted polymer waste. More research is needed on the behavior and degradation mechanisms of the blended materials, specifically how the polymer blend responds to different degradation temperatures and processes and how this influences larger waste management systems that spread the contamination. (Schyns and Shaver, 2021)

9.5.1 Complex Polymer/Filler Blends

During mechanical recycling, fillers can be used to lessen the effects of damage caused by chain degradation. Fillers can increase the recyclates' impact strength, elongation at break, and Young's modulus while decreasing the polymer's processability. Because of filler structure and thermodynamics, fillers in polymer blends migrate to the interfaces between the fractions. By reducing the energetic difference between particles in the bulk and at the interface, this lowers the interfacial tension between the two polymers. The chemical polarity of the polymer matrix must be considered when selecting fillers. The mechanical properties of the resulting material are reduced when filler and polymer are not properly matched because poor adhesion and poor stress and strain transfer across the polymer result from this. To increase the filler's and the polymer's adhesion, coupling agents like titanates or silanes can be added.

Fillers in recycling need to be used carefully, keeping processing in mind. Viscosity changes in processing can lead to excessive torques that damage processing equipment or necessitate recalibration of large industrial processes. Fillers alter the final extrudate's rheological behavior when they are added to recycles. Low shear rates yield low viscosity values because of brittle areas with high filler concentration. Viscous polymers, like those with a high filler content, require more energy to process, which raises the cost of processing. It's still difficult to strike a balance between increased recycle qualities, filler volume, and cost considerations.

Fillers made of naturally derived fiber are becoming more and more popular because they are renewable and may have less of an impact on the environment. Fiber fillers serve as physical compatibilizers in blends to reinforce polymers. Examples of these fillers include starch, cellulose, chitin, chitosan, soy, wheat gluten, gelatine, and lignin. Aguer et al., as well as Petchwattana et al., Both discuss how adding wood fibers to recycled materials has been successful in raising their quality over several processing cycles. The Petchwattana group. It should be noted that wood fiber filler and recycled PVC (rPVC) can withstand up to 5-7 cycles of use, according to Augier et al., Observe that the material's strength increases beyond 10–20 extrusion cycles without sacrificing its other mechanical qualities (Figure 9.13). PP and HDPE can also be filled with fibers that are naturally derived. Sadly, the degradation, disruption of the polymer matrix, and adhesion problems of these naturally derived fiber fillers necessitate the addition of additional coupling agent.

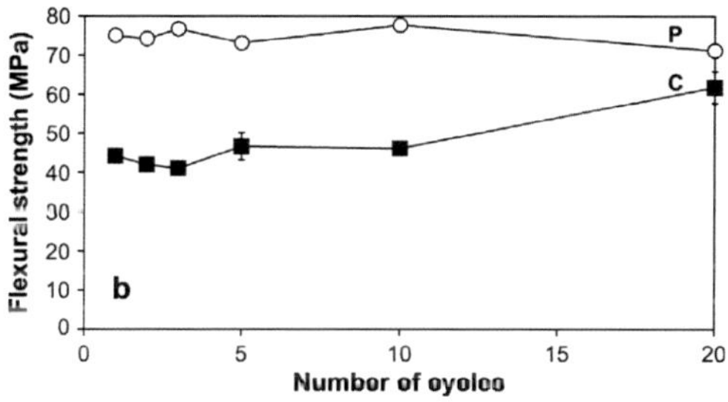

Figure 9.13. Flexural strength versus number of reprocessing cycles for PVC reinforced with wood fibers (C) and without fibers (P).

Source: *Z. O. G. Schyns, M. P. Shaver, Mechanical Recycling of Packaging Plastics: A Review. Macromol. Rapid Commun. 2021, 42, 2000415. https://doi.org/10.1002/marc.202000415*

Cheap composites with improved stiffness, creep resistance, and heat deflection are produced by using particle fillers like wollastonite, talc, $CaCO_3$, and glass fiber. Certain applications, such as replacing automotive parts, building heavy-duty bridges, and creating fire-retardant materials, call for the use of stiff composites. Glass fibers can even work in concert with another filler to further improve mechanical qualities and create materials with exceptional strength. Processing glass-fiber enforced polymers is extremely difficult, despite their potential benefits. Because glass-fiber stiffened polymers have higher viscosities, they degrade more quickly in polymer processing equipment.

Fillers are frequently customizable to meet the needs of particular products. Although they are not necessary in the majority of packaging applications, fire retardants are very important for safeguarding electronic equipment. Although high percentages of inorganic fillers like metal hydroxides, carbonates, or phosphates cause adhesion issues within the polymer, they can be used as flame retardants. Ammonium polyphosphate is used as a fire-retardant filler in flammable polypropylene to lessen the polymer's flammability; however, over time, the filler migrates out of the polymer matrix and breaks down inside the extruder. Aluminum trihydrate, magnesium hydroxide, zinc borate, melamine, graphite, titanium dioxide, flax, and other additives are also fire-retardant.

Fillers made of nanoparticles have been added in more recent innovations to increase the stability of polymer blends. Nanoparticle fillers lack specificity in contrast to polymeric compatibilizers. Taguet et al., suggest that when a nanoparticle's radius is noticeably greater than the gyrational radius of the polymers to be blended, the particle's stabilizing efficacy is maximized. The effectiveness of nanofiller is increased by minimizing interparticle interactions and having equal adsorption-affinities for both polymers. Blends of virgin and recycled polymers have been stabilized using a range of different organically coated nanofillers. The most sophisticated of these, rPET/HDPE, rPET/PP/HDPE, and rPET/rPP blends, have all been improved by using organic modified montmorillonite clay. Additionally, it has been reported that PP and PS's qualities are enhanced by montmorillonite clay. Research indicates that the presence of metals in the clay, surfactants, contaminants, or the collapse and re-aggregation of the clay into agglomerates may cause side reactions that lead to an enhancement of other degradation reactions.

It is also necessary to take into account the environmental effects of using compatibilizers and fillers in addition to other additives (Figure 9.14). The own recyclability of the fillers may differ from that of the host polymer matrix, necessitating an expensive separation step prior to the polymer's recycling. The extensive research on fillers has made it possible to quickly and affordably upgrade to mechanically recycled polymers (r-polymer), which allows polymers that might have ended up in a landfill to be used again for secondary applications where impact strength is more important than flexibility. (Schyns and Shaver, 2021)

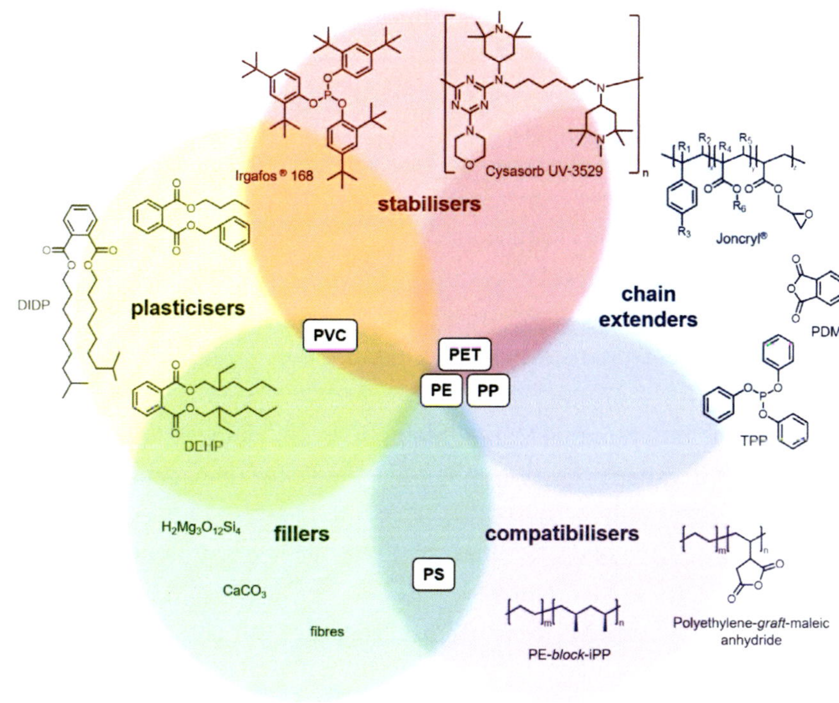

Figure 9.14. Common polymer additives used to improve polymer recyclates.

Source: *Z. O. G. Schyns, M. P. Shaver, Mechanical Recycling of Packaging Plastics: A Review. Macromol. Rapid Commun. 2021, 42, 2000415. https://doi.org/10.1002/marc.202000415*

REFERENCES

1. L. P. Silva, J. C. Prata, T. R. Walker, A. C. Duarte, W. Ouyang, D. Barcelò, T. Rocha-Santos, Chem. Eng. J. 2021, 405, 1266832.
2. Z. O. G. Schyns, M. P. Shaver, Mechanical Recycling of Packaging Plastics: A Review. *Macromol. Rapid Commun.* 2021, 42, 2000415. https://doi.org/10.1002/marc.202000415
3. Feil A, et al., Improvement of the recycling of plastics in lightweight packaging treatment plants by a process control concept. Waste Manag Res 2019;37(2):120e6.
4. Feil A, et al., Separate collection of plastic waste, better than technical sorting from municipal solid waste? Waste Manag Res 2017;35(2):172e80.
5. J. Kirchherr, L. Piscicelli, R. Bour, E. Kostense-Smit, J. Muller, A. Huibrechtse-Truijens, M. Hekkert, Ecol. Econ. 2018, 150, 264.
6. J. Maris, S. Bourdon, J. M. Brossard, L. Cauret, L. Fontaine, V. Montembault, Polym. Degrad. Stab. 2018, 147, 245.
7. Letcher TM, Vallero DA. Waste e a handbook for management. 2nd ed. Academic Press publications; 2019, ISBN 978-0-12-815060-3.
8. M. Avella, E. Bonadies, E. Martuscelli, R. Rimedio, Polym. Test. 2001, 20, 517.
9. Meddah, M.S. and M. Bencheikh, Properties of concrete reinforced with different kinds of industrial waste fibre materials. Construction and Building Materials, 2009. 23(10): p. 3196-3205.
10. PlasticsEurope, An analysis of European plastics production, demand and waste data for 2011. Plastics-the Facts 2012, 2012.
11. Pretz T, Feil A. Klassieren von Abfallgemischen. In: Thomé-Kozmiensky KJ, Goldmann D, editors. Recycling und Rohstoffe, Bd. 8. Neuruppin: TK Verlag; 2015. p. 577e85.
12. Pretz T, Feil A. Klassieren von Abfallgemischen. In: Thomé-Kozmiensky KJ, Goldmann D, editors. Recycling und Rohstoffe, Bd. 8. Neuruppin: TK Verlag; 2015. p. 577e85.
13. Pretz T. In: Wotruba H, Feil A, editors. 8th sensor-based sorting & control 2018. Aachen: Shaker Verlag; 2018. ISBN 3-8440-5805-2, 978-3-8440-5805-5.
14. Somani, R.H., et al.,, Flow-induced shish-kebab precursor structures in entangled polymer melts. Polymer, 2005. 46(20): p. 8587-8623.
15. Tabatabaei, S.H., P.J. Carreau, and A. Aji, Structure and properties of MDO stretched polypropylene. Polymer, 2009. 50(16): p. 3981-3989.

Chapter **10**

Applications and Future of Recycling and Recycled Plastics

Contents

INTRODUCTION

The applications and future of recycling and recycled plastics are rapidly evolving as the global community increasingly recognizes the need for sustainable solutions to plastic waste. Recycled plastics are now being integrated into a wide array of industries, from packaging and construction to automotive and consumer goods. These applications not only help reduce the demand for virgin plastics but also support the development of a circular economy, where materials are continuously reused rather than discarded.

In packaging, one of the largest consumers of plastics, recycled materials are being used to create bottles, containers, and films, meeting both environmental and regulatory demands. The construction industry is also embracing recycled plastics for producing durable materials such as composite lumber, insulation, and piping, which offer environmental benefits over traditional materials. In the automotive sector, manufacturers are increasingly using recycled plastics to produce interior components, bumpers, and other parts, contributing to lighter, more fuel-efficient vehicles.

The future of recycled plastics is promising, with ongoing innovations aimed at expanding their applications and improving their quality. Advances in chemical recycling technologies, for instance, hold the potential to break down complex plastic waste into high-quality raw materials that can be used to produce new, virgin-like plastics. Additionally, the development of biodegradable and compostable plastics is expected to complement recycling efforts, offering alternatives that reduce long-term environmental impact.

Moreover, there is a growing emphasis on designing products with recyclability in mind, known as "design for recycling." This approach is expected to play a crucial role in increasing the efficiency of recycling processes and the quality of recycled plastics. As governments, industries, and consumers become more committed to sustainability, the use of recycled plastics is set to expand significantly, paving the way for innovative applications and a more sustainable future.

10.1 RECYCLED PLASTIC

Plastic that has been reprocessed from waste or scrap plastic materials is referred to as recycled plastic. This plastic is collected, sorted, and cleaned before being melted down and made into new products, as opposed to being thrown away in landfills or the ocean. Recycling plastic helps to keep the environment free of pollution, conserves energy and natural resources, and lessens the amount of new plastic that is produced. Numerous goods, such as packaging, furniture, textiles, and even building materials, can be made from recycled plastic.

Recycled plastic is a fantastic substitute for new plastic because it has a number of advantages that draw in customers who care about the environment. Here are some of the reasons why we sometimes use recycled plastic in our products:

- **Fewer fossil fuels**: Using recycled plastic means that we are reducing our reliance on virgin plastic, which requires significant amounts of fossil fuels to produce. By using recycled plastic instead, we can conserve natural resources and reduce our carbon footprint.
- **Reuses scrap**: Recycling plastic waste helps to divert it from landfills or the ocean, where it would otherwise remain as litter or take hundreds of years to decompose. By reprocessing plastic waste into new products, we can reduce the amount of plastic waste that ends up in our environment and create a more circular economy.
- **Recyclable**: Products made from recycled plastic are themselves recyclable, creating a closed loop in which plastic waste can be continuously repurposed into new products. This helps to reduce the amount of new plastic that needs to be produced, as well as the amount of plastic waste that ends up in our environment.
- **Durable**: Recycled plastic is just as durable and strong as new plastic, making it a great material for a variety of products. In fact, recycled plastic is often even more durable than new plastic, as it has already undergone a previous life cycle and has proven to be able to withstand wear and tear.

10.1.1 Benefits of Using Recycled Plastic

Plastic should not be used at all, in our opinion, but if it must be used, it should only be recycled plastic because doing so will help to create a more sustainable future. Here are some of the benefits of using recycled plastic:

- **NO virgin plastic**: By using recycled plastic, we are reducing our reliance on new, or virgin, plastic. This helps to conserve natural resources and reduce the amount of plastic waste that ends up in our environment.
- **NO waste**: Recycling plastic waste into new products helps to divert it from landfills or the ocean, where it would otherwise remain as litter or take hundreds of years to decompose. This reduces the amount of plastic waste that ends up in our environment and creates a more circular economy.
- **NO virgin material**: In addition to reducing our reliance on virgin plastic, using recycled plastic also helps to reduce our reliance on

other virgin materials, such as wood or metal. This helps to conserve natural resources and reduce our impact on the environment.

- **NO fossil fuel extraction**: Producing new plastic requires significant amounts of fossil fuels, which contributes to climate change and other environmental issues. By using recycled plastic instead, we are reducing our reliance on fossil fuels and helping to create a more sustainable future.
- **YES recyclable**: Products made from recycled plastic are themselves recyclable, creating a closed loop in which plastic waste can be continuously repurposed into new products. This helps to reduce the amount of new plastic that needs to be produced, as well as the amount of plastic waste that ends up in our environment.
- **YES durable**: Recycled plastic is just as durable and strong as new plastic, making it a great material for a variety of products. In fact, recycled plastic is often even more durable than new plastic, as it has already undergone a previous life cycle and has proven to be able to withstand wear and tear.

10.1.2 Applications of Recycled Polyethylene

Because of its adaptability and capacity to lessen the environmental impact of many products, recycled polyethylene is a material that is being used more and more in a variety of industrial applications. We will define recycled polyethylene, discuss its uses, and go over the various varieties and purposes of polyethylene in the sections that follow.

Applications for recycled polyethylene are numerous and include consumer goods production, the building and automotive sectors, and more. We can list the following as some of its primary uses:

- Containers and packaging: It is used to make containers and packaging of all kinds, from bottles to bags and boxes. Due to its strength and light weight, it is a very popular material in the packaging industry.
- Construction: It is also used to make pipes, waterproofing sheets, geotextiles and other products used in construction. Its resistance to moisture and chemicals makes it ideal for these uses.
- Automotive: Recycled polyethylene is used in the manufacture of automotive parts such as door panels, bumpers and trim. Its light weight and impact resistance make it ideal for these uses.
- Textiles: Used to make textile fibers used in the manufacture of clothing, carpets and other textile products.

- Agriculture: Used in the manufacture of agricultural products such as greenhouse cover sheets, shade nets and irrigation pipes.

10.1.3 Recycled HDPE

One kind of recycled polyethylene that is obtained from the processing of bottles and containers holding liquids, such as milk, is called recycled HDPE. Compared to LDPE, this type of polyethylene is stronger and more resilient due to its higher density. Gathering the bottles and containers, crushing them, and cleaning them to get rid of residues and contaminants are the steps in the HDPE recycling process. After that, they are melted and molded to create pipes, sheets, and other items.

High-Density Recycled Polyethylene Uses

The main uses of HDPE include the following:

- Containers and packaging: Recycled HDPE is used to make bottles and containers for products such as detergents, chemicals, oils and lubricants. Its resistance to moisture and chemicals makes it ideal for these uses.
- Pipes: Recycled HDPE is used in the manufacture of pipes for drinking water, sewage systems and gas pipelines. Its abrasion and corrosion resistance make it ideal for these uses.
- Construction: Recycled HDPE is used in the manufacture of waterproofing sheets and geotextiles used in road and building construction. Its resistance to moisture and chemicals makes it ideal for these uses.
- Toys: Recycled HDPE is used in the manufacture of toys and other children's products because of its safety, strength and durability.

Low-Density Recycled Polyethylene Uses

Processing bags, films, and other low-density plastic products yield low-density polyethylene (LDPE). Due to its flexibility and low weight, LDPE is a material that finds extensive use. The main uses of recycled LDPE include the following:

- Bags and films: Recycled LDPE is used in the manufacture of bags and films for packaging of food, chemicals and other consumer goods.
- Bottles and containers: Recycled LDPE is used in the manufacture of bottles and containers for products such as milk, juices and other liquids. Due to its flexibility and impact resistance, it is a widely used material in the packaging industry.

- Insulation: Recycled LDPE is used in the manufacture of building insulation materials such as foils and foams. Its flexibility and moisture resistance make it ideal for these uses.
- Toys: Recycled LDPE is used in the manufacture of toys and other children's products due to its safety and flexibility.

Medium-Density Recycled Polyethylene Uses

After medium density plastic products like bags, films, and pipes are processed, recycled medium density polyethylene (MDPE) is produced. In terms of strength and flexibility, MDPE is in the middle of HDPE and LDPE.

Among the main uses of recycled MDPE, we can highlight the following:

- Pipes: Recycled MDPE is used in the manufacture of pipes for irrigation, gas distribution and pressurized liquid systems. Its resistance to abrasion and corrosion makes it ideal for these uses.
- Packaging films: Recycled MDPE is used in the manufacture of packaging films to protect delicate products such as fruit, vegetables and electronics. Its puncture and impact resistance make it ideal for these uses.
- Shopping bags: Recycled MDPE is used in the manufacture of reusable shopping bags due to its durability and strength. It is also a material used in the manufacture of rubbish bags.
- Containers and bottles: Recycled MDPE is used in the manufacture of containers and bottles for products such as shampoo, conditioner and cleaning products. Its resistance to chemicals makes it ideal for these uses.
- Toys: Recycled MDPE is used in the manufacture of toys and other children's products due to its safety and flexibility.

Applications for recycled polyethylene are numerous and it's a very adaptable material. Its use helps to save natural resources, lower greenhouse gas emissions, and lessen the quantity of plastic waste that ends up in the environment. A more sustainable circular economy must be achieved by promoting plastics' use in various industries and by encouraging their recycling.

10.2 APPLICATIONS OF RECYCLED PLASTICS

It is imperative that plastic waste that has outlived its usefulness be recycled for the sake of the environment and public health. One important material type that is crucial for minimizing environmental effects and conserving resources is recycled plastic. Plastic waste can be recycled in recycling facilities and used again in manufacturing. Plastic recycling, which primarily aims to lessen its impact on the environment, is an extremely important procedure for the health and well-being of the community. Plastic waste can be utilized in manufacturing processes after it has gone through recycling procedures. Recycled plastics can be used for a wide range of purposes.

Recycled plastics are widely used in manufacturing, helping to keep plastic waste out of the environment and promoting environmental sustainability in the industrial sector.

10.2.1 Packaging Materials

The packaging industry frequently uses recycled plastics. Products like grocery store bags, detergent bottles, food and beverage packaging, and packaging for cosmetics are made from recycled plastic.

10.2.2 Construction and Building Materials

Additionally, recycled plastics are highly favored in the building and construction industry. Generally speaking, recycled plastics can be used for pavement, flooring, fencing, floor coverings, and insulation. They are widely employed in the manufacturing of furniture and other building materials.

10.2.3 Automotive and Transportation

The automotive industry uses recycled plastics for parts like bumpers, panel coverings, seats, and interior parts that are used on both the outside and inside of cars. The purpose of this is to decrease the manufacturing of new plastic products.

10.2.4 Electronic Devices

Because they offer several benefits in the management of electronic waste, recycled plastic is frequently chosen for the casings and components of electronic devices.

10.2.5 Agriculture and Gardening

The agricultural and gardening industries frequently use recycled plastics that have undergone specific processing to create products like garden furniture, irrigation systems, greenhouse coverings, and ground coverings.

10.2.6 Healthcare and Medicine

Medical devices, packaging materials, and single-use medical supplies can all be made from recycled plastics in the healthcare industry.

10.2.7 Sports Equipment

Sports gear made of recycled plastic can be used for physical activities. Sports equipment, bicycle parts, outdoor furniture, and the surfacing and facing of sports fields are all made from recycled plastic.

10.3 APPLICATIONS OF DIFFERENT FORMS OF RECYCLED PLASTICS AS CONSTRUCTION MATERIALS

Plastics have developed into an essential resource for everyday human needs (Ma et al., 2021). The production of plastic materials has increased exponentially over the last few decades (Gilbert, 2017). By 2025, plastic production is predicted to reach more than 0.5 trillion tons. Approximately 60% of plastic waste is still not recycled (Huang et al., 2022). According to Tamoor et al (2022), Pakistan produced approximately 12 million tons of plastic waste in 2020, and by 2050, that amount might potentially reach 22 million tons. Plastics are widely used in a variety of industries, including electronics, construction, and packaging, because of their adaptability, durability, and low weight (Malkar, R.; et al., 2023). The production and accumulation of plastic materials that have reached the end of their useful lives is what causes the generation of plastic waste, which is a global environmental concern. Approximately 0.4 billion tons of plastic waste are produced worldwide (Tejaswini et al., 2022). Single-use plastics like bottles, straws, bags, and packaging are some of the main sources of plastic waste (Schnurr et al., 2018).

Thermoplastics and thermosetting plastics are the two different types of polymers, and they behave differently when heated. Because they are reversible, thermoplastics can be heated to take on any shape and are therefore referred to as recyclable materials. However, when thermosetting plastics are heated, a process known as polymerization takes place that causes chemical changes in the plastic that result in a hardened, rigid material. Their final form is the result of this irreversible behavior, which cannot be altered without seriously degrading the original form. Thermoplastics can soften or distort at high temperatures and typically have lower mechanical strength and heat resistance. On the other hand, thermosetting polymers are renowned for having exceptional heat resistance and high mechanical strength. Even at high temperatures, their structural integrity is preserved.

Because they are recyclable and reprocessable, thermoplastics are widely used in applications like consumer goods, automotive components, packaging, and 3D printing. In contrast, thermosetting plastics are used in electrical insulators, aerospace components, automotive parts, and composite materials because of their strength and high temperature resistance. E-wastes, also called electronic wastes, are used in the construction industry in a variety of ways and are derived from abandoned or outdated electrical and electronic equipment, devices, and components. These electronic wastes are frequently recycled and used as plastic aggregates in cementitious composites. Researchers studying flexible pavements have made extensive use of recycled plastic aggregates as an ingredient.

10.3.1 Generation of Waste Plastic

The amount of plastic waste generated varies by location but has been rising globally for some time, impacting both developed and developing nations. Because plastic waste clogs rivers and landscapes and adds to ocean pollution, it presents significant environmental problems.

Air, marine, and soil pollution are three ways that nonbiodegradable plastics harm the ecosystem. The majority of plastics cannot biodegrade and require centuries to transform into microplastics. These microplastics may also find their way into marine environments, and the ecosystem may be negatively impacted by their unpredictable nature. Their unpredictable nature is a result of the dangerous and carcinogenic chemicals that are added during processing to give the material the desired qualities. Furthermore, accurate forecasting of the generation of plastic waste is necessary for proper planning of waste management and waste plastic recycling.

The different kinds of thermoplastics and thermosetting plastics are displayed in Figure 10.1. Two of the thermoplastics that are produced most frequently worldwide are polyethylene and polypropylene. These polymers are widely used in consumer goods, automotive, packaging, and construction. Specialized thermoplastics, like polycarbonate, polyethylene terephthalate (PET), and polyurethane, are utilized in sectors like electronics, automotive, and aerospace in addition to commodity thermoplastics like PE, PP, and PVC. However, thermosetting polymers are essential to the creation of composite materials, such as carbon fiber, fiberglass, and aramid fiber composites, which find applications in the automotive, marine, and aerospace sectors (Bobade, SdotK. et al., 2016). The demand for and consumption of goods made of plastic have an impact on the amount of this kind of plastic waste that is produced globally.

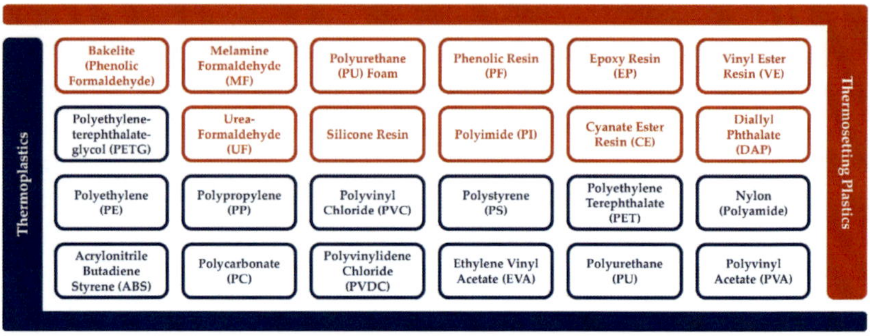

Figure 10.1. Types of thermoplastics and thermosetting plastics.

Source: Ahmed S, Ali M. Potential Applications of Different Forms of Recycled Plastics as Construction Materials—A Review. Engineering Proceedings. 2023; 53(1):5. https://doi.org/10.3390/IOCBD2023-15177

10.3.2 Waste Plastic Recycling Techniques

The circular economy is largely facilitated by the recycling of waste plastic. Various methods can be used to convert waste plastic into recycled plastic, depending on the type of conversion and the needs of the byproducts. Table 10.1 lists the most widely used methods for recycling waste plastics.

Table 10.1. *Plastic waste recycling techniques*

Approach	Process	Ref.
Mechanical Recycling	The sorting, cleaning, shredding, and then melting waste plastic.	(Schyns, Z.O.; et al., 2021)
Chemical Recycling	The breakdown of plastic waste into its constituent molecules through chemical processing.	(Rahimi, A.; et al., 2017)
Thermal Recycling	The breakdown of plastic waste into its constituent molecules through heating at a high temperature.	(Kijo-Kleczkowska, A.; et al., 2022)
Biological Recycling	The conversion of plastic waste into biomass or other useful products through microorganisms such as bacteria or fungi.	(Soong, Y.-H.V.; et al., 2022)

Source: Ahmed S, Ali M. Potential Applications of Different Forms of Recycled Plastics as Construction Materials—A Review. Engineering Proceedings. 2023; 53(1):5. https://doi.org/10.3390/IOCBD2023-15177

From the standpoint of the circular economy, the authors examined the procedures and difficulties involved in the mechanical recycling of PET, PE, PP, PS, and PVC. For plastics with comparable qualities, like clean and well-separated plastic containers, mechanical recycling works especially well. However, it also has drawbacks when it comes to recycling mixed plastics, contamination problems, and the possibility of material deterioration over several recycling cycles. Chemical recycling is advantageous because it doesn't require a lot of energy and is simple to do. This includes catalytic methods performed in mild temperatures. Furthermore, the method works well with mixed plastics, contaminated plastics, or plastics with intricate structures that are difficult to recycle mechanically (Rahimi, 2017).

The benefits of thermal recycling waste plastics include a decrease in the amount of plastic waste generated, energy recovery, and the potential to produce valuable fuels. For plastics with high calorific fractions, thermal recycling is advised. It uses plastic as a fuel or raw material while maintaining efficiency and environmental safety through appropriate design, oversight, and regulation. Soong et al., (2022) examined PET waste plastic recycling techniques, contrasting mechanical and chemical recycling methods with a biotechnological recycling approach that uses PET's enzymatic degradation. Both of the traditional approaches have negative financial and environmental effects in addition to producing recycled plastic of lower quality. However, the use of modified microorganisms in biotechnological processes can lead to the creation of useful byproducts. Tertiary recycling, also known as biological degradation, completely disassembles waste plastic into its constituent chemical elements. The type of plastic waste stream to be recycled determines the adoption of recycling processes based on the desired mechanical and durability properties.

10.3.3 Application of Recycled Plastic as a Construction Material

When it comes to building materials, most research on recycled plastic used it in aggregate form. Chen et al (2023) examined the development of environmentally friendly cement mortar by substituting 5% of the fine particles with thermoset plastic. It was found that the composition, which had a dense microstructure and a strong interfacial transition zone, produced the highest compressive strength. Authors Soni et alet al (2022) examined the combination of 50% high density polyethylene (HDPE) and 50% sand to create a sand-plastic composite meant for flooring applications. Maximum compressive and flexural strength was found to be the result of enhanced interfacial bonding caused by low water absorption and adequate adhesion. It was advised to use the composite as paving in public areas' non-traffic areas. A variety of plastic applications in the construction industry are displayed in Table 10.2.

Table 10.2. *Use of recycled plastic as a construction material*

Application	Composition	Modified Properties	Ref.
Eco-friendly cement mortar	5% thermoset waste as a replacement for fine particles	The highest compressive strength, a highly dense microstructure, and a robust interfacial transition zone.	(Chen, H.; et al., 2023)
Sand–plastic composites as floor tiles	50% sand and 50% HDPE	Maximum compressive and flexural strength due to sufficient adhesion, increased interfacial bonding, and low water absorption.	(Soni, A.; et al., 2022)
Wearing course of flexible pavement	Modified bitumen mix and processed waste plastic of about 5–10% by wt. of bitumen	Increased fatigue life and strength, ultimately enhancing the service life of pavement.	(Duggal, P.; et al., 2018)
Pavement construction	0.5–3% HDPE and PP by weight of graded bitumen at a temperature range from 160 °C to 170 °C	A successful blending-in of the polymer strands in bitumen was observed. Stable polymer-modified bitumen was obtained using PP for pavement construction.	(Appiah, J.K.; et al., 2017)

Source: Ahmed S, Ali M. Potential Applications of Different Forms of Recycled Plastics as Construction Materials—A Review. Engineering Proceedings. 2023; 53(1):5. https://doi.org/10.3390/IOCBD2023-15177

In the last ten years, plastic-based paver construction has also attracted attention. Duggal et al (2018) assessed the bitumen's performance in a wearing course for flexible pavement using a modified bitumen mix that included waste plastic addition at a rate of 5–10 percent by weight of bitumen. Better strength and fatigue life were found in the results, indicating a longer service life. An additional investigation was carried out to assess the impact of incorporating HDPE and PP into bitumen. In a temperature range of 160 °C to 170 °C, weight of graded bitumen was substituted for HDPE and PP at intervals of 0.05 percent to 3 percent. Consequently, the bitumen's polymer strands were successfully blended in. Furthermore, PP was used to create a stable bitumen that had been modified with a polymer for use in pavement construction.

10.4 APPLICATION OF RECYCLED PLASTICS IN THE AUTOMOTIVE INDUSTRY

Utilizing recycled plastics in the automotive sector is a challenging issue (the study by Maudet, 2006analyzes it analytically in : "A method for recycled plastic integration in the automotive industry"), which includes several risk factors, such as inquiries about the price and volume of the produced recycled materials (about 25% of the plastic materials used in the automotive industry are recycled plastics), as well as the quality of the recycled materials (in terms of the required properties). However, businesses have recently used recycled plastic to coat exterior and interior components.

10.4.1 Škoda

Application: PET bottles are used to make flooring, luggage compartments, and rug fibers (a Koda Scala, for instance, has 14 kg of recycled parts). Figure 10.2, the diagram below, provides an example of how recycled plastics are used.

For the interior seat coverings of the all-electric KODA ENYAQiV, the Czech automaker (Mladá Boleslav / Czech Republic, 2022) combines recycled polyester (PET bottles) with wool reinforcement.

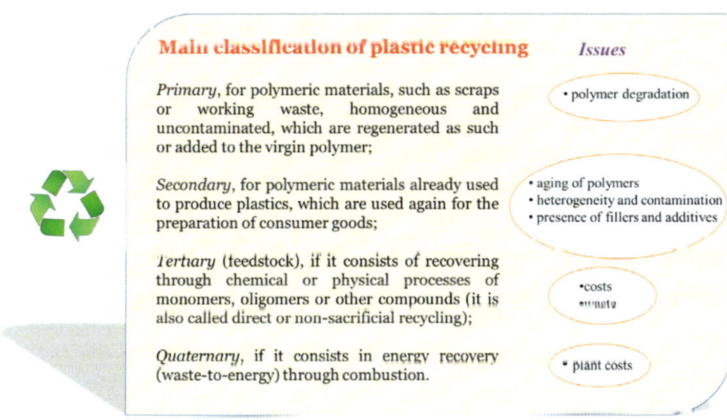

Figure 10.2. Recycling.

Source: Coltelli M-B, Gigante V, Aliotta L, Lazzeri A. Recyclability Perspectives of the Most Diffused Biobased and Biodegradable Plastic Materials. Macromol. 2024; 4(2):401-419. https://doi.org/10.3390/macromol4020023

10.4.2 Jaguar / Hyndai

The company created an extremely durable plastic used in the Jaguar I-Pace electric vehicle (in pilot cars) in collaboration with the German agrochemical company Badische Anilin and Soda Fabric, or BASF. This was done as part of the ChemCycling program. It is a material created by thermochemically converting waste plastic into pyrolysis oil. The material is a perfect sustainable option for designing the next generation of dashboards and external surfaces for Jaguar and Land Rover because it can be tempered and colored. As part of its Destination Zero vision, Jaguar also uses recycled PET bottles to make seats (minimum impact on the environment).

The aforementioned automakers, Jaguar and Hyundai, expanded their applied materials portfolio last year to include the so-called Econyl ® (yarn manufacturer: Aquafil). This fabric is created from recycled nylon found in the oceans, and it shares many of the original nylon's characteristics. Hyundai and the Healthy Seas Marine Conservation Organization have been collaborating for a considerable amount of time.

10.4.3 Volvo

Replacement of plastic parts (car: SUV plugin hybrid XC60 T8) with parts made of recycled plastic. An applied tunnel bracket made of recycled plastic and fibers from fishing nets and ship ropes is located inside the SUV XC60 T8. Research illustrates the commitment made by Volvo cars to use 25% recycled materials in plastics for all new Volvo models starting in 2025 and beyond (in collaboration with Axion Polymers).

10.4.4 Toyota Motor Corporation

The company created a unique recyclable plastic known as TSOP (Toyota Super Olefin Polymer. It was introduced in 1995 in various versions, including TSO P2, TSO P3, TSO P5, and TSO P6, each with unique properties, compounds, and uses. These are three additional resin compounds (talc + rubber) and "non-conventional high-performance PP.". According to a team of Toyota researchers, the new material performs similarly to TSOP but at a cost, specific gravity, injection moldability, economics, recyclability, and noise performance that is roughly 30% lower.

10.4.5 Ford

Many models of passenger seat cushions are made with upholstery made from recycled plastics (each Focus's seat fabric, for instance, is made from about 22 plastic water bottles). Over 23,000 tons of recycled plastic were recycled and repurposed by the company to make exterior parts for Ford vehicles produced in North America.

Table 10.3 shows that many additives can cause human exposure.

Table 10.3. *Migration risk and persistence in recycling, by analyte group, based on Danish EPA findings*

Analyte Group	Persistence in Recycling	Migration Risk
Heavy metals	Due to strong binding, expected to persist through mechanical recycling process. Mercury typically found in polyurethane, which cannot be mechanically recycled. The fate of mercury in feedstock recycling is not known, but most mercury is expected to have evaporated by that point.	Typically strongly bound, therefore not expected to migrate. As a result, the "exposure to consumers must therefore be considered low". Mercury an exception: not chemically bound, will migrate and evaporate, leading to some exposure risk. This risk is judged to be small.
Perfluorinated chemicals	Only used in certain types of plastics, and the fate of these substances by recycling is unknown. They suggest that "recycling is not normally practised".	These substances are not chemically bound, meaning there is a risk of migration.
Flame retardants	The fate in recycling depends on the plastic. Plastics which can be mechanically recycled (including PVC, PP, PS) will retain flame retardants during recycling. Newer, alternative flame retardants are less studied, characterized by "a lack of knowledge regarding both applications and fate in the products as well as by subsequent recycling activities".	Migration risk depends on the substance. Reactive flame retardants are chemically bound, and are considered of less risk. Additive flame retardants (such as most BFRs) are not chemically bound and will migrate easily, "and may thus result in significant exposure of consumers".
Phthalates	The migration rate is low enough to assume the main part of the plasticizer added to the product will remain in it until end of life. If mechanically recycled, they will "also be present in recycled materials".	Migration of plasticizers to food well studied. Generally, all plasticizers "must be anticipated to migrate and the use in plastics should thus be considered a source of exposure to consumers".

Analyte Group	Persistence in Recycling	Migration Risk
Bisphenols	They judge that if Bisphenol A is present in mechanical recycling, it will remain in the plastic.	Based on its physical properties, it should be regarded as a semi-volatile compound, able to migrate out of plastics. With time, "the major part of the substance will probably be released by leaching to the surface followed by evaporation or removal by washing".
Formaldehydes	In mechanical recycling, unreacted formaldehyde will likely evaporate due to its low boiling point and the high vapor pressure. As a result, "the substance will most likely not be present in recycled materials".	Its physical properties suggest it should migrate strongly. This strong evaporation could lead to occupational exposure.

Source: Lahl U, Zeschmar-Lahl B. Material Recycling of Plastics—A Challenge for Sustainability. Sustainability. 2024; 16(15):6630. https://doi.org/10.3390/su16156630s

10.4.6 Continental (Tire Manufacturer)

At the moment, Continental is incorporating recycled polyester made from PET bottles into the manufacturing of its tires. A passenger car tire contains roughly 40 grams of polyester fibers; 60 PET bottles are recycled to make one full set of tires. The fibers are extracted from waste bottles using a mechanical process, and then applied to the tire's structure.

Research on recycling specific plastics, creating compounds with fibers or reinforcements, evaluating mechanical properties, and other topics is still in progress. Particles that are appropriate, like glass fibers (GF) or basalt fibers, can be added to the final compound to change the properties of PET recycle.

Glass fibers are usually adopted as reinforcement due to their:

- remarkable mechanical and physical properties;
- rather low price

The use of rPET in combination with glass fibers (to produce a canopy strip installed at the vehicle's middle window portion) is described in the study. Three distinct sample types are tested, with the percentage of components varying. Before rPET was actually applied to the final composite, the bottles were recycled using the conventional method, which included sorting, crushing the bottles into flakes, pre-washing, friction washing, hot washing, cold washing, post-washing, magnetic separation, spin drying, and final flakes inspection. Samples: After extensive testing, it was discovered that 30% glass fiber and 70% rPET met the necessary specifications. Volpe, V. et al., also address a related topic regarding the application of recycled PET bottles into glass fiber reinforced composite material. Characterization of Recycled/ Virgin Polyethylene Terephthalate Composite Reinforced with Glass Fiber for Automotive Applications" is the title of their study. ". 2010 saw Mondadori et al., (2012) created composites with varying proportions of short glass fibers and recycled PET; the materials are showing improvements in mechanical performance as the glass fiber content rises.

Table 10.4 shows various microbes for PET degradation.

Table 10.4. *Summary of enzymatic degradation of PET with a different type of microorganism*

Type of Enzyme	Microorganism	Materials	Temperature (°C)	Product
Cbotu_EstA	*Clostridium botulinum ATCC3502*	PET film	50	TPA,MHET
Lipases	*Candida cylindracea*	PET nanoparticels	40	1,2-Ethandiol, TPA
Tcur0390	*Thermomonospora curvata DSM43183*	PET nanoparticels	50	NA
Hydrolase TfH	*Thermobifida fusca DSM43793*	PET pellete	44–55	TPA, EG
TfCut1	*Thermobifida fusca KW3*	PET nanospheres	55–65	HEB, MHET, BHET, Benzoic acid
Tha_Cut1	*Thermobifida alba DSM43185*	PET	50	TPA, Benzoic acid, HEB, MHET
MHETase and PETase	*Ideonella sakaiensis*	PET	30	MHET
BsEstB	*Bacillus subtilis 4P3-11*	PET	40–45	TPA, Benzoic acid, MHET

Type of Enzyme	**Microorganism**	**Materials**	**Temperature (°C)**	**Product**
LCC cutinase	*Leaf-branch compost*	Amorphous PET	50–70	MHET, TPA, Benzoic acid
Polymer polyhydroxyalkanoate	*Pseudomonas*	PET	70	EG,TPA
Hydroxyalkanoyloxyalkanoate	*Pseudomonas*	PET	70	EG,TPA
Thc_Cut2	*T. cellulosilytica*	PET films	50	MHET, TPA, BA, HEB
Arg29Asn	*Mutant The_Cut2*	PET films	50	MHET, TPA, BA, HEB
Lipases	*Aspergillus oryzae CCUG 33812*	PET fabric	55	BHET,TPA
Gln65Glu	*Mutant The_Cut2*	PET films	50	MHET, TPA, BA, HEB
TfCut2	*Thermobifida fusca KW3*	PET films	60	MHET
Thh_Est-esterase	*Thermobifida halotolerans*	3PET model substrate	50	MHET,TPA, BA, HEB

Source: Damayanti, Wu H-S. Strategic Possibility Routes of Recycled PET. Polymers. 2021; 13(9):1475. https://doi.org/10.3390/polym13091475

Table 10.5 below shows the kinetic degradation of PET as per select research findings.

Table 10.5. *Kinetic and modeling analysis of PET with different types of condition operation*

PET	Catalyst	Method	T (°C)	k	k_2	K	A	Ea (Kj·mol^{-1})	$Y_{B\text{-}HET}$, %	Y_{DMT}, %	Y_{TPA}, %
PET		P	373–443 448–503				6.4×10^{25}/min 9.31×10^{11}/min	347.4 172.6			
PET		P	390				48×10^{21}/min	323.8			
PET		P						202.1			
PET		P						198.6			
PET	H_2O	H	180	0.0037				378			
flake	H_2O	H	265	0.352 g PET/(mol min)		0.664		55.7			
PET	H_2@ZSM-5-25	H	230	0.1787				1.2			98.5
PET	H_2O without catalyst	H	230	0.0755				19.4			
flakes	NaOH	H	200	0.1900				99			97.9
flakes	KOH	H	160	2.24×10^{-6} L/(min cm^2)			419 L/(min cm^2)	69			90.9
PET	HNO_3	H	100	0.4800				101.3			

PET	Catalyst	Method	T (°C)	k	k_2	K	A	Ea ($Kj \cdot mol^{-1}$)	$Y_{B\text{-}HET}$, %	Y_{DMT}, %	Y_{TPA}, %
PET	HNO_3	H	88	1.9×10^{-2} L/(min cm^2)		0.042	659 L/(min cm^2)	88.3			
Flake < 2	sulfuric acid	H		2.0×10^{11} g.cm/(mol min)				99.7			90
PET	TBAB	Hydro-lysis	60–80	0.1100				75			99.0
flake	3Bu6DPB, NaOH	H	90	0.2 L/(mol h)			1.3×10^{9} L/(mol h)	68.2			>90
flake < 6	TOMAB NaOH	H	80					83			92
flake 3 × 5	marine water	H	205				5.33×10^{7}/h	75			96.0
flake	ZnAc	G	265	6.67×10^{3} L/mol/min				92.0			
pellet	Ionic liquid	G	190	0.16/min				51.6			
PET	EG	G					99783/min	46.2			
pellet 0.25	Na_2CO_3	G	196	3.767 L^2/mol^2/h		0.36 L/mol		185	80		

PET Catalyst		Method	T (°C)	k	k_2	K	A	Ea ($Kj \cdot mol^{-1}$)	$Y_{B\text{-}HET}$, %	Y_{DMT}, %	Y_{TPA}, %
fiber	Zn/Al	G	196					79.3	92		
PET	KOH	M	65	0.0132	0.0246			127.6	96.0		96.0
PET	CH_3OH	M	270	0.0033/s	0.0008/s						
BHET	CH_3OH	M	25	0.0017/day	0.0025/ day			56			
PET	CH_3OH	M	140	1.4 × 10^{-3} g PET/(mol min)			107.1 g PET/(mol min)	95.31			
PET	CH_3OH	M						66.5		80	

H: hydrolysis; M: methanolysis; P: pyrolysis; Flake 3 × 5: flakes of 3 mm × 5 mm, Ea: activation energy; A: preexponential factor; 3Bu6DPB: tributylhexadecylphosphonium bromide.

Source: Damayanti, Wu H-S. Strategic Possibility Routes of Recycled PET. Polymers. 2021; 13(9):1475. https://doi.org/10.3390/polym13091475

10.5 THE FUTURE OF PLASTIC RECYCLING

Since plastic is one of the materials that is used the most in our modern world, recycling plastic is more important than ever. It is an affordable, lightweight, and long-lasting material for both industrial and consumer use.

Because plastic is so widely used, we must handle it carefully after each use. The best method for extending the life of current materials is recycling. Plastic that cannot be recycled is either burned for energy or dumped in a landfill; in the worst case scenario, it may find its way into the environment and decompose into microplastic.

A complex system driven by markets, pricing strategies, and local laws governs the recycling of plastic. We are all expected to be a part of this system. Plastic recycling needs to be effective from all angles, whether you are a waste collector, consumer, or designer of plastic products.

10.5.1 Rising Needs To Enhance the Circular Economy by Recycling

The need to recycle plastic is growing because, once it is left in the wild, it poses a threat to the environment worldwide. By 2050, it is predicted that the oceans will contain more plastic than fish due to sustainable plastic recycling and increased plastic consumption. Additionally, the production of virgin plastic is expected to account for 15% of global greenhouse gas emissions.

The creation of a circular economy for plastics is one of the objectives of the European Union, which is the world's most advanced plastic recycler. The European Union's plastics strategy aims to have all plastic packaging composed of recyclable and reusable materials by the year 2030.

In the EU, numerous decisions have been made about the use and recycling of plastic. For instance, certain single-use items like plastic straws and deliberate microplastics that are easily disposed of in the ocean are prohibited. The use of innovative recycling techniques is encouraged for the member nations concurrently.

Overall, there is still a long way to go before plastic recycling is effective. Less than one-third of the approximately 26 million tonnes of plastic waste produced annually in the EU alone is recycled (estimate from 2018).

Plastics have a low recycling rate, which hurts businesses and the environment alike. Plastic packaging loses up to 95% of its value after just one use, which is an inefficient use of resources.

10.5.2 The Challenges and Possibilities of Plastic Recycling

1. Meeting Price and Quality

Consumer demand for recyclable products is more important than societal encouragement to recycle and participate in the circular economy. They can also exert pressure on businesses that meet the growing demand for environmentally friendly materials.

It is imperative that recycled plastic be regarded as elegant, high-quality, and long-lasting as the virgin plastic we are used to.

The quality of the final product is impacted by the recycling process's effectiveness. Sorting the recyclable plastic waste and separating the various plastic types from one another produces the best outcome.

For instance, the production of incredibly smooth and functional recycled materials is made possible by the effective recycling of plastic bottles.

2. Competitive Price

In terms of cost, recycled plastic is competitive with virgin plastic because the latter has a significant advantage in the global market. Virgin plastic is made from oils and has historically been less expensive than recycled plastic. When it comes to recycled materials, the price trend is less noticeable.

Enhancing plastic recycling's effectiveness is crucial for maintaining the industry's low and competitive prices. Reusing, sorting, and mechanically recycling materials are still frequently more expensive than creating new ones. Recycled plastics' reputation as an affordable alternative is continuously improved by regional incentives, regulations, and taxes.

3. Effective and Modern Recycling Infrastructure

It goes without saying that the world needs to increase its capacity for recycling plastic, as the need for recycled materials is a global megatrend. To ensure that plastic is recycled properly, it is estimated that the capacity for recycling plastics in Europe alone needs to be increased by at least a million tonnes annually.

One of the most important prerequisites is to make investments in cutting-edge recycling infrastructure. Recycling facilities that are highly automated and capable of expanding their capacity at a reasonable cost are required. When it comes to recycling plastic, you should automate everything you can because any additional downtime or curb will slow down the process and increase costs. For instance, at plastic recycling facilities, manually cutting plastic bales with wire is a hazardous, time-consuming task.

Notable is also the fact that recycling, when done correctly and in the modern era, can save more energy than is used in its various stages of operation. Recycling impure plastic can be very energy-intensive if waste transportation and washing water treatment are involved.

- It is up to consumers, businesses, and recycling plants to welcome recycled plastic as a new norm.
- The EU is setting new standards for a profitable and circular plastics economy. Also, North America is taking a lead in effective plastic recycling.
- Recycling infrastructure needs modernization as the demand for recycling is rising globally.
- Mechanical recycling is currently the most efficient way to recycle plastics. Chemical recycling of plastic, where plastic is processed thermochemically back into raw material, has not yet experienced its commercial breakthrough.

10.5.3 The Future of Plastics: Reclaim, Recycle, Repeat

An environment with less plastic waste could be made possible by new technologies in recycling processes. It's likely that you've seen the triangle of recycling: reduce, reuse, recycle. Plastic circularity represents the subsequent iteration of that notion.

Today: Mechanical Recycling

Plastics that can be recycled using current mechanical recycling technology are restricted in their types. The recycling of more plastic waste is therefore imperative to prevent possible long-term negative effects on ecosystems, the food supply, and human health.

"They continue to emerge from landfills and appear in our rivers and waterways.". We must find a solution to the waste plastic problem caused by these lightweight materials finding their way into the environment. ".

Nowadays, the majority of plastic is burned or dumped in landfills. Merely a small portion of the present plastic waste volumes are processed by mechanical recycling, which typically yields lower-quality plastic. For instance, your food packaging might not be converted back into food packaging, but rather into lawn furniture.

Because you're heating the product and weakening the molecular bonds, mechanical recycling frequently doesn't yield the same results. This implies that it is unable to re-enter the same category of demanding application. ..

The Upgrade: Chemical Recycling

Let's talk about chemical recycling, or advanced recycling. Plastic waste that is not suitable for mechanical recycling is converted chemically into a recycled polymer feedstock (RPF) through the process of chemical recycling. The production value chain for plastics can then use that feedstock to create new, superior plastics.

"You can increase the amount of plastic that is actually recycled by about three times over current levels with chemical recycling working in tandem with other recycling methods.". This permits the use of plastics in products such as food packaging, medications, and medical equipment, among others.

Reducing Waste – and Carbon Footprint

Plastics come in a wide variety of forms, such as PET, polypropylene, and polyethylene. Their options for dying differ greatly from one another. Comparing conventional incineration and landfilling to Honeywell UOP's UpCycle, plastic waste handling can typically reduce its carbon footprint by more than 50%.

Therefore, we're not only offering a fresh, dependable method of removing plastic waste from the environment and converting it back into usable plastics, but we're also assisting in reducing the global warming impact of these priceless materials ".

Table 10.6. *Properties and characteristics of widely used polymers in plastic packaging*

Polymer Type and Chemical Structure.	Resin Identification Code	General Application as Virgin Material	Properties	Toxic Chemical Present	Advantage	Challenges	Recycled Products
Polyethylene terephthalate (PET) 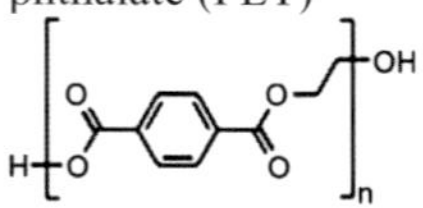	1 PETE	Bottle application because they are inexpensive, lightweight, and shatter-resistant	Clarity, strength, toughness, barrier to gas and moisture	BPA, leach antimony	More convenient (less energy intensive) for chemical recycling (catalytic methods) than thermolysis	Costly thermo-chemical process and hazardous byproduct generation, fiber quality is less purified than recycled material, toxic production process for nanoparticles	Mineral/drinking water bottles, cosmetic bottles
High-density polyethylene (HDPE, or PE-HD) 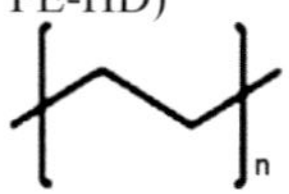	2 HDPE	Bottles, carry bags, milk pouches, recycle bins, etc.	Stiffness, strength, toughness, moisture resistance, permeability to gas	Unknown	Less extraction of fossil fuel, less dependence on petrochemical production, more convenient (less energy intensive) for chemical recycling (catalytic methods) than thermolysis	Expensive conversion process (isolation and purification), less effective mechanical properties than conventional recycled material	Tubes, sewer pipes, pallets, boxes, buckets, toys, bottles for detergents, construction, cable insulation, packaging of food products

Polymer Type and Chemical Structure.	**Resin Identification Code**	**General Application as Virgin Material**	**Properties**	**Toxic Chemical Present**	**Advantage**	**Challenges**	**Recycled Products**
Polyvinyl chloride (PVC) 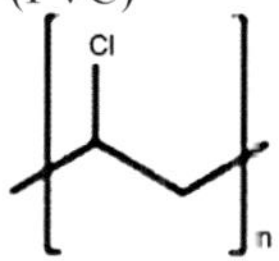	3	For pipes and fittings, tarpaulins, medical apps., etc.	Versatility, ease of blending, strength, toughness	Phthalates	Less extraction of fossil fuel, reduced volume to landfill, less dependence on petrochemical production, more convenient (less energy intensive) for chemical recycling (catalytic methods) than thermolysis	Fiber quality is less purified than recycled material, expensive conversion process (isolation and purification), less effective mechanical properties than conventional recycled material	Sewer pipes, window frames, construction, flooring, wallpaper, bottles, car interiors, medical products, planks, etc.)
Low/linear low-density polyethylene (LDPE/LLDPE) 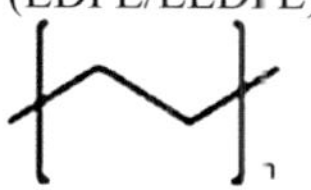	4 LDPE	Plastic bags, various containers, dispensing bottles, wash bottles, tubing, etc.	Ease of processing, strength, toughness, flexibility, ease of sealing, barrier to moisture	Unknown	Less extraction of fossil fuel, utilized as mixed plastic, ink affinity, hydrophobicity, durablility, less dependence on petrochemical production, more convenient (less energy intensive) for chemical recycling (catalytic methods) than thermolysis	Relatively complex process for the reduction in variability in thermal and mechanical properties, hazardous solvent utilization, expensive conversion process (isolation and purification), less effective mechanical properties than conventional recycled material	Flexible packaging, bin liners, carrier bags, tubes, agricultural mulch film, agricultural sheet, construction film, cling-film, heavy-duty sacks

Polymer Type and Chemical Structure.	Resin Identification Code	General Application as Virgin Material	Properties	Toxic Chemical Present	Advantage	Challenges	Recycled Products
Polypropylene/polypropene (PP) 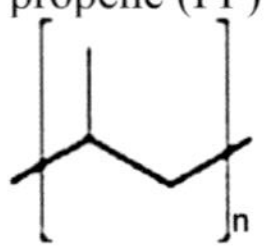	5 PP	Auto parts, industrial fibers, food containers, etc.	Strength, toughness, resistance to heat, chemicals, grease and oil, versatility, barrier to moisture	Unknown	Less extraction of fossil fuel, utilized as mixed plastic, reduced volume to landfill, convenient and less toxic component utilization in processing, lucrative applications in multiple advanced technologies	Relatively complex process for the reduction in variability in thermal and mechanical properties, fiber quality is less purified than recycled material, unwanted wax production affects equipment, toxic production process for nanoparticles	Pipes, pallets, boxes, furniture, car parts, pots of yogurt, buckets, butter, margarine, fibers, milk crates
Polystyrene (PS)/expended Polystyrene (PS-E) 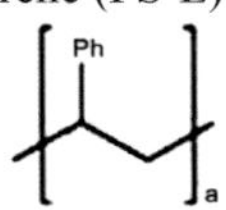	6 PS	Food service packaging, disposable cups, tray pitchers, refrigerators, liners, etc. It may also be used as cushioning materials for fresh produce, electronic or appliance industries, etc.	Versatility, clarity, easy formation	Leaks toxic chemicals when heated	Less extraction of fossil fuel, utilized as mixed plastic, less dependence on petrochemical production, more convenient (less energy intensive) for chemical recycling (catalytic methods) than thermolysis	Expensive conversion process (isolation and purification), less effective mechanical properties than conventional recycled material	Clothes hangers, park benches, flowerpots, toys, spoons, cutlery, picture frames, seeding containers

Polymer Type and Chemical Structure.	Resin Identification Code	General Application as Virgin Material	Properties	Toxic Chemical Present	Advantage	Challenges	Recycled Products
Thermoset plastics	7 Other	Thermoset plastics, multilayer and laminates, Bakelite, polycarbonate (PC), etc.	Stiffness, hardness, brittleness, chemical and stain resistance, good electrical insulators, non-remoldability	BPA	-	-	CDs, pallets, floors, roofs, furniture, sheeting, benches, shoe soles

Source: Hossain R, Islam MT, Shanker R, Khan D, Locock KES, Ghose A, Schandl H, Dhodapkar R, Sahajwalla V. Plastic Waste Management in India: Challenges, Opportunities and Roadmap for Circular Economy. Sustainability. 2022; 14(8):4425. https://doi.org/10.3390/su14084425

REFERENCES

1. Ma, Y.; Zhou, H.; Jiang, X.; Polaczyk, P.; Xiao, R.; Zhang, M.; Huang, B. The utilization of waste plastics in asphalt pavements: A review. *Clean. Mater.* 2021, *2*, 100031.
2. Gilbert, M. Plastics materials: Introduction and historical development. In *Brydson's Plastics Materials*; Elsevier: Amsterdam, The Netherlands, 2017; pp. 1–18.
3. Huang, S.; Wang, H.; Ahmad, W.; Ahmad, A.; Ivanovich Vatin, N.; Mohamed, A.M.; Deifalla, A.F.; Mehmood, I. Plastic waste management strategies and their environmental aspects: A scientometric analysis and comprehensive review. *Int. J. Environ. Res. Public Health* 2022, *19*, 4556.
4. Tamoor, M.; Samak, N.A.; Xing, J. Pakistan toward Achieving Net-Zero Emissions: Policy and Roadmap. *ACS Sustain. Chem. Eng.* 2022, *11*, 368–380.
5. Malkar, R.; Kagale, S.; Chavan, S.; Tiwari, M.; Patil, P. Applications of Bioplastics in Sports and Leisure. *Handb. Bioplastics Biocomposites Eng. Appl.* 2023, *2*, 299–315.
6. Tejaswini, M.; Pathak, P.; Ramkrishna, S.; Ganesh, P.S. A comprehensive review on integrative approach for sustainable management of plastic waste and its associated externalities. *Sci. Total Environ.* 2022, *825*, 153973.
7. Schnurr, R.E.; Alboiu, V.; Chaudhary, M.; Corbett, R.A.; Quanz, M.E.; Sankar, K.; Srain, H.S.; Thavarajah, V.; Xanthos, D.; Walker, T.R. Reducing marine pollution from single-use plastics (SUPs): A review. *Mar. Pollut. Bull.* 2018, *137*, 157–171.
8. Benin, M.A.; Retnam, B.S.J.; Ramachandran, M.; Sivapragash, M.; Dhas, J.E.R. Comparative study of tensile properties on Thermoplastic & Thermosetting polymer composites. *Int. J. Appl. Eng. Res.* 2015, *10*, 10109–10113.
9. Geyer, R. A brief history of plastics. *Mare Plast.-Plast. Sea Combat. Plast. Pollut. Through Sci. Art* 2020, 31–47.
10. Stieven Montagna, L.; Ferreira de Melo Morgado, G.; Lemes, A.P.; Roberto Passador, F.; Cerqueira Rezende, M. Recycling of carbon fiber-reinforced thermoplastic and thermoset composites: A review. *J. Thermoplast. Compos. Mater.* 2023, *36*, 3455–3480.
11. Zhou, P.; Tian, J.; Li, C.; Tang, Z. Comparative study of durability behaviors of thermoplastic polypropylene and thermosetting epoxy exposed to elevated temperature, water immersion and sustained bending

loading. *Polymers* 2022, *14*, 2953.

12. Liu, T.; Zhao, B.; Zhang, J. Recent development of repairable, malleable and recyclable thermosetting polymers through dynamic transesterification. *Polymer* 2020, *194*, 122392.
13. Bhagat, G.V.; Savoikar, P.P. Durability related properties of cement composites containing thermoplastic aggregates–A review. *J. Build. Eng.* 2022, *53*, 104565.
14. Huang, Y.S.; Zhou, Y.; Zeng, X.; Zhang, D.; Wu, S. Reversible Crosslinking of Commodity Polymers via Photocontrolled Metal–Ligand Coordination for High-Performance and Recyclable Thermoset Plastics. *Adv. Mater.* 2023, *35*, 2305517.
15. Oladele, I.O.; Okoro, C.J.; Taiwo, A.S.; Onuh, L.N.; Agbeboh, N.I.; Balogun, O.P.; Olubambi, P.A.; Lephuthing, S.S. Modern Trends in Recycling Waste Thermoplastics and Their Prospective Applications: A Review. *J. Compos. Sci.* 2023, *7*, 198.
16. Millet, H.; Vangheluwe, P.; Block, C.; Sevenster, A.; Garcia, L.; Antonopoulos, R. The nature of plastics and their societal usage. *Plast. Environ.* 2018, *2018*, 1–20.
17. Luhar, S.; Luhar, I. Potential application of E-wastes in construction industry: A review. *Constr. Build. Mater.* 2019, *203*, 222–240.
18. Nasier, S. Utilization of recycled form of concrete, E-wastes, glass, quarry rock dust and waste marble powder as reliable construction materials. *Mater. Today Proc.* 2021, *45*, 3231–3234.
19. Chen, H.; Qin, R.; Chow, C.L.; Lau, D. Recycling thermoset plastic waste for manufacturing green cement mortar. *Cem. Concr. Compos.* 2023, *137*, 104922.
20. Ma, J.; Nawarathna, H.M.; Hesp, S.A. On the sustainable use of recycled plastics in flexible asphalt pavements. *J. Clean. Prod.* 2022, *359*, 132081.
21. Singh, R.K.; Ruj, B. Time and temperature depended fuel gas generation from pyrolysis of real world municipal plastic waste. *Fuel* 2016, *174*, 164–171
22. Rajmohan, K.V.S.; Ramya, C.; Viswanathan, M.R.; Varjani, S. Plastic pollutants: Effective waste management for pollution control and abatement. *Curr. Opin. Environ. Sci. Health* 2019, *12*, 72–84.
23. Kibria, M.G.; Masuk, N.I.; Safayet, R.; Nguyen, H.Q.; Mourshed, M. Plastic waste: Challenges and opportunities to mitigate pollution and effective management. *Int. J. Environ. Res.* 2023, *17*, 20.
24. Silvarrey, L.D.; Phan, A. Kinetic study of municipal plastic waste. *Int. J.*

Hydrogen Energy 2016, *41*, 16352–16364.

25. Geyer, R.; Jambeck, J.R.; Law, K.L. Production, use, and fate of all plastics ever made. *Sci. Adv.* 2017, *3*, e1700782.
26. Pettipas, S.; Bernier, M.; Walker, T.R. A Canadian policy framework to mitigate plastic marine pollution. *Mar. Policy* 2016, *68*, 117–122.
27. Pivnenko, K.; Eriksen, M.K.; Martín-Fernández, J.A.; Eriksson, E.; Astrup, T.F. Recycling of plastic waste: Presence of phthalates in plastics from households and industry. *Waste Manag.* 2016, *54*, 44–52.
28. Eriksen, M.K.; Pivnenko, K.; Olsson, M.E.; Astrup, T.F. Contamination in plastic recycling: Influence of metals on the quality of reprocessed plastic. *Waste Manag.* 2018, *79*, 595–606.
29. Ghinea, C.; Drăgoi, E.N.; Comăniță, E.-D.; Gavrilescu, M.; Câmpean, T.; Curteanu, S.; Gavrilescu, M. Forecasting municipal solid waste generation using prognostic tools and regression analysis. *J. Environ. Manag.* 2016, *182*, 80–93.
30. Hamad, K.; Kaseem, M.; Deri, F. Recycling of waste from polymer materials: An overview of the recent works. *Polym. Degrad. Stab.* 2013, *98*, 2801–2812.
31. Maddah, H.A. Polypropylene as a promising plastic: A review. *Am. J. Polym. Sci* 2016, *6*, 1–11.
32. Amin, S.; Amin, M. Thermoplastic elastomeric (TPE) materials and their use in outdoor electrical insulation. *Rev. Adv. Mater. Sci* 2011, *29*, 15–30.
33. Bobade, S.K.; Paluvai, N.R.; Mohanty, S.; Nayak, S. Bio-based thermosetting resins for future generation: A review. *Polym.-Plast. Technol. Eng.* 2016, *55*, 1863–1896.
34. Shamsuyeva, M.; Endres, H.-J. Plastics in the context of the circular economy and sustainable plastics recycling: Comprehensive review on research development, standardization and market. *Compos. Part C Open Access* 2021, *6*, 100168.
35. Shen, L.; Worrell, E. Plastic recycling. In *Handbook of Recycling*; Elsevier: Amsterdam, The Netherlands, 2014; pp. 179–190.
36. Schyns, Z.O.; Shaver, M.P. Mechanical recycling of packaging plastics: A review. *Macromol. Rapid Commun.* 2021, *42*, 2000415.
37. Gu, F.; Guo, J.; Zhang, W.; Summers, P.A.; Hall, P. From waste plastics to industrial raw materials: A life cycle assessment of mechanical plastic recycling practice based on a real-world case study. *Sci. Total Environ.* 2017, *601*, 1192–1207.
38. Thiounn, T.; Smith, R.C. Advances and approaches for chemical recycling

of plastic waste. *J. Polym. Sci.* 2020, *58*, 1347–1364.

39. Rahimi, A.; García, J.M. Chemical recycling of waste plastics for new materials production. *Nat. Rev. Chem.* 2017, *1*, 0046.
40. Kijo-Kleczkowska, A.; Gnatowski, A. Recycling of plastic waste, with particular emphasis on thermal methods. *Energies* 2022, *15*, 2114.
41. Soong, Y.-H.V.; Sobkowicz, M.J.; Xie, D. Recent advances in biological recycling of polyethylene terephthalate (PET) plastic wastes. *Bioengineering* 2022, *9*, 98.
42. Zhuo, C.; Levendis, Y.A. Upcycling waste plastics into carbon nanomaterials: A review. *J. Appl. Polym. Sci.* 2014, *131*, 39931.
43. Lee, A.; Liew, M.S. Tertiary recycling of plastics waste: An analysis of feedstock, chemical and biological degradation methods. *J. Mater. Cycles Waste Manag.* 2021, *23*, 32 43.
44. Khalid, M.Y.; Arif, Z.U.; Ahmed, W.; Arshad, H. Recent trends in recycling and reusing techniques of different plastic polymers and their composite materials. *Sustain. Mater. Technol.* 2022, *31*, e00382.
45. Vollmer, I.; Jenks, M.J.; Roelands, M.C.; White, R.J.; van Harmelen, T.; de Wild, P.; van Der Laan, G.P.; Meirer, F.; Keurentjes, J.T.; Weckhuysen, B.M. Beyond mechanical recycling: Giving new life to plastic waste. *Angew. Chem. Int. Ed.* 2020, *59*, 15402–15423.
46. Soni, A.; Das, P.K.; Yusuf, M.; Kamyab, H.; Chelliapan, S. Development of sand-plastic composites as floor tiles using silica sand and recycled thermoplastics: A sustainable approach for cleaner production. *Sci. Rep.* 2022, *12*, 18921
47. Duggal, P.; Shisodia, A.S.; Havelia, S.; Jolly, K. Use of waste plastic in wearing course of flexible pavement. In Proceedings of the Advances in Structural Engineering and Rehabilitation: Select Proceedings of TRACE 2018, Greifswald, Germany, 24–27 April 2018; pp. 177–187.
48. Appiah, J.K.; Berko-Boateng, V.N.; Tagbor, T.A. Use of waste plastic materials for road construction in Ghana. *Case Stud. Constr. Mater.* 2017, *6*, 1–7.
49. American Chemistry Council (2017). Use of Recycled Plastics in Cars is Shifting into Overdrive. Retrieved from: https://www.plasticsmakeitpossible.com/plasticsrecycling/what-happens-to-recycled-plastics/use-ofrecycled-plastics-in-cars-is-shifting-into-overdrive/).
50. Plastic Europe. Enabling a sustainable future (2022). Retrieved from: https://plasticseurope.org/).

51. Maudet, C., Bertoluci, G., Froelich, D., & Viot, F. (2006). A method for recycled plastic integration in the automotive industry. In 13th CIRP: International Conference on Life cycle Engineering, Leuven.
52. Sláma, J. (2020). Use of recycled materials in ŠKODA cars: No Plastic Waste at All. Retrieved from: https://www.skoda-storyboard.com/en/skodaworld/innovation-and-technology/use-of-recycledmaterials-in-skoda-cars-no-plastic-waste-at-all/).
53. TASR (2022). Priemysel: Jaguar v spolupráci s BASF testuje nový recyklačný process. Retrieved from: https://www.engineering.sk/aktuality/8129-eko-srnitra-jaguar-basf-spolupraca-plasty-nrx (accessed: 22 August 2022).
54. Kadera, V. (2019). JLR a chemička BASF vyrábějí superpevný plast z odpadu. Již brzy se dostane do aut. Retrieved from: https://www.autorevue.cz/jlr-a-chemicka-basfvyrabeji-superpevny-plast-z-odpadu-jiz-brzy-sedostane-do-aut .
55. Plastics Recycling World, July/August. (2019). ELV bumpers can find a second life in the US. , 31-33. Retrieved from: https://issuu.com/plasticsrecyclingworldexpo/docs/pr w_july_august_2019_pdf_for_downlo).
56. Hyundai Motor NEWS (2022). Hyundai pokračuje v ochrane morí a rozširuje spoluprácu s Healthy Seas. Retrieved from: https://media.hyundai.sk/item/=eNortjIyslIqNU9Jq0uJ ck-xNTY3MC1OCM3ONLc1U7IGXDCMTAja (accessed: 25 July 2022).
57. Hyundaizubehoer.de - Hyundai (2022). Floor mats velour (ECONYL®), IONIQ 5 without sliding centre console. Retrieved from: https://hyundaizubehoer.de/en/ioniq-5/210446-floormats-velour-econyl-ioniq-5-without-sliding-centreconsole.html).
58. Potorčák, R. (2018). Recyklované plasty majú v nových autách Volvo tvoriť štvrtinu všetkých plastov. Retrieved from: https://www.odpadyportal.sk/Dokument/104132/recyklovane-plasty-majuv-novych-autach-volvo-tvorit-stvrtinu-vsetkychplastov.aspx, (accessed: 22 June 2022).
59. Renewable carbon (2018). PowerRibs™ enables ocean plastic in automotive interior parts. Retrieved fom: https://renewable-carbon.eu/news/powerribs-enablesocean-plastic-in-automotive-interior-parts/ (accessed: 23 August 2022).
60. Bcomp (2022). PowerRibs™ natural fibre reinforcement for high performance. Retrieved from: https://www.bcomp.ch/products/powerribs/ (accessed: 23 August 2022).
61. Kawamura, N., Nomura, T., & Nishio, T. (1996). Super olefin polymer

for material consolidation of automotive interior plastic parts. SAE transactions, 304-311.

62. Canadian Plastics (2003). New generation PP developed for IPs, bumpers. Retrieved from: https://www.canplastics.com/canplastics/newgeneration-pp-developed-for-ipsbumpers/1000008365).
63. Matsuda, M., Nomura, T., & Iwai, H. (1998). Material Consolidation for Automotive Interior and Exterior Parts through Development of a High Performance Material (No. 982410). SAE Technical Paper.
64. Automobiles review (2022). Toyota Auris 2008 - Environmental Care Retrieved from: https://www.automobilesreview.com/pictures/toyota/auris/wallpaper-28.html
65. Plastic expert (2022). How car manufacturers are using recycled plastic. Retrieved from: https://www.plasticexpert.co.uk/car-manufacturersusing-recycled-plastic/).
66. Continental AG. (2021). Spoločnosť Continental začne od roku 2022 používať pri výrobe pneumatík recyklované PET fľaše. Retrieved from: https://www.plasticportal.sk/sk/spolocnostcontinental-zacne-od-roku-2022-pouzivat-pri-vyrobepneumatik-recyklovane-pet-flase/c/7482/).
67. Mircea, C. (2022) Continental Rolls Out Its First Sustainable Tires Made From Recycled PET Bottles. Retrieved from: https://www.autoevolution.com/news/continentalrolls-out-its-first-sustainable-tires-made-fromrecycled-pet-bottles-185723.html#agal_2 (accessed: 22 June 2022).
68. Scelsi, L., Hodzic, A., Soutis, C., Hayes, S. A., Rajendran, S., AlMa'adeed, M. A., & Kahraman, R. (2011). A review on composite materials based on recycled thermoplastics and glass fibres. Plastics, Rubber and Composites, 40(1), 1-10.
69. Karthik, G., Balaji, K. V., Venkateshwara, R., & Rahul, B. (2015). Eco-Friendly Recycled PET (Polyethylene Terephthalate) Material for Automotive Canopy Strip Application (No. 2015-01-1304). SAE Technical Paper.
70. Volpe, V., Lanzillo, M. S., Molaro, A., Affinita, G., & Pantani, R. (2022). Characterization of Recycled/Virgin Polyethylene Terephthalate Composite Reinforced with Glass Fiber for Automotive Applications. Journal of Composites Science, 6(2), 59.
71. Mondadori, N. M. L., Nunes, R. C. R., Canto, L. B., & Zattera, A. J. (2012). Composites of recycled PET reinforced with short glass fiber. Journal of Thermoplastic Composite Materials, 25(6), 747-764.
72. HSM GmbH + Co. KG (2022). Sustainability through PET Recycling.

Retrieved from: https://eu.hsm.eu/en/know-how/compressing/petrecycling/ (accessed: 26 August 2022).

73. Koronis, G., & Silva, A. (2019). Eco-impact assessment of a hood made of a ramie reinforced composite. In Green Composites for Automotive Applications (pp. 99-114). Woodhead Publishing.

74. Lei, Y., Wu, Q., Yao, F., & Xu, Y. (2007). Preparation and properties of recycled HDPE/natural fiber composites. Composites Part A: applied science and manufacturing, 38(7), 1664-1674.

75. Oza, S., Wang, R., & Lu, N. (2011). Thermal and mechanical properties of recycled high density polyethylene/hemp fiber composites. International Journal of Applied Science and Technology, 1(5), 31- 36.

76. Adhikary, K. B. (2008). Development of wood flour recycled polymer composite panels as building materials. University of Canterbury, Christchurch, New Zealand. (PhD Thesis).

77. Pesante, P., Saavedra, K., & Pincheira, G. (2021). Mechanical Properties of a Wood Flour-PET Composite Through Computational Homogenisation. Cmc-Computers Materials & Continua, 67(3), 4061-4079.

INDEX

D

E

Q

R

S

Kolibri 360 GmbH
Ferdinand-Jühlke-Straße 7
99095 Erfurt, Deutschland
produktsicherheit@kolibri360.de